ALTERNATIVE CONSTRUCTION

ALTERNATIVE CONSTRUCTION

Contemporary Natural Building Methods

Edited by
LYNNE ELIZABETH
CASSANDRA ADAMS

WILEY

John Wiley & Sons, Inc.

New York • Chichester • Weinheim • Brisbane • Singapore • Toronto

This book is printed on acid-free paper. ∞

Copyright © 2005 by John Wiley & Sons, Inc. All rights reserved

Published by John Wiley & Sons, Inc., Hoboken, New Jersey

Published simultaneously in Canada

For general information on our other products and services or for technical support, please contact our Customer Care Department within the United States at 800-762-2974, outside the United States at (317) 572-3993 or fax (317) 572-4002.

Wiley also publishes its books in a variety of electronic formats. Some content that appears in print may not be available in electronic books. For more information about Wiley products, visit our web site at www.Wiley.com.

Library of Congress Cataloging-in-Publication Data

Alternative construction : contemporary natural building methods / edited by Lynne Elizabeth and Cassandra Adams.
  p. cm.
 Includes bibliographical references and index.
 ISBN 0-471-71938-2 (pbk)
  1. Building. 2. Building materials—Environmental aspects. 3. Green products. 4. Sustainable development. 5. Architecture and society. 6. Construction I. Elizabeth, Lynne. II. Adams, Cassandra.

 TH146.A48 2005
 691—dc22   2004061232

Printed in the United States of America

10 9 8 7 6 5 4 3 2

Contents

12.   Earthen Finishes

Part III.   APPLICATIONS

13. Integrated Systems with Rammed Earth

14. Straw, Clay, and Carrizo

15. Light-Clay House Additions

16. Variations on Earthbag

17. The Value of Indigenous Ways

Foreword

The year was 1968, and our world was exploding with technology. No longer satisfied with such things as neighborhood shops or electric fans, Americans were reinventing their lives with the help of tools that could keep their homes at a constant 70 degrees Fahrenheit and produce automobiles that could drive faster and farther than they had ever imagined. Creature comforts were the hallmark of life in America, and an envious world watched.

But while others were looking ahead, with their sights fixed on that next new shiny invention, Garrett Hardin was looking back. Harkening to the words of W. F. Lloyd's *Tragedy of the Commons*, Hardin made the point that the rational use of a common resource ultimately leads to its collapse. It was a hard sell in the 1960s.

It has been three decades since Hardin offered his prophetic thesis, and scientists worldwide have collected evidence supporting his case. Our population continues to grow while every living system on earth is in a state of decline. Even though knowledge of this sobering reality is spreading, as conflicts over water, fuel, and other resources escalate, the conviction that technology can fix the problem persists. We have been unable to discriminate between technologies that meet our needs sustainably and those that reduce the earth's carrying capacity. Citizens in the developing world imitate the American way of life, even though our sprawled communities lack the spirit and sense of place indigenous communities enjoy. Even worse, our model consumes resources at 20 times the rate of the world average.

In Lloyd's commons of the early 1800s, each herdsman was operating as a rational individual without benefit of scientific data or computer models to forecast the outcome of his grazing decisions. It is ironic that in spite of the vast amounts of information available to us, in the United States less than 4 percent of the earth's population holds all the records for consumption, waste, and pollution. Indeed, it is a continuing tragedy that we can be so engaged in seducing the rest of the world with our life-style and selling the latest products of comfort that we can ignore the logical outcome. The more successful we are with this marketing effort, the more we accelerate the collapse of the global commons.

The good news is that there is growing unrest in the United States—a discomfort in our own excess. This life-style doesn't feel as good as the advertisements had promised. It's like those new designer shoes we bought, even though they were a bit uncomfortable and far too expensive. After wearing them long enough to discover them to be truly painful, we also learned from Paul Hawken's *The Ecology of Commerce* that the leather was laced with residual chromium and that highly toxic lead was included in the polyvinyl chloride soles. Our definition of what constitutes good design has been expanded.

There is a growing awareness that by embracing technology as the ultimate solution, we have separated ourselves from nature. As Wendell Berry put it, "It is not only possible, but altogether probable, that by diminishing nature we diminish ourselves. . . . We seek some kind of peace, even an alliance between the domestic and the wild."

Daily, U.S. corporations and individual citizens are seeking designs for their homes, places of business, and communities that celebrate this alliance. Architects are working with these clients to integrate natural systems with elegant new technologies, creating environments that are resource-efficient and restorative. It is timely, then, that the experiences of the pioneers in the emerging movement of natural design are now available to us in a single textbook. This group of passionate visionaries has invested decades in understanding indigenous and low-impact construction techniques and in integrating these historic concepts into modern design standards of comfort, health, safety, and durability. We architects are indebted to the courage of these mavericks in alternative and ecological building systems. Their synthesis of natural flows and materials into more sustainable construction methods are transforming our design decisions.

This book informs and inspires us. It will, we hope, accelerate the creation of houses and communities that reconnect residents to the rhythms of nature and provide a model for reversing our patterns of consumption, waste, and pollution. If so, we may need a sequel to Hardin's treatise—perhaps *The Celebration of the Commons*!

—**Bob Berkebile**, FAIA

Since our emergence as a species, humankind has gone through three great cultural epochs, each with its prototypical forms of creating built environments and buildings. The first epoch was that of the pre-agricultural tribe of hunter-gatherers and primitive cultivators. The primary built environment was the village, and the primary building form was a dwelling constructed of natural materials gathered from the immediate environment. In what remains of indigenous tribal life around the world, we find the mud houses of the African *dogon*, the thatched and pole houses of the equatorial forest and flood plain, the bamboo structures of the Malay peninsula, the adobe of our Southwestern pueblos. These are the precursors of our own rediscovery of natural building materials and principles.

The simultaneous development of agriculture ten thousand years ago in various regions of the world gave birth to the second cultural epoch. Food surpluses allowed larger permanent settlements and gave rise to the great empires of the Mediterranean basin, Central America and Asia. In a real sense, agriculture is the mother of architecture, inasmuch as we make no large investment in buildings and cultural infrastructure until we have stable food supplies and the means to store and distribute them. Tribal forms gave way to hierarchical systems of organization and control, and the need to predict time and space gave rise to the first sciences and notational forms of recording time, quantity and events. Great sacred complexes were built during this period.

The third cultural epoch has its roots in the philosophy of the Greeks, the concept of the *demos*, or a society of equals. In the fifteenth and sixteenth centuries in Europe, an intellectual, artistic, and scientific flowering produced the key philosophic elements for the transition into the epoch of the corporate state. In earlier societal forms, the individual was first a member of a group. Enlightenment philosophy then proclaimed the individual as free and autonomous. Art, science, and religion were fused in earlier forms; now they became separate and differentiated. The new science of economics developed the idea of rational economic behavior based in free markets. Science, based on wrenching nature's secrets from her, began its long march toward domination and control over the natural world.

The dominant built environment and buildings of the third epoch follow an industrial model in which organic nature is disassembled into pieces and then reassembled into new mechanical entities. Living forests are stripped of trees and sawn into lumber. Gypsum is mined, combined with cellulose, and formed into gypsum board. New substances are made by rearranging hydrocarbon molecules obtained from petroleum feedstocks. Water is piped for use from far-off sources. Coal and oil are burned to make electricity to operate the machinery of our buildings. The living world and its people are resources with which to build and operate the economic system, of which culture is a part.

I believe we are at the edge of a new epoch. This will be an epoch in which the guiding principle is ecologic. Historically, each epoch incorporates and transcends the previous one. Thus, agricultural empires incorporated the oral traditions and sense of the sacred living world of the earlier tribal cultures. And the Secular Corporate Age incorporates the hierarchical organizational forms of the preceding agricultural empires. The Ecologic Epoch will see a resacralization of the living world in which life forms are more than resources—they are also our relatives. The new epoch will not reject science or technology but bring them into a context where phenomena are understood as parts of a systemic whole that includes the spirit of the whole.

That natural building principles, building systems and materials are the subject of this book is one of many signs that a new epoch is coming into being. Fifty years ago, anyone who proposed using the natural building materials discussed here would have been drummed out of the design and building professions. On a trip to South Dakota State University last year, I was shown an impressive rammed earth wall around the president's house. A bronze plaque read,

"EXPERIMENTAL PISÉ OR "RAMMED EARTH" WALL
built for the purpose of studying different methods
of bonding stucco to earth walls
by the Department of Agricultural Engineering
South Dakota State College Experiment Station
1934

Curious about the person who had initiated the project, I was told that his research career had been ended by this then unorthodox project, even though the experiment had been a complete success.

How does one explain the current growing interest among professionals and the public in natural building materials and systems? Perhaps the primary reason is the environmental benefits of using materials that are generally considered waste, such as low-grade soils and straw. The traditional North American light building material is wood, and, although wood remains an excellent building material, we are using it far faster than it is being replaced. Almost all the old growth forests in the United States have been logged, and we are seeing fewer primary wood products used in commercial housing.

Natural building materials appeal to people who are looking for alternatives to industrialized materials because of their environmental impact, their cost, and their appearance. In the 1960s, I often heard the song about, "little boxes made of ticky tacky . . . and they all look the same." The house, besides being our largest personal economic investment, is also our home— our symbol of self and our largest emotional investment. Not many people want to be fitted into some developer's, builder's, or market researcher's concept of who they are.

Finally, and most important, is the intangible quality of spirit we find in natural materials and systems. I believe people are drawn to natural materials because they stand in opposition to the industrial appetite for disassembling the organic, for destroying the soul and spirit inherent in living materials. For a long time, machine-made perfection was the aesthetic standard. Now that may be shifting to an appreciation for the irregularity, the softness, the handmade quality and workmanship of natural building materials. The shiny, smooth machine surfaces of mechanical materials do not age or wear well. They oxidize and stain. Their joints buckle and leak and deteriorate. Natural materials return to the earth more slowly and with more grace.

The use of natural building materials and systems is an important step in our move to an Ecologic Epoch, in shifting from a mechanical world-view to an ecologic world-view. We are slowly and painfully learning that the best path to a livable future requires that humans create ways to co-evolve with nature rather than destroy the living systems on which our life depends. Whether you are a building or design professional, a student, or an interested reader, the principles and technologies presented in this book can be a useful companion and guide on our common journey toward an ecologic future.

—Sim Van der Ryn

Preface

A mysterious and intangible force powers what has become known popularly in North America as "natural building." This indescribable energy motivates growing legions of urban architects to reassess their sophisticated habits, to turn off the computer, unlace their shoes, and stuff their daytimers with straw. The passion is essential, elementary, and basic as breath. It springs from subtle yearnings and a conscious desire to reconnect with nature.

Something feels so good about bare feet on sun-warmed adobe floors, so right about the shimmer of mica in a mud-based plaster. The gentle undulations of hand-shaped walls so sensuous and comforting, a bright mosaic on the kitchen splashboard, hearth-stones lugged from the creek—surely this is the way we were meant to live.

Once smitten, these people are never the same. They hunger for understanding of indigenous craft works, master building, and native ken. Beginning their practices all over again, they apprentice themselves to the earth. They start fresh, with the materials at hand—the clay of old puddles, straw from the field, reeds of the marsh. They look at their notions of how one should design, how one should build. What they once had to calculate, now they intuit. They feel with their feet, their fingers, their noses. The cob has enough sand. The slip is ready to pour. It is time to harvest the bamboo. They reconnect with the flow, the rhythm of nature. They rediscover their center, and smile at their handiwork.

This book on alternative construction methods was conceived in response to the growing need for a single text on the numerous and proliferating earth- and straw-based building systems in the United States. It provides a guide to the non-industrialized materials and assemblies currently evolving as ecological, low-impact options for residential and low-rise construction. *Alternative Construction* is not intended to replace the specialized technical manuals and books that have been published for some of these individual building systems, but rather to offer a full-breadth understanding of the various options for natural building in this country.

More than seven basic methods are addressed, with performance characteristics and construction techniques for each. The book includes, for each genre, a brief history of development, a review of environmental benefits, and considerations for application. For several systems the book also provides an analysis of bioregional appropriateness.

Each chapter has been written by an expert or team of experts—in some instances the original pioneers—in an individual building method. As such, this compendium work of more than two dozen contributors flows more like a natural builders' conference than a highly structured textbook. While each author covers the essential aspects of building with one particular system, each emphasizes what he or she thinks is most important, maintaining the color and richness of his or her unique perspective.

These alternatives to conventional wood-frame and steel-frame construction include adobe, cob, rammed earth, earthbag, light-clay, strawbale, and hybrid systems. A bonus chapter on bamboo construction suggests applications for its use in North American architecture.

The book opens with a report on recent reforms within international building codes that are moving toward more sustainable and performance-based building standards. This is followed by chapters on essential ecological design principles, including one on natural conditioning that features modeling and comparisons of the thermal performance of several earth, straw, and hybrid wall assemblies. The eight chapters of Part II, "Systems and Materials," constitute the heart of the book.

Part III presents case study applications in the United States and other countries, including projects that have synthesized hybrid methods. Several chapters address the cultural challenges of introducing alternative systems in other countries, in some cases newly engineered versions of indigenous systems that had been abandoned and, in others, systems that are entirely foreign to the local culture. Here, too, can be found practical principles for such technology transfer. The book closes with two appendixes: a substantial section on recommended references and a listing of domestic and international education centers for ecological and alternative building.

When the International Conference of Building Officials, source of the Uniform Building Code, devoted an entire issue (September-October 1998) of its national journal, *Building Standards*, to alternative building materials, it was an indication that these options were being taken seriously by the mainstream building industry. As this book goes to press, ICBO has released its January/February 2000 issue of *Building Standards* with no less than seven articles addressing alternative, natural construction technologies. Indeed, natural building systems offer practical solutions to many of the ecological, social, and economic problems looming at the opening of the twenty-first century.

On behalf of the contributors, we wish you good reading and hope you find this collective work both enlightening and useful.

Acknowledgments

This edited volume represents the collaborative effort of many—most notably its 32 contributing authors. We are deeply indebted to each of them for believing enough in the value of this project to set aside demanding careers to document what they practice. Trailblazers that they are, they have been an inspiration to us, as they are to one another, and we are all beneficiaries of the camaraderie and commitment to sharing knowledge that exists within the natural building community. The contributors have not been paid for their writing time, and because their regular work is often unfunded applied research driven by little more than visionary zeal, they deserve extra recognition for their generosity. In the same spirit, a percentage of the book's royalties will support a natural building alliance fund to further educational efforts, advocacy, and research in the field.

This book might not have been published without the support of Dan Sayre, original editor of the Wiley Series in Sustainable Design, who brought the book prospectus into contract. We are also grateful to Janet Feeney, another former editor at John Wiley & Sons, who shepherded the publishing process for several months. Senior Editor Amanda Miller must certainly be thanked for her wise guidance as well as her patience and faith, as we missed more than one manuscript deadline.

There were others besides the authors who gave unstintingly of their talents. Dietmar Lorenz translated the text of German author Frank Andresen. Sandra Leibowitz, compiler of the original guide to *Eco-Building Schools*, updated entries to the lists of resources Appendix B. David Kibbey, primary editor of the *West Coast Architectural Resource Guide*, published by

Architects, Designers, Planners for Social Responsibility (ADPSR), offered generous access to ADPSR's database of ecological building resources. Jeanette Owen-Kennedy helped verify addresses, and Greg Van Mechelen contributed numerous titles to the recommended references.

Photographs to illustrate the chapters were provided not only by the individual authors and their colleagues, but also by good friends and others who became friends in the process. All of these helpmates are extraordinary craftspeople who have made their own significant contributions to the natural building movement—Catherine Wanek, Robert Bolman, Kiko Denzer, Frank Meyer, Tom Wuelpern, Tara Teilmann-Way, Mike Carter, Christina Bertea, SunRay Kelley, Jon Hammond, David Bainbridge, David Arkin, Greg McMillan, Turko Semmes, John Swearingen, Thom Wheeler, C. E. Laird, and John Beck. We wish the book had room for their stories as well.

Coordinating contributors' chapters behind the scenes, editing text, and assembling photographs for their partners and workmates were Cynthia Wright, Margaret Caffey, Iliona Outram, and Alix Woolsey. We do appreciate their help.

Our special thanks go to Terry O'Keefe, who appeared like an angel in the last weeks of the manuscript's editing, to recraft uneven, if not confusing, parts of the text.

Many others who did not contribute directly should be recognized for their good advice and most especially for their expertise, from which the authors and we have drawn. These include Colombian bamboo architects Simón Vélez and Oscar Hidalgo, bamboo builder Kyle Young, underground architect Malcolm Wells, environmental architect Carol Venolia, Baubiologists Helmut Zcihc and Panther Wilde, and green building researchers Nadav Malin, Gail Vittore, Ann Edminster, Ray Cole, and Bill Browning. We appreciate Brady Williamson, civil engineering professor at the University of California at Berkeley, who generously lent his laboratory for two straw-bale wall fire tests.

In particular we are indebted to international earth building specialists Hugo Houben, Gernot Minke, and Franz Volhard for their valuable research, demonstration projects, and documentation. We are also indebted to this movement's grandfathers—Egyptial architect Hassan Fathy, who recognized a half-century ago the value of traditional construction methods to modern society; British-born architect Laurie Baker, who dedicated his life to the building of low-cost, earth-based housing for the poor in India; and the Aga Khan, supporter of Muslim architecture, some of the most majestic earth building in the world. We also thank Paul Oliver for his invaluable documentation of indigenous architecture, and offer our respect to the late Bernard Rudofsky for his appreciation of the vernacular.

Among the pioneers of alternative construction in the United States, we would like to recognize those teachers of the teachers, who have significantly expanded the boundaries of what is buildable—Pliny Fisk III, Nader Khalili, Judy Knox and Matts Myhrman, Ianto Evans and Linda Smiley, Paul McHenry Jr., and the late Ken Kern. May this book be a tribute to their vision and power to launch a movement.

To natural builders, throughout the land, who bring their skills to workshops, job sites, and wall raisings, we offer heartfelt thanks for the courage to build a sustainable future—your enthusiasm enlivens us all.

part I

DESIGN PRINCIPLES

1

Introduction

The Natural Building Movement

Lynne Elizabeth

Perhaps the soul could remember a little of its origination,
when people still belonged to the spirit of a place.

—Martín Prechtel, *Secrets of the Talking Jaguar*

Natural building in the United States is not just a phenomenon, it is a movement—a movement most visibly represented at the dawning of the twenty-first century by a particular set of non-industrialized construction technologies used primarily for residential applications. These include the traditional and modern earth- and straw-based building systems written about in this book, plus timber framing, stone masonry, and numerous indigenous forms.

Natural building is about far more than materials and wall assemblies, however. It encompasses a broad set of ethics, underpinned by a worldview

Alternative Construction: Contemporary Natural Building Methods, edited by Lynne Elizabeth
and Cassandra Adams ISBN 0-471-24951-3 © 2000 John Wiley & Sons, Inc.

that treats the earth as not only sacred, but alive. Its proponents concern themselves with what constitutes a healthy built environment, how to build with the least impact on the earth, and ways in which the built environment can nurture vibrant community. Natural building aligns itself with philosophies of holistic, integrated systems, such as Bill Mollison's Permaculture, Rudolf Steiner's Anthroposophy, or the German Bau-Biologie. Structures are understood not as isolated entities, but as parts of and within interdependent systems for providing shelter, food, clean water, energy, and waste recycling.

In contrast to a pervasive dependence on mechanical heating, cooling, and ventilation that consumes vast amounts of polluting energy, naturally conditioned buildings are designed with sensitivity to the site, the sun, prevailing winds, and the seasons. They offer healthful, inexpensive comfort and preserve the tranquility of our interior spaces. Daylighting is favored over artificial lighting, as is architecture that integrates buildings with their natural surroundings.

The movement is also imbued with an aesthetic appreciation of building materials in their unprocessed or minimally processed state—the beauty of raw earth, uncut stones, unmilled wood, and woven grasses. Architecture is inspired by natural flows, patterns, and an indefinable spirit of place. These values of harmonious ecological design have been popularized by the pictorially rich books of architect David Pearson, such as the *Natural House Book* and *Earth to Spirit*.

Many within the movement hesitate to call it "alternative," lest it be perceived as questionable or in any way be hampered from entering the mainstream. There is hope that architectural historians will look back at this time and note the widespread appearance of natural building as the beginning of a new construction era based on principles of ecological balance.

"Natural" as the norm may, indeed, not be far off. In October 1999, the American Institute of Architects Committee on the Environment held a conference in Chattanooga to explore "Mainstreaming Green." "Green" architecture has, in the last dozen years or so, grown from a minor pocket to an enormous presence. During this period it has defined itself largely as conventional construction that has been improved to meet higher environmental standards—standards that in the eyes of many natural builders often represent compromised solutions rather than those reflecting a deeper ecological awareness. All views considered, the territories of "green" and "natural" do overlap.

The 1990s have also witnessed the rise of the cult of "sustainability," which popped into popular parlance during that period and is now used to describe almost any enlightened response to environmental, economic, or social concerns. As much as it can be understood, sustainability is rapidly being adopted by nearly every civic institution as the ultimate policy. At its core lies a recognition that the prevailing operating system of our society is not capable of being maintained at its current pace or in its current form.

Some describe the problem as being out of step with nature; hence arose, also during the last decade, a deductive scientific movement from Sweden called the Natural Step. Its mission is to adjust the misalignments of our industrialized culture with natural operating systems.

The values of natural, green, and sustainable development, then, go hand in hand as guides for ecologically sound construction practices.

The Environmental Imperative

To baby-boomers, it is usually a surprise to learn that lightweight wood framing has become the predominant building method in the United States only since the end of World War II, when returning GIs and a flourishing economy latched onto it because it was expedient and cheap. Stick-frame, as it is popularly known in the United States, has only very recently been adopted to any extent in other parts of the world. At a time when forests throught the world are being clear-cut at unprecedented rates, it is tragic that wood framing should now take the fancy of builders in wood-poor countries—countries where masonry and other indigenous building systems have predominated within a more or less balanced ecology for centuries, if not millennia.

The ecosystems on our planet most discernibly threatened by human exploitation are the forests. We have lost nearly half (46 percent)—3 billion hectares—of the forests that originally blanketed the earth, and deforestation continues to expand and accelerate. Most of this forest cover was cleared during the twentieth century for timber or to convert land to other uses. Between 1980 and 1990 alone, 200 million hectares—together equivalent to an area larger than Mexico—were destroyed. The World Resources Institute has reported that only 22 percent remains of the world's irreplaceable "frontier forests"—areas of "large, ecologically intact, and relatively undisturbed natural forests." Within the temperate zones, that encompass much of the United States and Europe, the percentage of remaining frontier forests drops to 3.[1]

Ancient forests support roughly half the world's biodiversity; they also renew our air, stabilize our climate, and maintain our watersheds and soils. Most people take these and many other benefits of forests for granted; they consider trees valuable for fuel, construction, and paper.

Wood frame residential construction in the United States is a leading cause of global deforestation. Forty-five percent of all the wood harvested in the world in 1995 (3.33 billion cubic meters) was used for industrial round-wood—this is the wood that is used to make lumber, paper, plywood, and similar products. Nearly one-quarter of that roundwood is consumed in the United States, and 40 percent or more of this is used for construction. Ultimately, about 10 percent of the world's industrial roundwood is used by the U.S. construction industry, and most of that for residential buildings.[2]

Despite the critical need to stop this voracious forest consumption, the warning signs that filter into the construction market—diminished quality of lumber stock and higher prices—are minimal. They give little if any incentive for significantly changing building practices.

Organizations such as the National Association of Home Builders and the Natural Resources Defense Council have published recommendations for reducing wood demand, which include more efficient framing techniques and engineered wood products. Specifying lumber from sustainably managed forests is gaining more awareness as an important solution, as are salvage and recycling options. Considering, however, population growth and the fact that the size of the average single family home in the United States has more than doubled since 1950, all these measures for improving wood-frame building, even when combined, appear stop-gap at best.

Other insidious threats to health caused by industrialized construction include toxins emanating from buildings and pollution generated by the extraction and manufacturing of building materials. Transportation of the raw materials that go into building products and transport of the products themselves to construction sites are contributors to energy consumption and pollution of all kinds. These issues are well documented in a growing body of literature addressing the ecology of the built environment (see Appendix A).

Building with locally derived, unprocessed materials—materials as simple as the soil beneath our feet—is a natural response to this crisis. It significantly reduces the amounts of energy and secondary resources needed for extraction, processing, fabrication, and shipping. Rammed earth, adobe, cob, light-clay, and straw-bale wall systems can abate our demand for wood.

Coupled with vaulted, domed, or bamboo roof systems, these alternatives can significantly reduce reliance on wood. Designed with natural heating, cooling, ventilation, and lighting systems, such structures can substantially lessen our consumption of energy and resources and eliminate much pollution.

Reducing building size, designing with sensitivity to the site, and clustering development to preserve open space and lessen infrastructure demands are additional strategies for improving the ecology of our built environment. These are approaches the natural building movement has brought to the fore, but are also strategies that can be employed with any kind of development.

With so many obvious benefits for our local and global health, non-industrialized materials and systems are receiving wide recognition as solutions. Articles on the subject are appearing with increasing frequency in mainstream media such as the *New York Times*, the *Wall Street Journal, Good Morning America, National Geographic,* and *Metropolitan Home.* In addition, a rising number of trade periodicals cover alternative construction (see Appendix B), and a national consumer magazine was launched in 1999 called *Natural Home.* Architecture schools, too, are now beginning to teach alternatives that utilize earth-based or indigenous systems.

It also seems possible that the residential builders in this country, despite their vast numbers and entrenched habits, may smell the danger ahead and, like a herd of hoofed mammals all charging in the same direction, suddenly change course altogether. Perhaps, less dramatically, they will discover the greener pastures of natural building and migrate for many positive reasons. To some extent, the public is already demanding that they do so.

Building Craft, Building Community

Much of the natural building movement is underpinned by a renaissance of the blended trade and profession known as the master-builder. Christopher Alexander wrote about it more than a decade ago in *The Production of Houses*; essentially, it is a shift away from highly specialized roles—the architect as a conceptualizer and draftsperson and the contractor as a narrowly defined construction tradesperson—toward more overlapping, if not entirely enmeshed, roles. This holistic approach to design and building offers closer and more creative kinship with materials. The architect is not divorced from the medium and the contractor is not working under the restriction of faceless blueprints. It allows better response to subtleties of the site and the interface of building forms. It respects the unique talents of each participant in the building process and affords opportunity for greater self-expression in planning, execution, and embellishment. A master-building climate encourages innovation and is the ideal setting for the growth of alternative building methods. It has also spawned a revival of building as craft.

Pioneers of newly evolving alternative construction methods are conducting workshops and classes wherever interest springs up around the country. Timber framers, cobbers, thatchers, and many other experts in traditional building techniques have also been engaging in a great interchange of construction know-how here and abroad. Natural and traditional building schools have been proliferating (see Appendix B, "Alternative Construction Resource Centers"), and there is now an entire second generation of Americans trained to teach alternative methods.

Gatherings and conferences for the purpose of exchanging building technologies have also emerged in the last decade, most notably in the western states but now in the eastern states as well. The largest of these have become known as the Natural Building Colloquia, where during a week of long days the champions of all types of appropriate and intermediate technologies work together on experimental structures, teach newcomers, and share with colleagues what they have learned in the past year. A valuable cross-fertilization results, and several hybrid systems have been developed.

Probably nothing has nourished the growth of natural building more than the camaraderie and robust community spirit of these trailblazers, artisans, scientists, and seekers. Natural building attracts those wanting to build a healthy and healthful community. These new values include sharing the work with all, to cross gender, race, religion, age, skill-level, and just about any other social boundaries. The movement also supports self-help and community-supported building, which is sorely needed in a world of regulated, restricted, and exorbitantly expensive real estate development.

If non-industrialized building methods can be respected in this most industrialized of countries, vernacular methods stand a chance of being valued in other countries as well, and much of the beauty and wealth of human cultures can be maintained. Low-impact construction should not be associated with poverty; on the contrary, simple and regionally appropriate construction offers great freedom of expression and allows us to live closer to the riches of nature.

Mass-produced housing robs our neighborhoods of local color and our tradespeople of meaningful work. In contrast, supporting local building crafts enlivens the culture, and building with the materials and talents of the region strengthens local economies.

* * *

The job of shaping the built environment comes with a responsibility beyond the wants of the paying client, and beyond our personal wants as well. May the wisdom that we bring to our practice include an understanding of the effects of our building designs and materials choices on all beings now alive and their descendants.

Clearly, it is easy to be caught up in the concerns of the hour, the fashion of the year, and the powerful thrust of our cultural habits. To work from an awakened perspective is to feel the joy of being alive. May the world we build express that joy.

Notes

1. Janet N. Abramovitz, *Taking a Stand: Cultivating a New Relationship with the World's Forests* (Washington, D.C.: Worldwatch Institute, April 1998). Her primary sources were the report, "Frontier Forests," (Washington, D.C.: World Resources Institute, 1997) and data from the Food and Agriculture Organization of the United Nations.
2. Janet N. Abramovitz and Ashley T. Mattoon, "Reorienting the Forest Products Economy," *State of the World 1999* (New York: W. W. Norton, 1999).
3. Ibid.

The Realities of Specifying Environmental Building Materials

Cassandra Adams

The construction technologies being developed and refined by the architects, artists, owners, and builders featured in this book are their responses to environmental, ethical, and social issues surrounding the extraction of raw materials from nature and their use in construction of the built environment. Although these building materials and methods have traditionally been considered "primitive" and therefore inferior to more highly processed materials in terms of safety, durability, performance, occupant health, and comfort, the stories and photographs in this book provide convincing evidence otherwise.

With respect to environmental issues, consumption of building products and energy within the construction industry has created a significant demand for raw materials (both recycled and virgin) and for energy production, thereby contributing to the many environmental problems associated with the extraction processes (environmental degradation, loss of genetically diverse ecosystems, etc.) and with energy production (polluting by-products emitted into the air/water/soil, which become part of smog, acid rain, global warming, etc.). In addition, the toxic particulates and gases incorporated into building products (especially interior finishes and furnishings) during manufacture are emitted later, degrading interior air quality and contributing to health problems of those with environmental illnesses.

Ethical questions are raised by the fact that the average lifestyle of people in affluent nations directly impacts the lives of the world's poorest people, both to their benefit and detriment, by creating a demand for the export of their resources and agricultural products. In addition, the boom-and-bust type of economy that often accompanies timber and mineral extractive industries as they move from one site to the next is often disastrous for the stability of local communities, especially those that have traditionally depended on nearby forests for their livelihood. This condition occurs in industrialized and developing nations alike.

Social benefits accrue from the reaffirmation of communal bonds by those who participate in community construction projects (professionals as well as lay persons) or, as in Obregon, Mexico, where the process has led to improved economic opportunity. Another social benefit is the personal satisfaction associated with the experience of "making," the joy of working with one's hands and with sensual materials. Similarly, the aesthetic potential of these materials is considerable, varied, and unique to these materials; as is shown in the elegant simplicity of David Easton's structures, in the sensual shapes and textures found in Carole Crews' decorated walls, and in Simón Vélez' breathtaking bamboo cantilevers.

Strategies for reducing negative environmental impacts and for promoting positive impacts are not always stated explicitly by the architects, artists, owners, and builders in this book, but their presence can be seen in their work. Common to all the projects described in this book are the twin goals of broadening the "palette" of raw materials suitable for construction (thereby lessening the demand on existing supply sources) and the reduction of energy embodied in the production, manufacture, and transport of materials. Additionally, many of the buildings shown in these pages have passive heating, cooling, and daylighting strategies integrated into their design, and, in some, the yearly energy consumption is far below current energy-efficient design standards. Also important is the use of interior finishes that emit few (if any) volatile organic compounds (VOCs), although it should be noted that the emission of particulates from earthen finishes can sometimes be high.

Consumption Patterns

The construction industry's concern with energy and resource consumption is due to the fact that it has contributed significantly to overall consumption patterns. In 1997, according to the U.S. Commerce Department, about 36 percent of total energy use in the United States was consumed in the operation of commercial (16 percent) and residential (20 percent) buildings. This figure represents almost 9 percent of total worldwide energy use for that year and is close to the amount typically expended yearly for world cement production. For a comprehensive energy picture, one must also add the significant amount of energy expended for the construction process itself and for the production of other building products besides cement.

Materials consumption by the construction industry is even higher than its energy use. William Rees at the University of British Columbia estimates that 40 percent of materials consumption worldwide is for the construction and repair of the built environment. Table 1B-1 illustrates the magnitude of construction consumption of selected resources in the United States.

Resource consumption can also be examined from a land-use perspective, which expands our understanding of the broader environmental role of building materials. Rees developed a method to estimate the amount of land needed to support the lifestyles in various cultures. He incorporates energy into his calculations by balancing the carbon emissions from energy consumption with the hectares needed for an equivalent carbon sink (1 hectare = 2.47 acres).[1] The ecological footprints of selected countries are found in Table 1B-2. Note especially the ecological deficits in many of the larger and faster-growing nations.

Table 1B-1. Consumption of Selected Resources for Construction

Raw Materials	Recycled from Scrap	End Use	Consumption
Aluminum	20%	Transportation	36%
		Packaging	25%
		Construction and electrical	14%
		Electrical	8%
		Consumer durables and other	17%
Asbestos	Insignificant	Roofing products	48%
		Friction products	29%
		Gaskets	17%
		Other	6%
Cement	Small amount of concrete	Construction	100% (total)
		Readi-mix concrete	70%
		Concrete products	10%
		Road-paving contractors	10%
		Other construction	10%
Clays	Insignificant	Construction	55%
		Paper	13%
		Foundry and nonconstruction refractory	8%
		Other	24%
Copper	14%	Construction	42%
		Electric and electronic	25%
		Industrial and transportation	24%
		Consumer products	9%
Crushed stone	Insignificant	Construction	83%
		Chemical and metallurgical (includes cement and lime manufacture)	14%
		Agricultural and other	3%
Gypsum	Small amount	Construction (wallboard and cement)	81%
		Agricultural	10%
		Other	9%
Sand and gravel	Limited pavement recycling	Construction	97%
		Industrial	3%
Steel	61%	Warehouses and distributors	21%
		Construction	14%
		Transportation	13%
		Other	52%

Source: USGS.

Table 1B-2. Ecological Footprints of Selected Countries[2]

	Population in 1997	Ecological Footprint (in ha/cap)	Available Ecological Capacity (in ha/cap)	Ecological Deficit (in ha/cap)
		(All expressed in world averge productivity, 1993 data)		
WORLD	5,892,480,000	2.3	1.8	−0.5
Bangladesh	125,898,000	0.7	0.6	−0.1
Brazil	167,046,000	2.6	2.4	−0.1
Canada	30,101,000	7.0	8.5	1.5
China	1,247,315,000	1.2	1.3	0.1
Egypt	65,445,000	1.2	0.6	−0.5
Ethiopia	58,414,000	1.0	0.9	−0.1
Germany	81,845,000	4.6	2.1	−2.5
India	970,230,000	0.8	0.8	0.0
Indonesia	203,631,000	1.6	0.9	−0.7
Japan	125,672,000	6.3	1.7	−4.6
Mexico	97,245,000	2.3	1.4	−0.9
Netherlands	15,697,000	4.7	2.8	−1.9
New Zealand	3,654,000	9.8	14.3	4.5
Nigeria	118,369,000	1.7	0.8	−0.9
Russian Federation	146,381,000	6.0	3.9	−2.0
Thailand	60,046,000	2.8	1.3	−1.5
Turkey	64,293,000	1.9	1.6	−0.3
United Kingdom	58,587,000	4.6	1.8	−2.8
United States	268,189,000	8.4	6.2	−2.1

Note: Population figures are taken from the World Resources Institute, 1996. *World Resources 1996–1997 Database*, Washington, D.C.: WRI. file "hd16101.wk1".

Environmental Assessments

As noted by some of the authors in this book, the environmentally conscious building material specification process is more complex than simply making decisions to incorporate recycled and low-embodied-energy materials or to use materials obtained locally. This is due to the fact that every building material, every building system, and every construction practice impacts the natural environment in numerous ways at every stage of its life cycle, beginning with resource extraction and ending with building demolition and recycling of the debris. Every design decision involves an environmental compromise, thereby requiring the designer or builder to evaluate and compare the environmental impacts that occur throughout all the life cycle phases. The necessity of having to make choices is directly related to the fact that "environmentally conscious" design is *not* the same as "sustainable" design.

Currently, the most widely used method for evaluating a building's environmental impact is to conduct an environmental assessment of its life cycle, where the inputs and outputs of energy and resources are identified and quantified for each phase. The phases typically considered include raw materials extraction, processing and manufacture (this may involve several steps), onsite construction, occupancy, demolition, and debris disposal or recycling. Inputs occur during each phase and include all materials and all the process and transport energy. Outputs also occur in each phase, and besides the "product" itself, they include waste energy (such as heat or noise) and by-products (both polluting and nonpolluting). Life cycle assessment methodologies and software are in development for evaluating the environmental impacts embodied in a building's materials and/or for assessing the environmental quality of a building. Some in use include the U.S. Green Building Council's LEED Program, BEES 2.0, Athena, BREEAM (UK), and EcoHomes (UK).

One might argue that houses should be exempted from this lengthy and time-consuming environmental evaluation process due to the fact that they are much smaller and less resource and energy intensive than larger structures. However true this might be, the fact is that residential construction comprises 40 to 60 percent of construction expenditures in the United States (depending on the economy) and residential buildings (in aggregate) consume 30 percent more energy per year than do commercial buildings. These figures suggest that this evaluation process should not be waived.

Many of the existing assessment programs have been developed to the point where they are now (or soon will be) able to estimate *quantifiable* environmental impacts related to energy-consumption, resource quantities, and carbon-cycle effects. However, some important environmental impacts are not quantifiable in terms of dollars, energy, carbon, etc., and/or are difficult to compare. How does one value the impacts on human life and health, on loss of genetic diversity, on ecosystem degradation or destruction, or on the effect of climate change on agricultural patterns? Some assessment programs address these issues in terms of checklists or sliding-scale rating systems, but work continues on the development of more sophisticated assessment methods that address more of this complex mix of variables.

In addition to the quantitative and qualitative issues described above, there are some other issues that should be considered. These are discussed below.

PRIORITIZATION OF ENVIRONMENTAL IMPACTS Environmental impacts are *not* all equal. Some are more critical than others and should be given more weight. Important prioritizing factors include:

- ❖ *Sphere of influence.* Some impacts have a more widespread area of influence than others (global warming versus streams siltation).
- ❖ *Duration.* Some impacts last only a few months or years while others continue forever (nuclear waste dumps versus patchwork clearcutting).
- ❖ *Magnitude of risk to human or ecosystem health.* Some impacts have severe consequences for human or ecosystem health while others have little or no effect (toxic waste dumps versus well-managed municipal landfills). Sometimes the same behavior will have different effects (the reaction to VOC emissions by healthy people versus those with environmental illnesses).
- ❖ *Reversibility.* Some impacts are irreversible while others are technologically possible to repair (destruction of a genetically diverse ecosystem versus reclamation of a former strip mine site).

IDENTIFICATION OF CAUSAL FACTORS It is important to identify which relationships between environmental problems and a particular design tradition (or construction behavior) are causal relationships; that is, where the environmentally destructive behavior is driven by construction industry demand and where the discontinuation of the practice will improve the environmental situation. If there is a direct causal relationship, then the practice should be avoided or mitigated. However if the driving force comes from some other segment of society and the construction industry is only making efficient use of leftover wastes, then a potentially bad environmental practice becomes environmentally beneficial. An example illustrating this point would be the factors underlying the destruction of world forests. In those regions where agricultural conversion is the primary motivating factor behind forest destruction (such as in some tropical forests), it would be better to use the timber than to burn it. This is the reasoning that supports the broadening of tropical species utilization. In the United States, where demand for construction lumber is the motivating factor for forest destruction (such as in Pacific Northwest forests), the appropriate environmental response is to reduce wood consumption in housing.

RESOURCE INTENSITY Some building products consume less raw material than others in fulfilling the same use. The weight of a 2 × 14 joist required to span a given distance is greater than an engineered-wood I-joist used for the same purpose, therefore some preference should be given to the product that uses raw materials more efficiently. This consideration is more important for products whose raw materials are in limited supply than for abundantly available materials or materials found on site. However, it should be noted that even soil can be in limited supply, as is the case in China, for example, in some of its agricultural regions.

ASSESSMENT BOUNDARIES An environmental assessment must be comprehensive and inclusive, because a single-issue environmental decision can conceivably be worse for the environment than the "standard" practice. Furthermore, the scope of the assessment must be meaningful with respect to its greater context. Energy consumption is a case in point, because this is one place where environmentally conscious design practice can fall short. Currently, the scope of energy assessments is typically restricted to the traditional scope of the design professions; that is, within the building envelope and on the building site. The problem here is that every bit of energy savings designed into a project by careful materials selection (that reduces embodied energy) and by careful design of thermal conditioning systems (that reduce operational energy) can be easily and quickly overridden by a poor choice made during the site selection process and the resulting energy-related transportation issues.

The Issues There remains a fundamental limitation in the methodology used for assessing the "environmental impacts" of building materials. Current assessment practices work on a building-by-building basis with the implicit goal of the assessment process being an improvement over previous buildings. This is essentially the same method that is used for energy consumption, where the ultimate goal of "sustainable energy use" would be for the building to be off the grid or even to sell energy back to the electric utility. The analogous case for building materials would be for the building to be equipped with the means to extract all its raw materials on site. Obviously this will only be the case for a very few privileged buildings in rural or campuslike settings, which means that existing mines and forests must be shared. The "sustainability" of building materials, then, cannot be determined from an environmental assessment. It can only be determined on an industrywide basis, where the total demand for resources can be balanced with the available supply sources and maximum allocations made to each economic sector. This ideal would require resource management and cooperation among industries of a kind that is almost totally lacking. At present, there is one notable exception to this. Within the U.S. timber industry, the nonprofit Forest Stewardship Council has begun to certify forests that are managed by their owners in a sustainable manner (in which annual growth exceeds annual harvest) and with environmentally sound practices. Certified sustainable forest management currently is being practiced in only a small percentage of U.S. forests, so its importance (so far) is an example of what truly sustainable management practices can look like.

Finally, a word of caution to our readers. It is important to emphasize that the construction methods described herein vary in terms of their development to meet current standards of health, safety, and performance. David Easton, for example, has spent several decades developing rammed earth and PISÉ construction in California to the point where his local building officials feel comfortable issuing permits, a condition that also applies to a few other locales in this country, Europe, Australia, and elsewhere. The same situation also exists for adobe and straw-bale construction in some locales.

Other construction methods, however, are presented in earlier stages of their development, so the performance of these materials and methods over longer periods of time, in all climates, and for all structural conditions is still not completely understood nor worked out. Several have not yet been adopted into the building codes, but they hold promise—and it is our hope that their presence in this book will inspire some readers to contribute to their further development. The unknowns about the performance of bamboo, straw walls, long bags, tires, composite wood-cement blocks, etc. will undoubtedly become known and design standards and construction methods will eventually be agreed upon. Meanwhile, in a litigious society like ours, it is important for all parties (designers, builders, *and* owners) to clearly communicate to each other what the unknowns and the risks are and what the implications are for building durability and performance.

Notes

1. Wackernagle, Mathis and William Rees, *Our Ecological Footprint* (Philadelphia: New Society Publishers, 1996).
2. Wackernagle, Mathis, Larry Onisto, Alejandro Callejas Linares, et al. "Ecological Footprints of Nations: How Much Nature Do They Use? How Much Nature Do They Have?" (prepared for the 1997 Rio +5 Forum, The Earth Council, 1997).

A New Context for Building Codes and Regulation

David Eisenberg

The strong and growing interest in natural and alternative building methods today has generated a corresponding need for strategies to gain code approval for these alternatives. This chapter addresses big-picture strategies involving systemic change in the regulatory system, as well as specific strategies for individual projects.

To effectively deal with building codes, code organizations, and code officials, it helps to know a bit of historical background. Building codes were not brought down from the mountaintop on stone tablets, but their origins are thousands of years old. For several millennia, people have been devising rules to ensure that those who design and build for others will be held accountable for their work.

Hammurabi enacted the first known written building code in Babylon in 1758 B.C. This code was stringent by today's standards—not because it provided rigorous detail on proper building methods, which were not covered at all, but because of the penalties for the failure to build well. The code

Alternative Construction: Contemporary Natural Building Methods, edited by Lynne Elizabeth and Cassandra Adams ISBN 0-471-24951-3 © 2000 John Wiley & Sons, Inc.

mandated, "If a builder has built a house for a man and his work is not strong, and if the house he has built falls in and kills the householder, that builder shall be slain." The code further stipulated that if a house fails because of poor workmanship, the builder is responsible for correcting the bad work and paying for it out of his own pocket. One imagines that, given the penalties, builders paid attention to the quality of their work without a need for licensing, certification, or permitting. Of course, those penalties were probably also an impediment to innovation.

There were many intermediate steps between Babylon and the International Building Code. With the growth of cities came a need to regulate safer building practices, especially where the size or close proximity of buildings presented dangers to surrounding buildings and people. Preventing recurrences of major fires was the focus of most early codes, such as that enacted in A.D. 1189 in London, requiring official approval of common walls between buildings, and others banning dangerous construction like wood chimneys. Eventually, problems related to existing buildings resulted in codes to regulate systems such as fire escapes, ventilation, water supply, toilets, and stair railings. In 1905 in the United States, the National Board of Fire Underwriters, an insurance industry group, wrote the National Building Code. This code, devised to minimize the risks to property as well as to the occupants of buildings, succeeded in raising interest and support for the regulation of building construction nationwide and subsequently led to the formation of organizations of building officials.

By 1940, three major model code organizations had been established in the United States, each with its own code. The Building Officials and Code Administrators International, Inc. (BOCA, covering the northeastern United States), produced the BOCA National Building Code (NBC); the International Conference of Building Officials (ICBO, covering the western half of the country), developed the Uniform Building Code (UBC); and the Southern Building Code Congress International (SBCCI, covering the southeastern United States), published the Standard Building Code (SBC). Between them, these codes covered almost the entire country. The three codes, although very similar, had differences that were problematic for anyone designing or building in more than one region or manufacturing materials or equipment for national distribution.

In an effort to simplify the process and gain national consistency, the three model building code organizations formed the Council of American Building Officials (CABO) to coordinate changes and additions to their codes and to eliminate conflicts. CABO eventually developed the One and Two Family Dwelling Code and the Model Energy Code (MEC). For various reasons, BOCA, ICBO, and SBCCI formed a new organization in 1994, the International Code Council (ICC), to supersede CABO. The ICC was created to facilitate the development of a single, national set of building codes, the International Codes (I-Codes), the first edition of which was published in 2000. The I-Codes replace the individual codes that BOCA, SBCCI, ICBO, and CABO formerly produced. This set of codes includes the International Building Code, the International Residential Code (replacing the CABO One and Two Family Dwelling Code), the International Mechanical Code, the International Plumbing Code, the International Fire Code, and the International Energy Conservation Code (replacing the CABO Model Energy Code), among others.

The development of the I-Codes does not fully consolidate all of the codes or regulations of the United States. Many other organizations are involved with both these and other codes, such as the National Fire Protection Association (NFPA), which produces the National Electric Code (NEC) and other fire codes. Still other organizations develop standards for materials, equipment, methods, and testing that are referenced by the codes.

Creating the I-Codes also did not change how codes are adopted and enforced in the United States. Unlike countries with codes that are developed and enforced nationally, the United States continues to have local, county, or state code adoption. Now there is a single set of codes available for adoption nationwide, instead of the former regional codes, though a few states and local jurisdictions maintain their own, unique codes. Most jurisdictions retain the right to amend the codes and append local chapters or sections as they see fit. This maintains the opportunity to make codes responsive to local or regional materials, methods, and other considerations.

Performance Codes

An important initiative of the ICC code effort was to develop and publish the International Performance Code. This code provides an alternative path for approval based on performance criteria rather than on the prescriptive criteria found in the other codes. Prescriptive codes dictate in detail what must be done and how in the construction of a building and all of its systems. Performance codes describe what a building, material, component, system, or design must accomplish, rather than how it will accomplish it. The performance approach allows greater freedom in designing a system as long as the system meets the performance criteria. The prescriptive method limits variation and impedes innovation.

Each approach has both advantages and drawbacks. Performance codes, in providing much greater flexibility, also increase the burden on the designer, building official, and, frequently, the owner. The designer or proponent must demonstrate, through testing results, calculations, or other means, that the proposed approach is capable of meeting the requirements of the code. Usually, a licensed design or engineering professional must verify the adequacy of the method and perform any necessary calculations which adds to the cost of the project. The performance approach also requires the building department to interpret a set of criteria much different from those used with the prescriptive method.

For natural building proponents to use the performance path, the critical requirement will be to present credible information, such as test results, on which to base the calculations and other claims of conformance with the codes. The use of performance codes that require new test data can be very expensive and time-consuming. Because the prescriptive approach is faster and often less expensive, proponents of innovative systems, once they have gained wider acceptance through alternative or performance provisions, tend to develop prescriptive provisions to facilitate cost-effectiveness. Thus, for individual projects, the process for gaining approval through performance provisions is not necessarily less expensive or time-consuming than using the current alternative materials and methods provisions that exist in all the model codes. However, as the regulatory sector becomes more familiar with the performance basis for approval, gaining acceptance for alternatives through either method should become easier.

The Unintended Consequences of the Industrial Building Model and the Need for Change

It is estimated that more than 40 percent of the material resources entering the global economy today are related to the building industry, with the developed countries responsible for an enormously disproportionate share.[1] Buildings in the United States alone account for 10 percent of the world's energy usage.[2] The impact of these buildings is staggering and global. Current building regulatory practice creates risk at the highest, or global, level to avoid it at the lowest, or local, level. Few other human activities have this much negative impact on the planet and, today, more than at any other time in history, what happens in the most developed countries strongly influences what happens in the developing world.

While we are raising awareness about the need to make our buildings energy-efficient, we must not overlook the energy consumed and pollution generated to manufacture our building materials. Production of the 100 million tons of cement used annually in the United States emits an equivalent 100 million tons of CO_2. (Photo by Robert Bolman.)

Open-pit copper mine near Silver City, New Mexico. Our patterns of consumption and industrial processes create an enormous demand for natural resources and raw materials. Consider the investment of energy alone required to extract and process these resources. (Photo by David Eisenberg.)

Industrial building replaces labor intensity with resource intensity. In developing countries, where typically there is an abundance of low-cost labor, where resources and technology tend to be scarce and expensive, and where manual crafts and skills are often highly evolved, the trend toward industrialized construction is a recipe for economic, social, and environmental disaster. Yet indigenous methods are stigmatized by the perception that they are obsolete or used only by an impoverished society. Abetted by efforts to market modern building technologies in these countries, this attitude presents the largest barrier to the continued use of traditional construction.

The replacement of traditional, low-impact building methods with higher-impact industrial approaches in developing countries is a trend that has both benefits and dangers. The phenomenal growth of interest in natural building in the high-end custom housing market in the United States, coupled with appropriate improvements in traditional methods, can help to reverse this trend by demonstrating that those people who can afford to build anything, regardless of cost, are choosing natural building methods. Knowing that acceptance and widespread use of these methods in the United States can slow their rejection in developing countries and result in

Luxurious, multilevel, solar-oriented adobe residence in Albuquerque, New Mexico, by designer/builder C. E. Laird. (Photo by C. E. Laird.)

greatly improved lower-impact structures everywhere provides further impetus to support their acceptance and use.

At the dawn of the twenty-first century only one-third of the 6 billion people on earth live in buildings of manufactured or industrially processed components, 2 billion more live in earthen structures, and the other 2 billion live in other vernacular types of buildings or no buildings at all.[3] Coupled with projections that world population will be reaching 9 to 10 billion by the year 2050,[4] it becomes easy to see why maintaining the viability of low-impact alternatives is so crucial. Within the United States alone, at current building rates it will take less than 40 years to build half again as many buildings as exist in the country today.[5]

The sheer number of people on the planet and the intensity of the impacts of our buildings are forcing us to confront a very different set of issues than have been considered in the past. These broader issues are rarely raised within the regulatory community and, often, not even recognized.

Framing of typical code-approved house in the United States. The components of this building— almost exclusively processed or manufactured materials—have high levels of embodied energy, generate waste, use resources unsustainably, and often contain toxic compounds that outgas or leach out over time. Building regulations are beginning to recognize the need to ensure healthful indoor air quality. (Photo by David Eisenberg.)

Yet they are central to the task of finding sustainable strategies for building and development today, and they will remain the most critical issues for many generations to come.

Using Non-Industrial Building Methods in an Industrial Environment

The desire to use natural building materials and methods of construction in today's building projects often evokes resistance from a regulatory community that has little experience with non-industrial building systems. These systems, many of which have been in continuous use around the world for centuries, have nonetheless been judged inferior to their industrial counterparts and are considered obsolete in most of the developed world.

The rejection of non-industrial, nonproprietary materials and methods in developed countries has meant that they have been largely ignored in modern building codes. They have had no constituency in the code development process, so few formalized codes or standards have been written for them. However, a growing awareness of both the negative impacts of today's conventional building practices and the benefits of lower-impact alternatives has created a rapid increase in interest and sparked numerous grassroots efforts to gain acceptance for them.

Yet building codes alone do not control the building industry, nor do they determine what is designed or built. They are, however, the rules by which the design and construction industry must operate. The greatest

Interior of a prototype house using earthbag and soil-cement vault construction at Cal-Earth Institute, Hesperia, California. Nader Khalili's efforts over many years resulted in the development of earthbag structures, and then earth-tube structures and soil-cement vaults and domes. These buildings have now undergone structural testing, demonstrating that they surpass even California's high seismic structural requirements. (Photo by David Eisenberg.)

influence on building codes is the building industry itself. Industry proposes the majority of code changes, pays for the research and testing to back its claims, spends the most time and money supporting those code changes through the process of development, adoption, and enforcement, and works the hardest to make building codes and standards conform to its particular interests and needs. Building code organizations and officials are generally diligent in pursuit of their mission to protect the public from harm caused by the built environment. They strive to maintain fair and open processes that minimize the influence of special interests, yet they remain, nonetheless, responsive to and strongly influenced by the building industry.

Thus, the greatest challenge to gaining regulatory and institutional acceptance for natural alternatives lies in the industrial context in which building is viewed today. The industrial paradigm frames the whole field, demanding a high level of uniformity and predictability in buildings and their components. At the same time, this worldview creates an illusory set of expectations of universality and interchangeability that foster the belief that location, environmental conditions, and external or cumulative impacts need not be considered.

This industrial view requires that natural components mimic and exhibit the same degree of uniformity as industrial materials and methods. It is worth noting that the opposite was the case not long ago. As new industrial and synthetic materials and systems were developed and introduced, their proponents had to prove that they were not inferior to the familiar, age-old non-industrial resources. These industrial replacements have typically been developed to reduce the labor and costs of construction, not necessarily to improve the quality or safety of buildings. In fact, much of what has been added to building regulations in the twentieth century has been a direct response to this situation. Our comfort with many of today's materials and systems is the result of the regulatory response to their repeated failures and problems, leading eventually to defined limitations and accepted standards of practice. Their widespread use today is more a testament to the investment of money and effort in understanding and managing these problems than to their perfection or superiority over their non-industrial predecessors.

The Challenge for Current Practice

There are two primary avenues for introducing natural and alternative building methods to standard practice. The first is short-term and project-

Residence of Virginia Carabelli, built in Tesuque, New Mexico, in 1991. This was the first straw-bale structure in the United States to be granted permits. It was also the first to be bank-financed and insured. (Photo by David Eisenberg.)

specific, and the second involves change in the larger regulatory system. In both cases, gaining acceptance should be seen as a process, not an event. This process usually relies on presenting authoritative and accurate information that addresses the advantages and problems of both the proposed alternative and the current practice it would replace.

The goals of those promoting sustainability are not contrary to those of the code officials. In fact, they are largely congruent. The basic difference is that the code community has yet to fully acknowledge the larger, more pervasive safety issues that the sustainability community takes quite seriously. Vital to overcoming resistance is a shared intention to create regulations that result in structures that are safe for their occupants and the other inhabitants of earth, both current and future.

The establishment of a cooperative atmosphere is vital for acceptance of any alternative approach. The educational process requires time, and it should not be expected that a code official will approve an unfamiliar alternative without adequate information and time to study it. Building on common concerns for safety is key to positive relationships and having plans approved.

Expanding the awareness of the larger context will vary with the alternatives being proposed, the attitudes and knowledge of the officials involved, and the quality, quantity, and sources of the information and documentation submitted on behalf of the alternatives. The less authoritative and extensive the specifics, the more important it will be to establish the larger context.

The nonproprietary nature of alternatives means that there have rarely been adequate financial resources to carry out full-scale research, testing,

Post-and-beam straw-bale residence in Santa Fe, New Mexico. This house with radiant-heated earthen floors and earthen plasters demonstrates that natural materials can result in high-quality buildings of exceptional value. (Photo by David Eisenberg.)

and development. The inability to supply certification or recognized test results for alternatives is often viewed as evidence of their inability to comply with the requirements of the code or an attempt to get away with using substandard materials or methods because of cost or for other reasons. Code officials that request proponents to go through the evaluation and certification processes offered by the model code organizations often do not recognize the enormous time and expense—far beyond the scope of almost any construction project budget or schedule—required. Constructively making the distinction with building officials that the proposed

Rammed earth house in Tucson, Arizona. Rammed earth is an ancient technique of building walls by compacting earth into forms. It is essentially a rapid sedimentary process. (Photo by David Eisenberg.)

system is non-proprietary, without such recognized testing, and then establishing that it is viable and safe, is a formidable task.

Life cycle analysis and full-cost economics for building systems will go a long way toward leveling the playing field, as will broad-based educational programs designed to facilitate better understanding of the larger issues at stake. To that end, the Development Center for Appropriate Technology, a nonprofit organization in Tucson, Arizona, is developing a program called "Building Sustainability into the Codes." The first such effort in the United States, it seeks to form a partnership between the model code organizations and those committed to sustainable building and development. Early efforts are focused on raising awareness of broader issues of health and safety in the regulatory community and on fostering the direct, effective involvement of ecological building proponents in the code development processes.

Part of the educational strategy is to reveal how focusing on details tends to obscure the bigger picture. To focus only on the myriad details detached from their cumulative effects, whether in the wording of a paragraph of a subsection of the code, the design of a structural connection, mechanical system, or a whole building, is scale-of-one thinking. And that scale-of-one thinking is the basis for decision making for the design and construction of millions of buildings.

Clearly, responsibility for change does not lie solely at the doorstep of building regulation, but must be shared among all stakeholders in the process, whether occupants, designers, engineers, builders, or code officials.

Getting Permission to Build

In seeking a permit for a project incorporating natural or alternative materials or methods of construction, often the first task is to raise the comfort level of the building official through education about the viability of the proposed alternatives. There is a great deal of variability in the regulatory system from one jurisdiction to the next. This can be advantageous when a sympathetic official is open to alternative approaches, and even more helpful if he or she is also willing to communicate with other officials.

Building officials have authority to approve whatever they deem to be adequate to meet the intention of the codes. Their authority for approving alternatives comes from provisions within the codes that state the criteria required to make such approvals. The 2003 edition of the International Building Code includes the following:

The light-clay Zucher residence near Austin, Texas, is a wonderful example of human-scale architecture. It was hand-built using many native and natural materials, such as stone for foundations, fireplaces, and chimneys; local timber for the roof framing, and light-straw-clay for the infill walls. The walls were finished with lime and sand plasters. (Photo by David Eisenberg.)

104.11 *Alternative materials, design and methods of construction and equipment.* The provisions of this code are not intended to prevent the installation of any material or to prohibit any design or method of construction not specifically prescribed by this code, provided that any such alternative has been approved. An alternative material, design or method of construction shall be approved where the building official finds that the proposed design is satisfactory and complies with the intent of the provisions of this code, and that the material, method or work offered is, for the purpose intended, at least the equivalent of that prescribed in this code in quality, strength, effectiveness, fire resistance, durability and safety.

104.11.1 *Research reports.* Supporting data, where necessary to assist in the approval of materials or assemblies not specifically provided for in this code, shall consist of valid research reports from approved sources.

104.11.2 *Tests.* Whenever there is insufficient evidence of compliance with the provisions of this code, or evidence that a material or method does not conform to the requirements of this code, or in order to substantiate claims for alternative materials or methods, the building official shall have the authority to require tests as evidence of compliance to be made at no expense to the jurisdiction. Test methods shall be as specified in this code or by other recognized test standards.

These provisions are somewhat more stringent than earlier versions of such provisions, yet they still allow for discretion on the part of the building official.

Frequently, as stipulated in the code, a building department requires that a test or set of tests be carried out to demonstrate the adequacy of a particular material or method to meet the requirements of the code. As mentioned earlier, if the only tests acceptable to the jurisdiction are those performed in certified labs and carried out in accordance with all accepted standards, gaining acceptance can become extremely difficult if not impossible. This is a point at which a supportive building official from another jurisdiction may be able to intervene.

It is also not unusual for the code jurisdiction to decide that it is not qualified to determine the adequacy of a material or method, or of the testing and other supporting documentation submitted. It may refer the review process to the regional or central office technical staff of its model code organization. This organization is basically a consulting service, however; it has no jurisdictional authority. The authority for approval still lies fully with the building official in the jurisdiction in which the project is located. It is usually a hard

Low-cost straw-bale house designed for Habitat for Humanity, New Mexico. (Photo by Tara Teilman-Way.)

sell to get a local jurisdiction to deviate far from the recommendations of its regional or central technical staff. Thus, if the plans are sent for review, the focus of attention, in terms of providing supportive information, should also shift to members of that staff. Once their determination has been made, it is difficult to return to them for another interpretation.

An approach that has had some success is to offer to give the jurisdiction a letter or legal document that holds it harmless from all responsibility for the alternative materials and methods used. This offers an avenue for approval, assuming the owner is willing to take that responsibility, that minimizes the risk for the building department. In projects for which an architect or an engineer has stamped the plans, the argument can also be raised that this practitioner has already taken some legal responsibility by designing the project and sealing the plans.

Finally, there are two other avenues for getting approval, both of which have been used successfully many times. One is the local appeals process. At the request of an applicant who has been refused a permit or denied ability to use a particular alternative, the building department must convene an appeals board meeting. A selected group of local or regional building professionals

This center for environmental education and the Saguaro Chapter Girl Scouts in Tucson, Arizona, was built of post-and-beam straw-bale in accordance with the Pima County code. (Photo by David Eisenberg.)

hears the applicant's request and supportive testimony, as well as that of the building department, and makes a ruling on whether to back or overrule the decision of the building official. Occasionally, a building official will even request this process and join the applicant in support of the alternative in order to set a precedent and have wider backing for the decision.

The second tactic is the application of political pressure on the building department through elected officials of the jurisdiction (i.e., mayor and city council, county supervisors or commissioners, or state-level elected officials if the state has jurisdiction). Lobbying or applying pressure, including media coverage, can be an effective tool for change. It is also possible to have codes or regulations developed and adopted totally through the political or legislative process, bypassing the regulatory community entirely. These last two avenues should be pursued carefully, and only as a course of last resort, because they can generate a great deal of resentment in the long run within the regulatory community. The creation of code through legislative process can best be accomplished cooperatively, inviting supportive code officials to assist in the development of the proposed legislation.

Guidelines for the Process

The process of acquiring permits for building with alternative materials can be made easier by heeding the following recommendations. In some cases, the preparation, attitude, and follow-through described here will make the difference between rejection and acceptance.

START EARLY Define as many of the nonstandard aspects of the project as early as possible, to give the longest lead time. Building officials, like everyone else, resist being hurried or pressured and, in this situation, may believe that there is a hidden reason they are being rushed through the approval process. Give them time to absorb and respond to the material presented in support of the alternative, and allow for a number of exchanges with them as they raise objections and concerns and consider the information provided.

GATHER INFORMATION The initial step is to learn both about the jurisdiction in which the project is located and about the alternatives being proposed. Find out who has jurisdiction and which codes are enforced at the project location. Get a copy of the permit process requirements and, if possible, a current version of the code that is in force. Study the sections related to the alternative approaches that will be included in the project and identify areas of concern.

Gather as much background as possible on the building methods to be used, including test results and historic and recent precedents, both locally and in other jurisdictions. Seek out and communicate with knowledgeable resource people, including sympathetic code officials if any can be identified, to ensure that the information is as complete as possible. Anticipate objections that may be raised.

Provide the building department with the best available reference materials—books, the best or most authoritative publications, videos, test results, and documentation of the successful use and approval of the alternative in other places. Buying these resources for the building department, rather than lending them, is a good overall strategy because it demonstrates a serious commitment to working through the process. These materials become part of the department's resource library, where they may have an influence on future permit applications. If high-quality newsletters or magazines are available, consider giving the code jurisdiction a subscription as early as possible to increase the level of awareness in the department before a permit request is ever submitted.

Such investments are minor in relation to the time and cost of a typical building project and can yield high returns in ease of process. Moreover, educational materials circulated in the building department may fall into the hands of a potential advocate within the department.

In choosing the supportive materials to submit, quality is more important than quantity. When there is a lack of authoritative or high-quality information, it is sometimes effective to submit a broader sampling of media that document the widespread and accepted use of the alternative elsewhere; however, it is difficult to succeed with this type of information alone.

Where relevant, supporting material should address situations as regionally, climatically, seismically, or otherwise similar to the local circumstances as possible. It is reasonable for building officials to question the assumption that because something works in one region or climate it will automatically work in another. Often, in natural building, the materials are local and the methods are traditional, with a local or regional historical precedent that can be cited. Universities can sometimes be good resources for documentation of the viability and durability of traditional building materials and methods.

DEVELOP GOOD RELATIONSHIPS Because this is partially a process of creating trust, having already established a good relationship with the building department can be of great help. Lacking such a relationship does not doom the effort to failure, but typically lengthens it. If there is no relationship or some bad history, it is often beneficial to enlist the help of someone who has a good working relationship with the department.

Strive to maintain a cooperative, open-minded, and positive attitude, with the expectation of a successful outcome. Be flexible and consider the objections and concerns from the building official's point of view. It should always be acceptable to challenge unfounded assumptions and to take the conversation beyond the letter of the code in seeking the basis and intent that led to the code's being as it is. Yet one should be careful to avoid creating an adversarial or antagonistic relationship with building department staff, because that will always be counterproductive.

View building officials as resources, rather than opponents. Enlist their help in finding the path to approval, instead of asking whether it is possible to use an alternative. Openly acknowledge the extra effort required by the building department, as this helps create better rapport and demonstrates an understanding of the process.

Finally, remember that being a pioneer includes some measure of responsibility for those who may follow, to make their path easier. Although it is not always easy to remember in the face of what may seem to be unreasonable requirements or frustrating setbacks, those who go first often strongly influence the experiences of those who come next.

MEET AND SHARE INFORMATION WITH BUILDING OFFICIALS When the project is well enough defined, have an initial meeting to discuss the project and the proposed alternatives informally. If there are known sympathetic officials or inspectors, talk to them first. Have copies of the resource materials ready to leave with the code officials. Give them enough time to absorb what has been provided.

GET SPECIFIC FEEDBACK FROM BUILDING OFFICIALS Both in preliminary discussions and when the plans have been submitted, if the building department says no, get it to list its specific objections in writing if possible. If it will not, make a list of the concerns and objections and review it with the department. The more specific the response to its objections the better, because this creates a finite list of concerns needing to be addressed.

ADDRESS CONCERNS WITH REASONABLE AND FACTUAL RESPONSES It is important to address the concerns of the building department with reasonable, factual responses. This step is often a repeat of the preceding steps, with a progressively narrower focus on specific issues each time through. Again, recognize the official's perspective and the responsibilities and constraints that go with his or her job. Demonstrate that you understand and respect the merits and limitations of the building system or alternative, and that what you plan to do is safe and reasonable. Set up another meeting with the official to address his or her concerns, and in the meantime do more homework. This is an area in which the influence of another code official who is familiar with, and supportive of, the proposed alternative can be of enormous benefit.

NETWORK WITH OTHERS WHO HAVE HAD SIMILAR EXPERIENCES There are lessons to be learned from the experiences of others who have gone through this process. On the Internet or via other avenues, seek out knowledgeable organizations, groups, and individuals and study their successful approaches. The most valuable contacts are often experienced building officials who have approved and worked with the materials or methods in question, or who are open-minded and receptive to alternatives.

Conclusion

Building codes will play a crucial role in the future of natural and alternative building in the United States, either by inhibiting or by helping these materials and methods to gain broader usage. For these alternatives to achieve mainstream acceptance, their adequacy to meet the primary criteria in the

Interior of a cob house near Austin, Texas. Cob is an example of a simple, natural building material, used successfully for centuries by people all over the world. Because of the nature of the material and the building process, cob buildings generally have a human scale and are as unique as the people who build them. (Photo by David Eisenberg.)

building codes for safety, health, and durability must be demonstrated. Success in reaching that goal will depend, in part, on allowing them to be fairly viewed not only in relationship to their industrial competitors, but also in relationship to the overall consequences of building with either the traditional or the new approach.

The moral authority of building codes is based on the need to protect the public welfare from threats resulting from the built environment. Until now this concern has been expressed almost exclusively in terms of safety in specific built structures. Codes have rarely addressed aggregated impacts, other than in the relatively recent introduction of energy codes. At this enlightened time, we are able to see what we build in its larger environmental and economic context. It is imperative for those who understand the cumulative planetary impacts of construction, now and into the future, to bring this larger vision to the regulatory community as an issue of risk avoidance of the highest importance.

Notes

1. David Malin Roodman and Nicholas Lenssen, *A Building Revolution: How Ecology and Health Concerns Are Transforming Construction,* Worldwatch Paper #124 (Washington, D.C.: Worldwatch Institute, March 1995), 5.

2. The United States uses 25 percent of the world's energy (International Energy Annual 1997, Appendix E—"World Energy Consumption [Btu], 1988–1997," and Washington, D.C.: U.S. Department of Energy, Energy Information Administration, document DOE/EIA-0219, 1997) and buildings in the United States account for 40 percent of the U.S. energy consumption (same as Note 1. Roodman and Lenssen), thus U.S. buildings use approximately 10 percent of world energy.

3. David Malin Roodman and Nicholas Lenssen, *A Building Revolution,* 7 and 28, and Hugo Houben and Hubert Guillaud, *Earth Construction: A Comprehensive Guide* (London: Intermediate Technology Publications, 1994), 6.

4. *World Population Projections to 2150* (New York: Department of Economic and Social Affairs, Population Division, United Nations Secretariat , February 1998).

5. Michael Myers, *Sustainable Communities: Green Buildings* (Washington, D.C.: U.S. Department of Energy, January 1997), 1, and personal communication July 6, 1998, with Michael Myers.

Natural Conditioning of Buildings

Ken Haggard, Polly Cooper, and Jennifer Rennick
with assistance from Phil Niles

Natural conditioning provides for the passive heating, cooling, lighting, and ventilation of buildings and is dependent on neither mechanical systems nor imported energy. It is acheived through holistic architectural systems that respond to chaotic conditions on-site. These conditions include climate, human use, and the microclimatic effects of winds, topography, and the building itself.

Achieving comfort and delight by utilizing on-site energies requires creativity by the designer. The design goal is to organize a whole, made up of integrated multifunctional components. For example, building walls are space definers, but they can also create thermal barriers, direct breezes, reflect or absorb sunlight, and act as thermal mass. With the help of computer simulation tools, natural conditioning optimizes the visual and thermal environment to the point where mechanical conditioning is needed only as backup for the climatic extremes, thus providing comfort while using only 5 to 30 percent of the imported energy consumed by standard industrial-era buildings. If done well, the result is superior comfort, minimization of expensive infrastructure, and contribution to a healthier planet. Natural conditioning of buildings is part of our cultural transformation from the present highly developed but depletive industrial era to the evolving regenerative sustainable era of human development.

Alternative Construction: Contemporary Natural Building Methods, edited by Lynne Elizabeth and Cassandra Adams ISBN 0-471-24951-3 © 2000 John Wiley & Sons, Inc.

History

The vast majority of ancient buildings were passively heated, cooled, lit, and ventilated. There were a few exceptions—for instance, the baths of Rome and the hypocaust systems of Korea—but essentially historical buildings were passive buildings. Although standards of comfort were far different then, many of these buildings were amazingly effective for their period and culture—heating in the Mongolian *ger* (yurt) or the vernacular structures of the Bernese Alps, cooling in the traditional architecture of Iran, and the natural ventilation of classic Japanese architecture, for example. These architectural examples testify to humankind's ability to integrate art, function, technology, and culture in building, a legacy that should inspire us today.

In addition to the inspiration that can be drawn from ancient history, an understanding of recent history can help us avoid reinventing the already known. Ignorance of collective knowledge in this field can waste valuable creativity and exuberance. Moreover, historical perspective can surmount the problem of cynicism—cynicism that results from an inability to perceive progress and a failure to appreciate the uneven flow of evolution, as it stops and starts and occasionally doubles back on itself.

True to this pattern, the recent history of natural conditioning of buildings has occurred in four waves of intense activity followed by troughs of relative neglect.

WAVE 1   1880–1914: EARLY INDUSTRIAL OPTIMISM

In 1881, Edward Morse of Lowell, Massachusetts, patented a solar heating system consisting of a glazed masonry south wall, what we have come to call a trombe wall. At about the same time, Clarence Kemp designed and marketed a thermo-siphon water heater, which dominated the roofs of buildings in Los Angeles before a widespread utility grid was established. A similar use of passive water heaters occurred in Texas and Florida. In the 1900s, the California Bungalow, a building style that emphasized natural cooling, was developed and mass marketed. Bungalows had large roof overhangs, well-ventilated attics, natural ventilation vegetable coolers, and covered porches facing the street.

In established architectural circles at the time, Frank Lloyd Wright was designing architecture that emphasized beautiful natural lighting, and Victor Horta and others were integrating "wintergardens"—essentially greenhouses—into their Art Nouveau versions of urban row housing in Northern Europe.

This optimistic era ended with the catastrophic destruction of World War I and the following periods of economic boom and bust.

WAVE 2   1945–1955: POST–WORLD WAR II Optimism about technology at the end of World War II stimulated the feeling that research and better materials could create more naturally conditioned buildings. In 1947, Libby-Owens-Ford started marketing double-pane glazing, and to promote this product selected 48 architects to design solar homes, one for each state in the union. The results were published in the book *Your Solar Home*. Most of these buildings were not optimized for performance because sizing techniques were nonexistent. From 1947 to 1952, the Massachusetts Institute of Technology (MIT) constructed a series of test cells and full-scale solar test buildings to attempt to get numerical data on technique and performance. The second of these test facilities used a passive approach employing water walls and movable insulation. The state of insulation technology caused MIT to abandon this approach—prematurely, as history now shows. *House Beautiful* also published a remarkable series of excellent technical articles on climate and design in the early 1950s.

The Texas A&M Experiment Station produced a series of research reports on natural ventilation and natural lighting that utilized low-speed wind tunnels and lighting models to formalize important principles of airflow and lighting effects in regard to architectural form. As a result, there were many naturally lit and ventilated elementary schools built in the southwestern part of the United States in the early 1950s.

In 1952, in Chicago, the Keck Brothers developed a solar housing tract utilizing south-facing facades of double-pane glass and optimized roof overhangs. Frank Lloyd Wright designed "the solar hemisphere" in Wisconsin, a partly earth-bermed house with large expanses of south-facing glass and some thermal mass. In Europe, Le Corbusier was building very effective sun control facades for sunny climates and Alvar Aalto was designing buildings for overcast climates that provided elegant natural lighting.

However, by the mid-1950s mechanical air-conditioning became widespread, artificial lighting was improved, and new standards began to increase lighting levels to excessive heights. Energy from fossil fuels seemed cheap and inexhaustible, and suburban sprawl was rapidly becoming the norm. In this period the country evolved toward the windowless grade school, suburban planning, massive oil imports from the Middle East, and even more massive carbon dioxide exports to the atmosphere. Natural conditioning was addressed by Van Dresser in New Mexico and Nieberger in California, and by Victor and Aladar Olgay in New Jersey with their book *Design with Climate*. These efforts, however, were generally ignored until the next wave of development.

WAVE 3   1975–1980:
THE PASSIVE SOLAR
MOVEMENT

Advances in the use of solar heating and cooling for architecture by Hay, Yellott, Baer, David Wright, and others, coupled with the energy crisis of 1973, stimulated much interest. In 1972, Phil Niles created the first computer simulation model of a passively conditioned building, thereby achieving the original holistic quantification of buildings. With the first national passive solar conference in 1976, put together by Doug Balcomb of Los Alamos Labs, there developed a nationwide and then worldwide movement. Research expanded with federal support during the Carter Administration, thermal prediction models were improved, books on the subject were published, and significant construction and product development were carried out. Wave 3 essentially overcame the problems of the first two periods—the lack of appropriate materials and an inability to quantify the use of such materials in a holistic system.

Wave 3 did so much so fast that the following trough was immense. Most of the research activity was cut by the Reagan Administration, oil prices dropped considerably, and the United States drifted back toward its old wasteful ways. In architectural circles, postmodernism became the rage. This style, which developed architectural form from largely historical metaphor, played down social, functional, and technical relevance. Integration was, for a few years, considered an inappropriate architectural endeavor.

WAVE 4   1995 TO THE
PRESENT:
THE SUSTAINABILITY
MOVEMENT

Slowly, concerns with planetary resources and the worldwide effects of industrial wastefulness became prevalent. Myriad concerns about health, from the planetary scale to the individual level, became widespread. These concerns, along with the growing awareness of the ozone hole and global warming, coalesced into the budding sustainability movement. The now matured technological knowledge about passive space heating, cooling, and lighting began to expand to other concerns—the embodied energy of materials, biologically integrated waste systems, health implications of built environments, and regional effects of building practices. "Green architecture" and "sustainable architecture" became common terms to describe architecture that addressed these wider concerns.

Basic Principles

This section presents the technical terms and processes that are basic to passive heating, cooling, lighting, and ventilation of buildings.

ON-SITE SOURCES WE CAN USE FOR PASSIVE HEATING	ON SITE SINKS WE CAN USE FOR PASSIVE COOLING

ON-SITE SOURCES WE CAN USE FOR PASSIVE HEATING

1. SOLAR RADIATION

Solar radiation can heat a building through direct and diffuse radiation. Diffuse radiation is the component of sunlight which reaches the ground after scattering and reflection from the earth's atmosphere.

2. OUTSIDE AIR

Outside air when it's warmer than about 75°F can act as a heat source for most buildings.

3. INTERNAL METABOLISM

People, lights, and equipment all add heat to the interior of a building and can thus be considered on-site heat sources.

ON SITE SINKS WE CAN USE FOR PASSIVE COOLING

1. SKY AND SPACE

Some heat is always radiated from a building to the sky. Under certain circumstances such as clear nights and low humidity the upper atmosphere and space can act as a appreciable sink.

2. OUTSIDE AIR

Outside air that is cooler than about 75°F act as a heat sink. Cool night air can be utilized as an effective sink via night ventilation.

3. WET OR DAMP SURFACES

Wet surfaces provide on-site sinks because heat is absorbed when water evaporates.

On-site thermal sources and sinks. (Illustration courtesy of San Luis Sustainability Group.)

HEAT TRANSFER BASICS Heat can be transferred via three processes—conduction, convection, and radiation. Conduction transfers heat by molecular movement within a material, or from material to material if in direct contact. Convection transfers heat by warming a fluid which, when warmed, rises, taking heat with it. Radiation is the transfer of heat from warmer surfaces to cooler surfaces by electomagnetic radiation. This radiation does not warm the transparent material (usually air) between the two surfaces.

THERMAL SOURCES AND SINKS The siting of most buildings can provide three on-site thermal sources for heating and three on-site thermal sinks for cooling. If a building is properly designed, this is enough energy to provide most of the thermal conditioning needed for comfort.

ILLUMINATION SOURCES There are three on-site illumination sources—direct sunlight, sunlight reflected from the ground or nearby objects, and sunlight diffused and reflected by the sky. There was, until recently, a bias toward using only the light from an overcast sky—the traditional north light with its minimal heat and glare. That bias arose because the early scientific application of natural lighting was done in northern Europe, where overcast conditions predominated. However, with a more holistic look at passive building, lighting is seen as part of the total flow of energy available for use at a particular site. With this approach, direct sunlight—especially from the south—serves as a

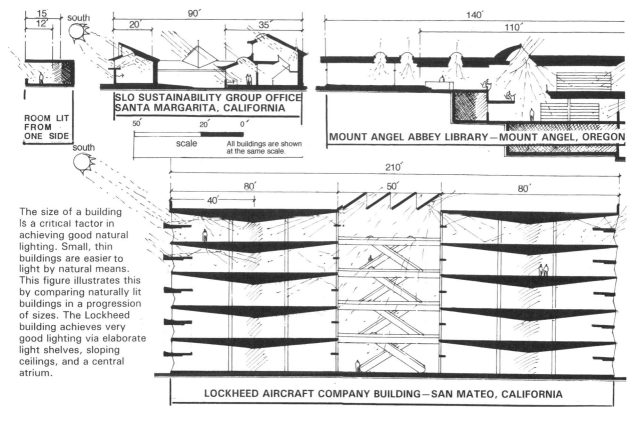

The size of a building is a critical factor in achieving good natural lighting. Small, thin buildings are easier to light by natural means. This figure illustrates this by comparing naturally lit buildings in a progression of sizes. The Lockheed building achieves very good lighting via elaborate light shelves, sloping ceilings, and a central atrium.

ROOM LIT FROM ONE SIDE

SLO SUSTAINABILITY GROUP OFFICE SANTA MARGARITA, CALIFORNIA

scale — All buildings are shown at the same scale.

MOUNT ANGEL ABBEY LIBRARY—MOUNT ANGEL, OREGON

LOCKHEED AIRCRAFT COMPANY BUILDING—SAN MATEO, CALIFORNIA

A spectrum of approaches to natural lighting. (Illustration courtesy of San Luis Sustainability Group.)

common natural lighting source, and control of heat and glare is handled architecturally.

Under normal conditions, natural daylight will penetrate 12 to 14 feet horizontally into a space. Thus, one of the main objectives of natural lighting is to increase the reach of daylight. The second major consideration in daylighting is to provide light from several directions in order to reduce glare. The result of both of these considerations is relatively thin buildings or large buildings with atria or courtyards.

BUILDING SCALE, METABOLISM, AND FORM

On a particular site, illumination sources and thermal sources and sinks tell how much energy is available for use in a building. Thus, it may seem that once the building's use is determined, it should be relatively easy to determine the energy needed—the building's load. Determining load, however, is one of the more subtle aspects of natural conditioning, since loads are a product of the complex relationships of the building's scale, metabolism, and form. This is a complex problem, because the relationship of these fac-

Assume $X = 1$
$A = 1 \times 1 \times 6 = 6$
$V = 1 \times 1 \times 1 = 1$
Area to volume ratio $= 6/1 = 6$

$A = X \times X \times 6 = X^2 \times 6$ $V = X \times X \times X = X^3$

Assume $X = 300$
$A = 90,000 \times 6 = 540,000$
$V = 300 \times 300 \times 300 = 27,000,000$
Area to volume ratio
$= 540,000/27,000,000$
$= .02$

Thermal loads of hummingbirds and elephants. (Illustration courtesy of San Luis Sustainability Group.)

tors is not linear. To understand this, consider the thermal loads of two different animals, a hummingbird and an elephant.

Animals, like buildings, need to maintain relatively constant interior temperature. Heat to maintain this temperature is provided by the animal's metabolism, and heat loss occurs mostly through the animal's skin. Heat loss is therefore related to the skin's surface area. A hummingbird is a very small animal, but the relationship of its skin area to the volume of its body is relatively large. Thus, the hummingbird loses heat very rapidly and must have a high corresponding metabolism to maintain interior temperature. Hummingbirds feed on the nectar of flowers, one of the most concentrated foods available in the natural world, and must eat often. In contrast, elephants, although vastly larger, have a lower skin-area-to-volume ratio, with correspondingly proportionate lower heat loss. Thus, elephants can survive with occasional meals of low-energy food like plant roughage.

Buildings act similarly. There are buildings, like the hummingbird, with a large skin-to-volume ratio. These are called "skin-dominated" buildings, that is, buildings in which the building envelope dominates the thermal loads. Corresponding to the elephant are the "interior load-dominated" buildings. These are buildings that are large enough that the internal loads dominate their thermal character. Knowing these rather simple relationships destroys one of the myths of passive heating: that it is easier to heat small buildings than large buildings. Actually, in most temperate zones the reverse is true. Heating larger buildings is usually easier because they produce enough internal heat and their skin-to-volume ratio is small. Thus, if properly designed, they can often heat themselves by their own internal metabolism—metabolism being defined as heat generated inside the building by lighting, equipment, and people. Cooling, however, presents another set of problems.

In most temperate zones, heating, cooling, or combinations of the two are the limiting design factors for skin-dominated buildings. In interior load-dominated buildings, heating is generally easier to accomplish than cooling, particularly if artificial lighting is not overused. The reason is that artificial lighting produces a great deal of heat that is added to the cooling load. In many large, thick industrial-era buildings, lack of natural lighting is usually the limiting condition.

The recent bias has generally been toward interior load-dominated buildings because they lend themselves to a single purpose—maximum rental area for the lowest first cost. However, from an aesthetic, social, and environmental viewpoint the typical interior load-dominated building of the industrial era is fairly disastrous. This bias should change to a more rational approach. Skin-dominated buildings are usually more appropriate in most temperate zones, because balanced natural heating, cooling, and lighting are easier to achieve within these buildings. In addition, social relationships are better when users feel less cut off from the exterior environment. The perceived density advantages of big, interior load-dominated buildings proves not to be as obvious as assumed, except in exceptionally large buildings.

COMFORT STANDARDS There is an unspoken assumption that passively conditioned buildings will be less comfortable thermally than mechanically conditioned buildings, and that we should build them only for environmental reasons. In reality, the opposite is true. A properly designed passive building should inherently be

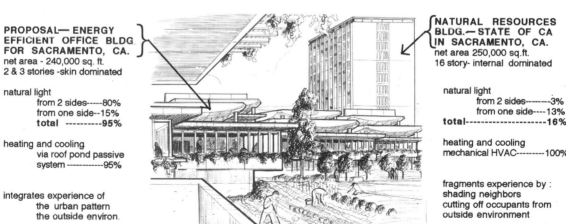

PROPOSAL— ENERGY EFFICIENT OFFICE BLDG. FOR SACRAMENTO, CA.
net area - 240,000 sq. ft.
2 & 3 stories -skin dominated

natural light
 from 2 sides-----80%
 from one side--15%
 total ---------95%

heating and cooling
 via roof pond passive
 system ------------95%

integrates experience of
 the urban pattern
 the outside environ.

NATURAL RESOURCES BLDG.— STATE OF CA IN SACRAMENTO, CA.
net area 250,000 sq.ft.
16 story- internal dominated

natural light
 from 2 sides--------3%
 from one side----13%
 total--------------------16%

heating and cooling
 mechanical HVAC---------100%

fragments experience by :
shading neighbors
cutting off occupants from
outside environment

Comparison of a naturally conditioned skin-dominated building to a mechanically conditioned interior load-dominated building. (Illustration courtesy of San Luis Sustainability Group. Data taken from K. Haggard, J. Pohl, and P. Cooper, "An Office Building for an Era of Transitions," **Proceedings of the Second Annual Passive Solar Conference.** *Philadelphia: American Solar Energy Society, 1978.)*

more comfortable than a properly designed mechanical building. The reasons are described as follows.

For a given clothing and activity level, the main determinants of comfort are the temperatures of the air and surrounding surfaces. These are of roughly equal importance. Raising the surrounding surface temperature by one degree allows the lowering of air temperature by one degree without any change in comfort sensation. This means that if a person feels comfortable in a building with air temperature at 72 degrees Fahrenheit and the floor and walls at 64 degrees (typical of conventional buildings using forced-air systems), then that person will feel equally comfortable in a building with the air temperature lowered to 68 degrees if the interior room surfaces are also at 68 degrees (typical of a passive building). Because the air inside passive buildings is usually in thermal equilibrium with the interior surfaces, the mean interior surface temperatures are close to the mean air temperature. Besides allowing people to be comfortable at cooler air temperatures, another result of passive buildings is the sensation of a uniform temperature throughout the interior, in contrast to the usual sensing of cold spots and drafts typical of mechanical systems. The advantage of not having to rely on forced streams of air, in which the humidity has been uncomfortably altered mechanically, is fewer colds and less respiratory discomfort during winters and summers.

A common concern about naturally conditioned buildings is the possible discomfort resulting from the relatively large indoor temperature swings inherent in the operation of passive systems. Most performance simulation modeling of passive systems assumes that the indoor temperature is allowed to vary between 65 and 80 degrees Fahrenheit. The illustration below shows that this 15-degree band is almost always acceptable to occupants engaged in normal activity. This assumes that occupants can take

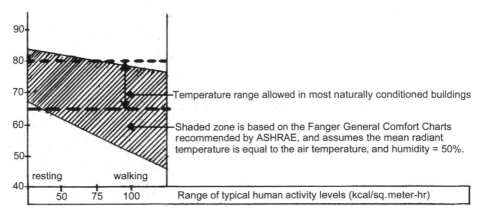

Chart of comfort zone for typical human activity. (Illustration courtesy of San Luis Sustainability Group. Data taken from P. O. Fanger, Comfort: Analysis and Applications in Environmental Engineering. *New York: McGraw-Hill, 1970.)*

Temperature range allowed in most naturally conditioned buildings

Shaded zone is based on the Fanger General Comfort Charts recommended by ASHRAE, and assumes the mean radiant temperature is equal to the air temperature, and humidity = 50%.

resting   walking

Range of typical human activity levels (kcal/sq. meter-hr)

advantage of a slight clothing change and can tolerate temperature swings sensed by the average person as ranging from slightly warm to slightly cool. Passive buildings therefore rely on mechanical systems only as backup to prevent temperatures from ranging beyond this comfort zone. The size and energy use of such mechanical systems are usually much smaller than those of mechanical systems used to totally condition a building.

In summary, the subject is two different kinds of comfort. The first kind is disconnected from the natural environment and based on an ideal constant air temperature. This approach has high maintenance requirements and is expensive to humans and to the planet. In contrast, natural conditioning is a comfort system connected to the environment in which chaotic temperature swings occur within the prescribed limits of the comfort zone. Here we rely on mechanical systems only when the temperature strays beyond the comfort zone. With natural conditioning, operating costs and maintenance requirements are less and impact on the planet is less.

THERMAL MASS A factor that distinguishes many passive buildings is the availability of thermal mass—material that has the ability to hold heat or coolness. Recent industrial building techniques produce very little thermal mass. In fact, lightness was considered a virtue in industrial-era buildings (the legacy of Buckminster Fuller, Mies van der Rohe, and others). Because light buildings cannot hold heat or coolness very long, they are generally poor passive buildings. In temperate climates a good passive building usually needs a good amount of thermal mass.

Common materials that are nontoxic and can hold heat or coolness are grouped in several categories:

1. Masonry materials such as brick, concrete, stucco, adobe, earth, and the stucco skin of straw-bale construction
2. Eutectic materials, such as Glauber's salt (sodium sulfate), that store heat upon change of state from solid to liquid or from different states of crystallization
3. Water

Each of these has characteristics that must be considered in architectural applications.

Masonry is commonly available and provides good thermal mass, but in this material absorption of heat is relatively slow. In theory, phase-change materials hold great promise for high-efficiency thermal mass, and a great deal of work was done with them in Waves 2 and 3 of recent history. These are materials that change state at appropriate temperatures and, in the process, release or absorb relatively large amounts of heat. At present, however, they are not used much, because of problems with long-term stability and the need for large-scale production to be economically feasible.

Water is excellent thermal mass, with roughly two times by volume and four times by weight more thermal storage than masonry materials. Life is possible on earth because it is a water-covered planet that provides enough thermal mass to moderate temperature extremes on the surface. Our bodies, which are also mostly water, act similarly. However, water used as thermal mass must be contained and the potential for leakage must be dealt with.

Architectural uses of these thermal mass materials are categorized in two general applications: concentrated mass, which most often uses water and masonry, and distributed mass, which usually consists of thin surfaces of masonry. Each has its advantages in regard to direct-gain heating and night-vent cooling in dry temperate climates.

Concentrated mass, as the name implies, is concentrated in specific areas in a building. It is most often more efficient for heating than cooling in direct-gain heating and night-vent cooling systems. This is because the surface area for transfer of coolness from night air to the mass is small relative to the amount of storage involved. Water is a very good material for such application, as it stirs itself to distribute the heat and has such high heat storage capacity. Illustrated on the next page are several concentrated mass direct-gain components and a chart illustrating optimal sizing for one particular climate.

Distributed mass is mass that is spread over large surfaces. It is good for night-vent cooling, because large surface areas provide opportunities for the transfer of coolness from night air to the interior of the building, which can be used to override heat gains during a following hot day. For most temperate climates where buildings are responding to diurnal patterns, the optimal thickness for masonry used as distributed mass is about 2 inches. The heating performance of distributed mass systems, although generally not as efficient as concentrated mass systems, is relatively good. Therefore, this approach is appropriate for conditions where the heating and cooling loads, as well as thermal sources and sinks, are relatively balanced.

Of course, these two approaches to mass can be used together within the same building. The advantage of this approach can be seen in a structure

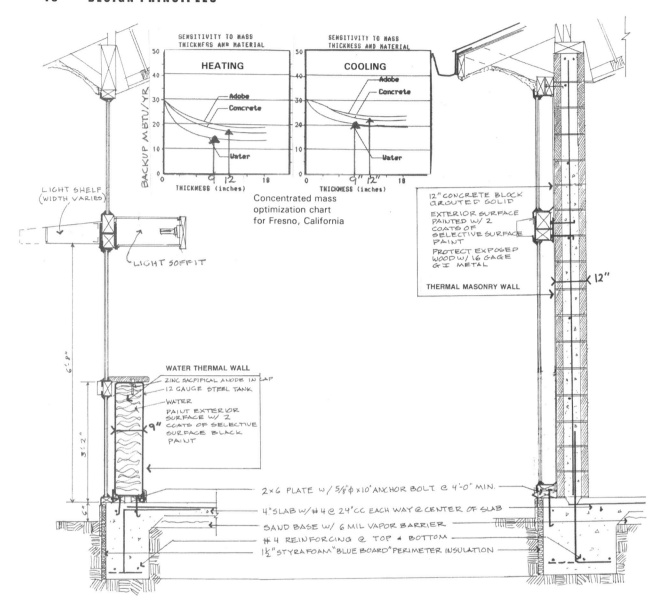

Several concentrated mass components and their optimization. [Illustration courtesy of San Luis Sustainability Group. Data taken from K. Haggard and P. Niles, **Passive Solar Handbook for California.** *Sacramento: California Energy Commission, 1980, 119 ("Distributed Mass Performance Sensitivity") and 171 ("Concentrated Mass Performance Sensitivity").]*

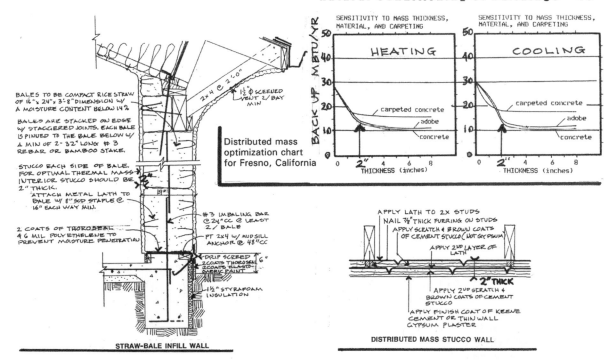

BALES TO BE COMPACT RICE STRAW OF 16"x 24"x 3'-8" DIMENSION W/ A MOISTURE CONTENT BELOW 14%

BALES ARE STACKED ON EDGE W/ STAGGERED JOINTS. EACH BALE IS PINNED TO THE BALE BELOW W/ A MIN OF 2- 32" LONG # 3 REBAR OR BAMBOO STAKE.

STUCCO EACH SIDE OF BALE. FOR OPTIMAL THERMAL MASS, INTERIOR STUCCO SHOULD BE 2" THICK.
ATTACH METAL LATH TO BALE W/ 8" SOD STAPLE @ 16" EACH WAY MIN.

2 COATS OF THOROSEAL & 6 MIL POLYETHELENE TO PREVENT MOISTURE PENETRATION

2+4 @ 2'-0"

1½" Ø SCREENED VENT 2/BAY MIN

19"

#3 LM BALING BAR @ 24"CC @ LEAST 2/BALE

PT 2x4 W/ MUDSILL ANCHOR @ 48"CC

DRIP SCREED
2 COATS THOROSEAL
2 COATS ELASTOMERIC PAINT
6"

1½" STYRAFOAM INSULATION

Distributed mass optimization chart for Fresno, California

BACK UP MBTU/YR

SENSITIVITY TO MASS THICKNESS, MATERIAL, AND CARPETING

HEATING

carpeted concrete
adobe
concrete

THICKNESS (inches)

SENSITIVITY TO MASS THICKNESS, MATERIAL, AND CARPETING

COOLING

carpeted concrete
adobe
concrete

THICKNESS (inches)

APPLY LATH TO 2X STUDS
NAIL ⅞" THICK FURRING ON STUDS
APPLY SCRATCH & BROWN COATS OF CEMENT STUCCO (NOT GYPSUM)
APPLY 2ND LAYER OF LATH

2" THICK

APPLY 2ND SCRATCH & BROWN COATS OF CEMENT STUCCO

APPLY FINISH COAT OF KEENE CEMENT OR THIN WALL GYPSUM PLASTER

STRAW-BALE INFILL WALL

DISTRIBUTED MASS STUCCO WALL

Several distributed mass components and their optimization. [Illustration courtesy of San Luis Sustainability Group. Data taken from K. Haggard and P. Niles, Passive Solar Handbook for California. Sacramento: California Energy Commission, 1980, 119 ("Distributed Mass Performance Sensitivity") and 171 ("Concentrated Mass Performance Sensitivity").]

that needs a large amount of thermal mass for the particular climate in which it is located. Integrated multiple systems afford a wider range of options for including thermal mass throughout the building, increasing the amount of mass that can be architecturally accommodated.

ARCHITECTURAL CHARACTERISTICS OF RADIANT TRANSFER

Some of the mistakes made in the mid-1970s can clarify important principles in regard to radiant transfer in architecture. One example concerns the color of distributed thermal mass. Many early designs used dark colors on the surfaces of interior masses, based on the assumption that more heat would be absorbed. This idea proved to be simplistic. Masonry, even with a dark surface, absorbs heat somewhat slowly. Thus, the surface temperature can get too hot and overheat the adjacent air, creating too large a temperature swing within the room while adding very little extra heat to storage. Dark surfaces have also created glare problems in buildings that use natural lighting, by producing too much contrast with the windows. Therefore, it was found that with distributed mass, it is generally better to keep the surfaces light in color. Enough direct sunlight is still absorbed to build up storage, and the excess is reflected to other mass that is not in direct sun at that particular time. In addition, natural lighting works significantly better.

In contrast, the water tank detail used to illustrate concentrated mass should certainly be as dark as possible on the sun side for short wave radiation absorption, even to the point of using a selective surface. A selective surface has higher absorptivity than emissivity, so it can act as a one-way thermal valve, providing more absorption and less emission back to the receiving side of the concentrated mass. However, this concentrated mass need not be dark on the interior side, where heat is being radiated from the mass to the room. The reason is that this transfer of heat is in the form of long wave radiation, and it makes very little difference at these temperatures what color this surface is as long as it is not bare metal, which is a very poor radiating surface. Once again, natural lighting works far better with light colors on the interior face of the mass.

RELATIONSHIP BETWEEN INSULATION AND THERMAL MASS

For most passive buildings there are two simultaneous needs. The first is the need to improve the building envelope in regard to thermal transfer so that the relatively diffuse on-site energies can be used effectively. The second need is to get interior temperature swings within the comfort zone by providing enough thermal mass. Providing one without the other is generally not very effective, as illustrated by the adobe wall shown on the facing page, which provides wonderful interior temperature damping but poor insulation. Usually, the optimal condition is achieved by combining materials so that there is good insulative material toward the outside of the building and good temperature-damping material on the interior surface.

SOLAR GEOMETRY AND ARCHITECTURE

The angles of the sun on a site and building are critical to heating, cooling, and lighting, and the designer must take them into account. The use of descriptive geometry, accurate sections, computer analysis, or physical models can determine whether an area will be shaded for a particular orientation and time on a particular site.

With the use of sun angle information, the building's facade can be designed to accept sunlight in winter, to be shaded in summer, or both. In temperate zones, underheating in early spring and overheating in late fall pose a greater design challenge, because the sun angles are the same for these two very different loading conditions. It is here that designers some-

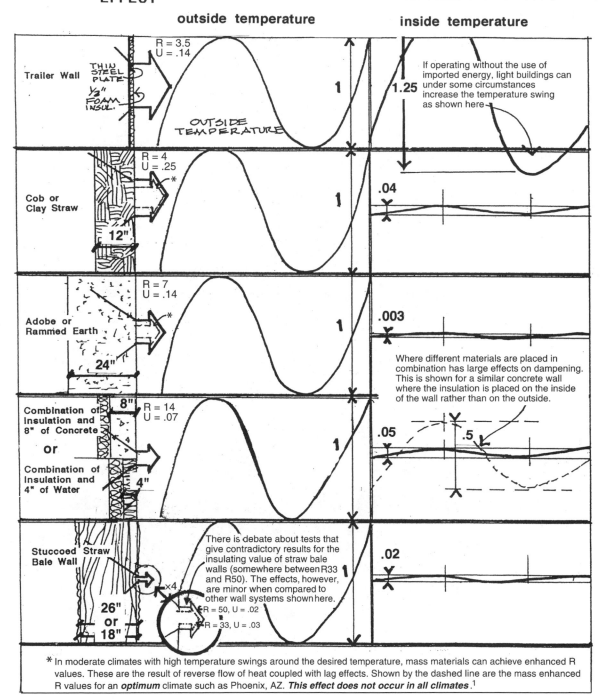

INSULATING EFFECT DAMPENING EFFECT

outside temperature   inside temperature

Trailer Wall — THIN STEEL PLATE, 1/3" FOAM INSUL. — R = 3.5, U = .14 — OUTSIDE TEMPERATURE — 1 — 1.25

If operating without the use of imported energy, light buildings can under some circumstances increase the temperature swing as shown here

Cob or Clay Straw — R = 4, U = .25 * — 12" — 1 — .04

Adobe or Rammed Earth — R = 7, U = .14 * — 24" — 1 — .003

Where different materials are placed in combination has large effects on dampening. This is shown for a similar concrete wall where the insulation is placed on the inside of the wall rather than on the outside.

Combination of Insulation and 8" of Concrete — 8" — R = 14, U = .07 — 1 — .05 — .5

or

Combination of Insulation and 4" of Water — 4"

Stuccoed Straw Bale Wall — 26" or 18" — ×4 — R = 50, U = .02 — R = 33, U = .03 — 1 — .02

There is debate about tests that give contradictory results for the insulating value of straw bale walls (somewhere between R33 and R50). The effects, however, are minor when compared to other wall systems shown here.

* In moderate climates with high temperature swings around the desired temperature, mass materials can achieve enhanced R values. These are the result of reverse flow of heat coupled with lag effects. Shown by the dashed line are the mass enhanced R values for an *optimum* climate such as Phoenix, AZ. **This effect does not occur in all climates**.[1]

Insulation and dampening effect on interior temperature swings by various materials and combinations. The damping effect of high-mass materials on exterior walls can generally be beneficial only when the outdoor temperature cycles around the comfort zone. However, interior mass, with adequate insulation to the outside, can also be helpful by storing heat from the sun or coolness from cool night air or, in some cases, the night sky radiation, to make the interior temperatures comfortable. (Illustration courtesy of San Luis Sustainability Group. Data taken from C. G. Ramsey and H. Sleeper, Architectural Graphics Standards, 8th Ed. New York: John Wiley & Sons, 1988, 730–733.)

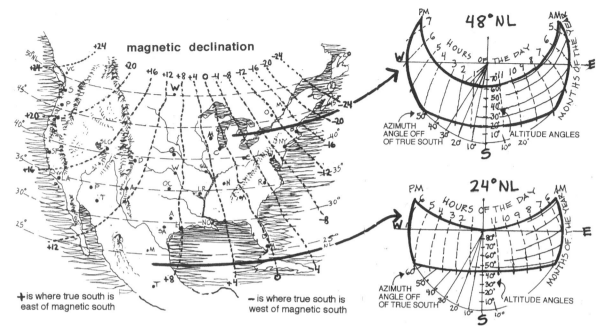

Sun angle charts. (Illustration courtesy of San Luis Sustainability Group. Data taken from C. G. Ramsey and H. Sleeper, **Architectural Graphics Standards,** *8th Ed. New York: John Wiley & Sons, 1988, 730–733.)*

WINTER

SUMMER

Illustration of solar acceptance and control by architectural design. (Photos courtesy of San Luis Sustainability Group.)

times resort to movable control devices (see "Appropriate Passive Strategies," page 56) to further fine-tune solar utilization and control.

AIRFLOW PATTERNS AND ARCHITECTURE Wind-induced airflow patterns are complex and often chaotic, difficult to predict exactly but easy to understand by examining patterns. The basics of these patterns are shown on the next page in a brief digest of technical reports from the Texas Engineering Experimental Station published in the mid-to-late 1950s.[2]

MODELING TECHNIQUES Because a passive building is a complex holistic entity, optimizing performance generally requires computer simulation. Good mathematical models were developed during the passive solar era in the late 1970s and early 1980s. Since then they have been validated, improved, and made relatively user-friendly. For our "Comparative Analysis for Six Different Climates" (shown later in this chapter), we used CALPAS, a relatively early modeling program, which is a predecessor to Energy 10 developed by Doug Balcomb's group at the National Renewable Energy Laboratory and marketed by the Passive Solar Energy Industries Association.

TRUE COSTS What does natural conditioning cost? First we must recognize that we live in a highly subsidized, fairly artificial economy. Therefore, what we pay for some things often does not reflect their actual cost. Amory Lovins of the Rocky Mountain Institute and others have pointed out, for example, that oil is subsidized about $150 per barrel if the cost of defending our tenuous supply from the Middle East is included in the calculation. Even this amount does not take into account the damage caused by the by-products of burning this oil, in the form of contributions to the greenhouse effect, acid rain, air pollution, and so on.

One can ask the value of personal health, the value of community health, or the value of healthy ecosystems. These values are very meaningful, yet difficult to quantify. In an evolving sustainable era we seek to create economic systems in which things are priced more closely to their actual cost to all of us. Working toward true-cost accounting entails complex and challenging, yet critically important, cultural changes.

The costs of passive systems can be small because, by their very nature, such systems are integrated. They can also offer high value in relation to their cost. A key value in naturally conditioned buildings is greater comfort. In addition, if each building is part of a larger whole that provides more

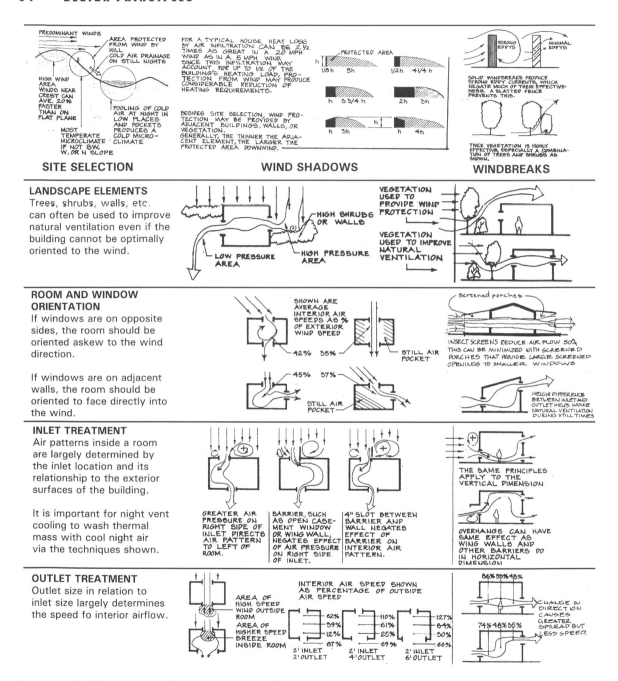

SITE SELECTION

PREDOMINANT WINDS

AREA PROTECTED FROM WIND BY HILL
COLD AIR DRAINAGE ON STILL NIGHTS

HIGH WIND AREA. WINDS NEAR CREST CAN AVE. 20% FASTER THAN ON FLAT PLANE

POOLING OF COLD AIR AT NIGHT IN LOW PLACES AND POCKETS PRODUCES A COLD MICRO-CLIMATE

MOST TEMPERATE MICROCLIMATE IF NOT SW, W, OR N SLOPE

FOR A TYPICAL HOUSE HEAT LOSS BY AIR INFILTRATION CAN BE 2½ TIMES AS GREAT IN A 20 MPH WIND AS IN A 5 MPH WIND. SINCE THIS INFILTRATION MAY ACCOUNT FOR UP TO ½ OF THE BUILDING'S HEATING LOAD, PROTECTION FROM WIND MAY PRODUCE CONSIDERABLE REDUCTION OF HEATING REQUIREMENTS.

BESIDES SITE SELECTION, WIND PROTECTION MAY BE PROVIDED BY ADJACENT BUILDINGS, WALLS, OR VEGETATION. GENERALLY, THE THINNER THE ADJACENT ELEMENT, THE LARGER THE PROTECTED AREA DOWNWIND.

WIND SHADOWS

PROTECTED AREA
1/8 h   5h   1/2 h   4 1/4 h
h   3 3/4 h   2h   3h
h   3h   h   4h

WINDBREAKS

STRONG EDDYS   MINIMAL EDDYS

SOLID WINDBREAKS PRODUCE STRONG EDDY CURRENTS, WHICH NEGATE MUCH OF THEIR EFFECTIVENESS. A SLATTED FENCE PREVENTS THIS.

THICK VEGETATION IS HIGHLY EFFECTIVE, ESPECIALLY A COMBINATION OF TREES AND SHRUBS AS SHOWN.

LANDSCAPE ELEMENTS

Trees, shrubs, walls, etc. can often be used to improve natural ventilation even if the building cannot be optimally oriented to the wind.

HIGH SHRUBS OR WALLS

LOW PRESSURE AREA

HIGH PRESSURE AREA

VEGETATION USED TO PROVIDE WIND PROTECTION

VEGETATION USED TO IMPROVE NATURAL VENTILATION

ROOM AND WINDOW ORIENTATION

If windows are on opposite sides, the room should be oriented askew to the wind direction.

If windows are on adjacent walls, the room should be oriented to face directly into the wind.

SHOWN ARE AVERAGE INTERIOR AIR SPEEDS AS % OF EXTERIOR WIND SPEED

42%   35%   STILL AIR POCKET

45%   37%   STILL AIR POCKET

Screened porches

INSECT SCREENS REDUCE AIR FLOW 50%. THIS CAN BE MINIMIZED WITH SCREENED PORCHES THAT PROVIDE LARGE SCREENED OPENINGS TO SMALLER WINDOWS

HEIGHT DIFFERENCE BETWEEN INLET AND OUTLET HELPS INDUCE NATURAL VENTILATION DURING STILL TIMES

INLET TREATMENT

Air patterns inside a room are largely determined by the inlet location and its relationship to the exterior surfaces of the building.

It is important for night vent cooling to wash thermal mass with cool night air via the techniques shown.

GREATER AIR PRESSURE ON RIGHT SIDE OF INLET DIRECTS AIR PATTERN TO LEFT OF ROOM.

BARRIER, SUCH AS OPEN CASEMENT WINDOW OR WING WALL, NEGATES EFFECT OF AIR PRESSURE ON RIGHT SIDE OF INLET.

4" SLOT BETWEEN BARRIER AND WALL NEGATES EFFECT OF BARRIER ON INTERIOR AIR PATTERN.

THE SAME PRINCIPLES APPLY TO THE VERTICAL DIMENSION

OVERHANGS CAN HAVE SAME EFFECT AS WING WALLS AND OTHER BARRIERS DO IN HORIZONTAL DIMENSION

OUTLET TREATMENT

Outlet size in relation to inlet size largely determines the speed fo interior airflow.

AREA OF HIGH SPEED WIND OUTSIDE ROOM

AREA OF HIGHER SPEED BREEZE INSIDE ROOM

INTERIOR AIR SPEED SHOWN AS PERCENTAGE OF OUTSIDE AIR SPEED

62%   110%   127%
59%   61%   84%
12%   25%   30%
87%   69%   66%
2' INLET 2' OUTLET   2' INLET 4' OUTLET   2' INLET 6' OUTLET

86% 59% 48%   CHANGE IN DIRECTION CAUSES GREATER SPREAD BUT LESS SPEED.
74% 48% 35%

Summary of airflow patterns. (Illustration courtesy of San Luis Sustainability Group.)

value (local jobs, equity, health, and efficiency), we have the beginnings of a more balanced and sustainable economy.

Nested Design Considerations

Architectural design involves a series of studies and decisions about specific aspects of a building, which must eventually merge to create a unified whole. Thermal and illumination needs must be given the same design priority they once held, before we became overreliant on mechanical systems. It is critically important that these considerations receive attention at the beginning of the design process, because early decisions often have the largest effects.

SITING, ORIENTATION, AND FORM

There are few rules in architectural design that are as consistently true and as important as that which requires a building to be oriented toward the equator of our planet in temperate zones (south orientation in the Northern Hemisphere and north in the Southern Hemisphere). This is an immutable relationship, not to be discounted without undesirable results. There are many reasons for this dictum. In the Northern Hemisphere, for example, south sun can be available over a whole winter's day and can be easily shaded in summer. East and west have only a half day's orientation in each season, and the low sun angles they intercept are difficult to control, thus increasing unwanted thermal loads and creating severe glare problems for natural lighting. There is some flexibility to this rule for small-angle variations. Such variation, however, should not diverge more than 15 to 20 degrees from true south, depending upon the specific site conditions.

In addition to solar orientation, siting must take into effect airflows, not only for ventilation and passive cooling potentials but also to minimize infiltration losses. All of these factors are related to vegetation and topography, so a thorough site analysis is a prerequisite for any passive building design.

Another early design consideration should be the form of the building in regard to scale and thermal type (see "Building Scale, Metabolism, and Form" under "Basic Principles,"earlier in this chapter). These factors affect the ease or difficulty of achieving a synthesis of passive heating, cooling, and lighting. Equally important are the social effects. Huge, interior load–dominated buildings are more difficult to integrate into the landscape, urban setting, or social fabric. In the emerging sustainable era, buildings must be considered systems within larger systems, not just maximum rental areas or monumental icons.

BUILDING ENVELOPE The envelope of a passive building is a human-designed boundary that optimizes through flows by multiple architectural elements. The incoming flows of consequence include the following:

1. Air via infiltration and ventilation. Ventilation can be enhanced mechanically by passive means, such as by chimney effects or by photovoltaic-powered fans.
2. Heat via radiation, convection, and conduction. Radiation is a more frequently used heat transfer mechanism in passive buildings than in mechanically conditioned buildings, which generally rely on convection.
3. Sunlight—either direct or diffuse.

Outflows of consequence generally include the following:

1. Air exfiltration and ventilation
2. Heat either by radiation, convection, or conduction

APPROPRIATE PASSIVE STRATEGIES Once the appropriate flows through the envelope are developed, the energy in the interior space of the building can be further optimized by

1. Using thermal mass to hold heat or coolness during periods of need
2. Arranging various surfaces to enhance energy transfers
3. Designing spatial arrangements that can transfer energy by convection
4. Making use of shades, shutters, blinds, and reflectors to fine-tune transfers of energy and light

| | HEATING | | COOLING hot, humid climate | | COOLING hot, dry climate | | | |
| | | | | | day | | night | |
	IN	OUT	IN	OUT	IN	OUT	IN	OUT
AIR FLOW	X*	X*	✓	✓	X	X	✓	✓
HEAT FLOW	✓	X	X	✓	X	X	X	✓
FLOW OF LIGHT	modified to be visually functional		modified to be visually functional and comfortable with minimal associated heat gain					

*except for ventilation required for a healthy interior

Envelope flow strategies for several climate needs. (Illustration courtesy of San Luis Sustainability Group.)

The various passive strategies are characterized by the type of flow through the envelope, major thermal function, and type of mass. Thus, for a situation in which sunlight enters directly into the interior space to heat thermal mass that consists of thin stuccoed interior walls, there is a *direct gain/distributed mass* passive approach to heating. There are also *direct loss/distributed mass* applications for cooling by night ventilation. Because most passive buildings need both heating and cooling, the designer usually attempts to combine functions using the same elements—in this case, a *direct gain/loss–distributed mass* system. There are also *direct gain/loss–concentrated mass* approaches and, of course, combinations of thermal mass applications.

If thermal mass is integrated within the envelope so that the interior space is heated or cooled by the surfaces of the envelope, the approach is called *indirect gain* in the case of heating and *indirect loss* in the case of cooling. Indirect strategies can be applied to walls, called thermal walls, or to roofs, where such strategies can take the form of roof ponds. Thus, there can be *indirect gain/loss–wall mass* and *indirect gain/loss–roof mass,* as in the case of roof ponds.

generic base	basic passive types	variations and enhancements
DIRECT SYSTEMS	direct gain/loss-distributed mass	• integral reflector • direct gain roof • solar induced ventilation • thin eutectic storage • blinds, shutters, and shades
	direct gain/loss-concentrated mass	• focusing aperture • movable insulation • eutectic storage
INDIRECT SYSTEMS	thermal wall heating and cooling	• vented thermal wall • reflectors • movable insulation
	roof pond heating and cooling	• roof enclosed for heating assist • fan coil assist for multistory • evaporative cooling assist • stepped insulation with reflectors
ISOLATED SYSTEMS	thermosiphon loop heating sun-room heating cooling tower	

Summary of passive strategies. (Illustration courtesy of San Luis Sustainability Group.)

A third approach to passive strategies is to have a separate architectural element perform each thermal function required. This approach is classified as an *isolated system.* An isolated system can consist of a greenhouse or sun-room, thermo-siphon air heaters for heating, or a cooling tower for cooling. *Isolated gain–distributed mass,* for example, would describe a thermo-siphon air heater that heats a thin, high-mass plenum under a floor.

FINE-TUNING OF PASSIVE SYSTEMS Finally, there is a more detailed level of thermal control for different seasons, different rooms, and different times of the day or night, accomplished by movable insulation, insulating curtains, reflectors, blinds, shades, and shutters. The manufacture and availability of these devices has been greatly improved. As a result, there are much better off-the-shelf components available for these functions than ever before.

Comparative Analysis for Six Different Climates This section introduces thermal sizing and calculations via holistic computer modeling. It compares the performance of several of the alternative material assemblies that are the subject of this book for very simple, skin-dominated, direct gain/loss buildings. In addition, there are excursions into other passive strategies in which these assemblies may provide better results for a particular climate.

(a)   *(b)*   *(c)*

Fine-tuning of a direct gain system. (a) Tuning of light penetration with movable insulated louvers. (b) Control of fall and spring solar gain with movable shades that can reverse adjust. (c) Specialty shades for harsh late-afternoon sun control. (Photos courtesy of San Luis Sustainability Group.)

Several building schemes were modeled comparing the thermal performance of different wall assemblies. The building schemes utilize a direct gain/loss distributed mass passive system with slab-on-grade and wood frame roof.[3] A skin-dominated direct-gain system for a midsized residence was chosen so that the wall assemblies could be easily compared without inadvertently comparing different types of passive heating strategies. Six U.S. temperate climate zones were chosen: Denver, Colorado, and Helena, Montana, for cold climates; Fresno, California, and Philadelphia, Pennsylvania, for moderate climates; Tampa, Florida, and El Centro, California, for hot climates. For each climate zone two sets of buildings were analyzed thermally. The first (Iteration 1) models a base case building with stick-frame construction. The second (Iteration 2) models a modified version of the original base case building (the floor plan and elevations are conceptually the same) that improves passive performance by better orientation, sun control, glazing, and use of thermal mass. For each iteration several wall assemblies were substituted for the stick-frame construction of the base case building.

These new building configurations were modeled for stick-frame construction with mass, uninsulated adobe, insulated adobe, straw-bale construction, and a combination of adobe and straw-bale. For three of the climate zones a third set (Iteration 3) of buildings was modeled. This third set was chosen where a passive system other than direct gain–distributed mass would improve the building's performance. The results of the computer simulations, expressed as annual heating and cooling loads in Btu/square foot of building area, are shown on the accompanying charts.

In each of the climate zones the performances are different, as may be expected. It is worth noting that there are no hard-and-fast rules, such as "Straw-bale is better than adobe or rammed earth" or "Adobe and rammed earth do not need exterior insulation." This study shows, for example, that in situations similar to those found in the first set of modeled buildings for Fresno, an uninsulated 24-inch adobe or rammed earth building performs better than a code-complying stick-frame building typically found in the area. This does not tell the full story, however. The second iteration stick-frame, with some added mass, greatly outperforms the uninsulated 24-inch adobe. The straw-bale outperforms both, but when insulation is added to the adobe, this new combination becomes the best performer. Thus, for this particular climate, adobe needs insulation.

For each of the six climate zones the conclusions have one conceptual similarity: *The best-performing wall assemblies are those composed of both mass and insulation*—the insulation applied to the exterior and the mass to the interior. The proportion of the mass to the insulation depends on the particular climate. It is important to note that the effectiveness of the mass and the insulation is not fully realized until the efficiency of the building as a whole is optimized. For example, adding mass to an Iteration 1 scheme may not make as great an improvement as possible until the building is better insulated and the size and type of the windows are adjusted as well. For earthen and straw structures the goal should be to choose an appropriate insulative material and an appropriate massive material to complement the original material of choice. Then, with the help of computer thermal optimization, the amounts of insulation and mass can be determined.

DENVER, COLORADO Denver has mild summers and cold, but often sunny, winters. Denver and other similarly cold climates have a primary need for heat conservation, and thus insulation, and a secondary, but very important, need for thermal mass. The full potential of the mass and insulation cannot be realized until the building has been designed with the climate in mind, as in Iteration 2.

The importance of insulation is clearly demonstrated in Iteration 1, in which the highly insulative straw-bale version is the best choice. Uninsulated adobe or brick is not appropriate for this climate.

The performance of the original buildings can be greatly improved by decreasing the amounts of north, east, and west glass, improving the window glazing R-value with low-E argon, decreasing the overhangs, and similar measures (Iteration 2). The stick-frame building can be dramatically improved by performing all of these steps and by adding 1,500 square feet of 2-inch mass (2.1). If the same techniques are employed, the 24-inch uninsulated adobe (2.2) will use nearly the same amount of energy for heating, but much less for cooling.

Although performance is improved, two problems are being ignored, underinsulating and/or undermassing. The stick-frame is underinsulated and undermassed, and the 24-inch adobe is underinsulated. The best choice in Iteration 2 is a hybrid version: the envelope wall composed of straw bales on the exterior and 24-inch adobe on the interior, which could also be achieved by using straw-bale for the exterior walls and the adobe for interior walls (2.5). This scheme requires only a fraction of the energy that the original base case required. A close second is the 24-inch adobe with an equiva-

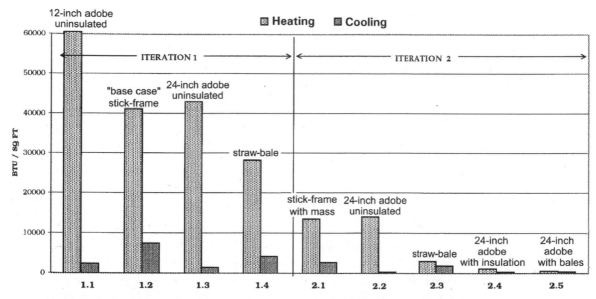

Performance comparison, Denver, Colorado. (Illustration courtesy of San Luis Sustainability Group.)

24" Adobe with R-21, Interior Float Temperature, and Exterior Temperature Range

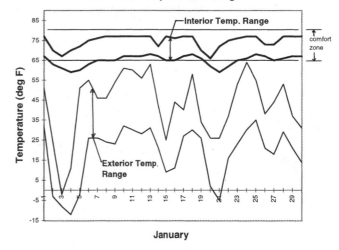

lent R-21 stud wall to the exterior (2.4), and third is the straw-bale with 2 inches of interior cement stucco (2.3). The latter three all have insulation and mass. When the interior temperature of the 24-inch insulated adobe (2.4) is allowed to float without mechanical backup heating, this scheme very nearly requires no backup energy. If it is totally independent, there would be a few nights below comfort zone during the winter and a few days above comfort zone during summer, perhaps uncomfortable for a short time, but certainly not life-threatening.

HELENA, MONTANA   Helena has very cold winters and mild-to-warm summers. The primary concern in Helena is heating and energy conservation. Even more critical than in Denver's situation, a high level of insulation is of tremendous importance, and, to a lesser but significant degree, so is mass.

For Iteration 1, the best solution is to follow energy-conserving building practice for cold climates (1.3),[4] then replace stud walls with straw bales (1.4). The performance of a building can be greatly improved by making subtle changes in the design to work better in the climate, as in Iteration 2. Once this is done, the second and third best performers are the 24-inch adobe interior with straw-bale exterior (2.4) and the straw-bale with 2-inch interior cement stucco (2.3), respectively. The best-performing scheme uses walls two bales thick (2.5). This is perhaps overkill by current building standards, but for a cold climate where energy resources may not be available, there may be a place for this practice. Low-insulated, low-massed structures, such as the stick-frame with mass, or uninsulated high-mass structures, such as 24-inch uninsulated adobe (not shown in Iteration 2 because it is such a poor performer), are not appropriate.

Iteration 3 achieves improved performance by excursions into other passive systems. The energy performance, and hence the comfort of these schemes, may be improved by adding an attached sun space on the south.

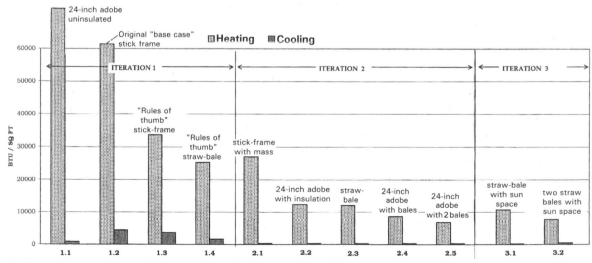

Performance comparison, Helena, Montana. (Illustration courtesy of San Luis Sustainability Group.)

The sun space can allow for extended outdoor use, as well as act as a buffer zone to the building. With the addition of a sun space to the straw-bale version (2.3) of Iteration 2, the heating and cooling performance is predicted to improve by 11 percent. The sun space covers two-thirds of the south elevation and is thermally isolated. With the adding of the same sun space to a superinsulated building, such as the two-bale-wide exterior wall with interior walls of adobe (2.5), the heating energy load improves by less than only 1 percent. The temperature of the sun space fluctuates wildly, but there are many days each month in winter when one can use the sun space comfortably. This is an incredible advantage over other homes that do not offer the additional weather-protected area. Moreover, the windows and walls that adjoin the sun space will not feel as cold as those that have direct contact with the extremely cold exterior temperatures. However, if one made the mistake of trying to keep the sun space within the comfort zone year-round, energy use would skyrocket.

FRESNO, CALIFORNIA Fresno has cool, foggy winters, with temperatures in the 30s to 50s (degrees Fahrenheit) and hot summers, with temperatures at night in the 50s, in the day in the day 90s+ (degrees). This climate exhibits a need for both heating and cooling, and for both insulation and mass.

The interesting discovery is that a highly massed building and a moderately insulated, highly massed building perform quite similarly in terms of overall energy use. In other words, the highly insulative straw-bale with 1½- to 2-inch interior cement stucco (2.3) and the 24-inch adobe with the equivalent of R-13 stud wall to the exterior (2.4) have nearly equal overall energy use. The straw-bale version is better in the heating mode; the adobe version is better in the cooling mode. Typically, but not surprising, neither situation is built in practice. To improve the performance of heating and cooling for these buildings, add the equivalent of R-21 to the exterior of the adobe (2.5) and add more interior mass, either distributed or concentrated, to the straw-bale (not shown).

Uninsulated high-mass structures, such as 24-inch uninsulated adobe, and low-insulated low-mass structures, such as the stick-frame with mass, are not advisable. Equally inadvisable is straw-bale with gypsum plaster or no other interior mass (2.4).

PHILADELPHIA, PENNSYLVANIA Philadelphia has cool-to-cold winters, in the 20s to 40s (degrees Fahrenheit), and warm summers, in the 70s to 90s. This location has a primary need for heating and some cooling. Thus, insulation and heat/energy con-

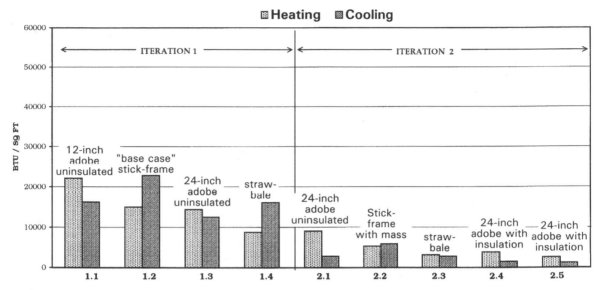

Performance comparison, Fresno, California. (Illustration courtesy of San Luis Sustainability Group.)

servation are of key importance. Thermal mass plays a significant role in cooling once it is beyond 2 inches thick. High mass, equivalent to 12- or 24-inch adobe, can play an important part once it has exterior insulation beyond that of an equivalent R-21 stud wall. Uninsulated adobe brick is not appropriate.

The best performer for Iteration 1 is clearly the straw-bale version. Insulation is a necessity for cold climates, which can also be seen in the Iteration 2 buildings. The best performer for Iteration 2 is a 24-inch adobe with straw bales to the exterior (2.6). Second best, requiring twice the energy of the prior situation, is a 24-inch adobe with equivalent R-21 to the exterior (2.4). An adobe with less than an equivalent R-21 yields poorer performance than the straw-bale version with its 1½- to 2-inch interior stucco.

TAMPA, FLORIDA   Tampa has mild winters, in the 50s to 70s (degrees Fahrenheit), and warm, humid summers, in the 70s to 90s. Tampa clearly has a primary need for cooling. The first investigation (Iteration 1) reveals that the 24-inch uninsulated adobe wall is the best choice, implying that high mass is of major importance and insulation not much of a issue.

However, this is not true once the building is efficiency optimized for the climate (Iteration 2). The straw-bale with 2 inches of interior cement stucco (2.4) outperforms the 24-inch uninsulated adobe version (2.3). Both

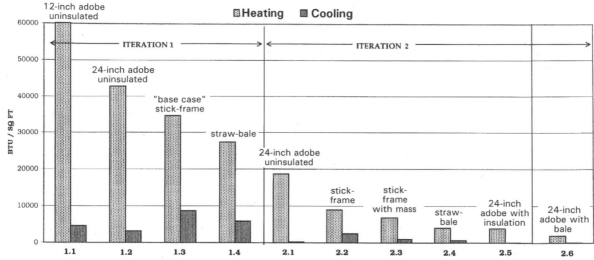

Performance comparison, Philadelphia, Pennsylvania. (Illustration courtesy of San Luis Sustainability Group.)

are outperformed, however, by adding the equivalent of R-13 wall to the exterior of the 24-inch adobe wall (2.5). Note that the latter two buildings both have insulation, straw-bale or equivalent R-13, and mass, 2-inch stucco or 24-inch adobe. *This implies that the question is not simply which material is better than another, but what combination of massive and insulative materials gives the best performance.* Although mass may be of primary importance, insu-

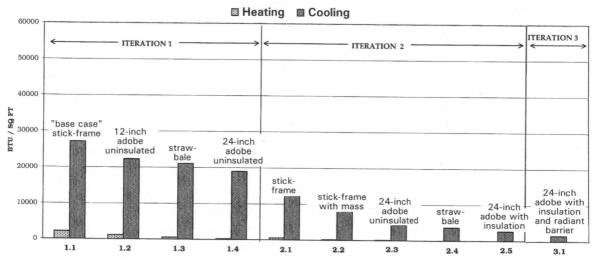

Performance comparison, Tampa, Florida. (Illustration courtesy of San Luis Sustainability Group.)

lation can improve the ability of mass to cool. This can be seen in the 38 percent improvement in cooling from the 24-inch uninsulated (1.4) to the 24-inch insulated adobe version (2.5).

However, results can be still better. Other cooling strategies are available to the designer: greater shading and ventilation techniques such as those found in the "dogtrot" style house; double roof systems; and the use of radiant barriers in walls and roofs.[5] Thermal performance improves when the 24-inch insulated adobe is modeled with a radiant barrier in both the walls and the roof and when stack ventilation is possible by utilizing the full height of the two-story structure (3.1).

A 42 percent improvement can be achieved using these techniques with exterior straw-bales and 4-inch adobe or cob interior walls, a scenario that can easily be built.

EL CENTRO, CALIFORNIA El Centro has cool-to-mild winters, in the 30s (degrees Fahrenheit), and very hot summers, at night in the 60s, during the day at 110+ degrees. This low desert community has some need for heating and great need for cooling; thus, it is a candidate for both insulation and mass. The extreme climate, at a minimum, warrants high insulation with moderate mass, or high mass with moderate insulation. The nearly equal need for mass and insulation is seen in Iteration 1, as the 24-inch uninsulated adobe and the straw-bale building perform equally in energy use.

After the building is designed to work within the climate, the simulation shows that the straw-bale assembly with interior cement stucco (2.2) significantly outperforms the uninsulated adobe (2.1), presumably because of the integrated mass and insulation. Both the straw-bale with interior cement stucco (2.2) and the 24-inch adobe with the equivalent of R-13 stud wall exterior insulation (2.3) require 46 percent less energy to cool and 90 to 100 percent less energy to heat than the uninsulated adobe structure in Iteration 1.

Adding mass to the straw-bale structure improves its heating and cooling performance, and adding more insulation to the 24-inch insulated adobe structure improves its performance. Continuing to add mass or insulation works only to a certain point, after which increased performance is achieved only with other types of cooling systems.

For this type of climate one should investigate other passive cooling strategies such as a high-mass roof system with movable insulation, like the roof pond system (Iteration 3). For the El Centro climate and other low-

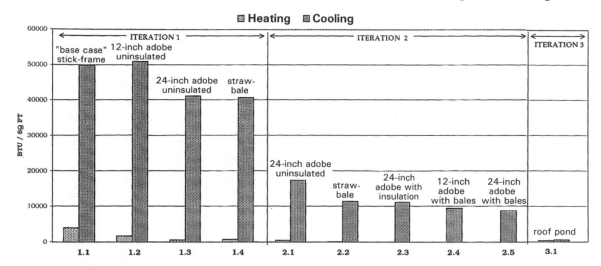

Performance comparison, El Centro, California. (Illustration courtesy of San Luis Sustainability Group.)

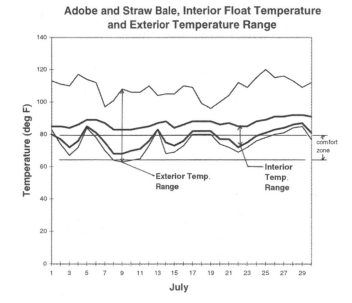

elevation deserts, the roof pond system has potential to greatly outperform the direct gain/loss–distributed mass system modeled in Iterations 1 and 2.[6] A roof pond system with 900 square feet of pond area 6 inches deep, R-6 movable insulation, and R-19 walls is predicted to use less than 5 percent of the cooling energy used by the straw-bale or 24-inch insulated adobe versions of Iteration 2. The heating load increases slightly, presumably because the walls are only R-19. Slab on grade with perimeter insulation, exterior building color, and glazing areas remain constant from Iteration 2 to Iteration 3.

Conclusions

MATERIALS

For each of the six temperate climate zones modeled, particular conclusions were drawn as to which material combination and proportions perform best. A common finding is that no material will perform as well as it can without the implementation of sound passive solar and natural cooling design strategies. The argument within the sustainability/earthen materials realm should address the question of what proportions of mass and insulation are best suited for a particular job. It should include prior discussion about which passive strategies are appropriate and what materials are available within the region. Nearly all the changes from Iteration 1 to Iteration 2 were subtle: increasing or decreasing overhangs, using heat-accepting or heat-rejecting low-E glass, decreasing non-south glass areas, adding fixed vertical or seasonal shading devices, and so forth. None of these strategies is new.

What is new is applying these strategies in conjunction with various *alternative* wall assemblies. It is unfortunate, but true, that many enthusiasts of alternative building materials, or straw-bale and earthen materials, see each material as a pure thing in itself. This perspective ignores one of the fundamental principles of sustainability and nature: *All things must be seen in conjunction and in cooperation with the whole. Nothing stands on its own.*

The best performers of all six climate zones were the hybrid schemes incorporating both mass and insulation. How much mass and how much insulation depend on the climate, the target comfort zone, and the amount of outside energy one wishes to waste or conserve.

PASSIVE CAPABILITIES

To achieve the level of performance discussed throughout this chapter requires analysis beyond the designer's intuitive ability. Holistic computer modeling permits designers to extend their intuition and significantly improve performance, even allowing in some cases stand-alone naturally conditioned buildings and, in most cases, reducing mechanical systems to the status of backup systems.

Notes

1. "Thermal Mass and R Value," *Environmental Building News* 7, no. 4 (April 1998).
2. B. Evans, *Natural Air Flow Around Buildings,* Research Report No. 59 (College Station: Texas Engineering and Experiment Station, Texas A&M College System, 1954).

3. P. Cooper and J. Rennick, " How to Make Your Straw Bale Oven," *Proceedings of the 22nd Annual Passive Solar Conference* (Washington, D.C.: American Solar Energy Society, 1997). The tongue-in-cheek title of this publication refers to poor solar design—not ovens.

4. J. Balcomb, R. Jones, R. McFarland, and W. Wray, *Passive Solar Heating Analysis: A Design Manual* (Atlanta: American Society of Heating, Cooling, Refrigerating, and Air-Conditioning Engineers, 1984).

5. P. Fairey, *Radiant Energy Transfer and Radiant Barrier Systems in Buildings,* FSEC-DN-6 (Cape Canaveral, FL: Florida Solar Energy Center, 1984).

6. K. Haggard and P. Niles, "El Centro Comparison of Passive Types," *Passive Solar Handbook for California,* 47 (Sacramento: California Energy Commission, 1980).

Structural Properties of Alternative Building Materials

Bruce King, P. E.

The increasing popularity of the so-called alternative materials in the industrialized nations presents an engaging and worthy challenge to the structural engineering community. Schooled to work almost exclusively with the "Big Four"—concrete, masonry, steel, and wood—many engineers are quite cautious or reluctant to work with materials they never read about in their textbooks. This reluctance can reach ironic proportions when, for example, an engineer freshly trained in America or Europe returns to his or her home in the Middle East or Africa but is unwilling to work with the vernacular sun-dried brick or adobe architecture that has been used within that culture for several millennia and is still, generally, all that the local population can afford. Upwards of 30 percent of the world's population is estimated to live in earth (chiefly adobe) housing, yet there is astonishingly little mention, much less study, of earth construction in engineering literature or building codes.[1]

In the 10,000 years or so that human beings have been erecting shelter, it has only been within the last two hundred years that there has been anything but anecdotal knowledge; the success or failure of structures depended entirely on the experience, judgment, and intuition of builders. With the refined application of the scientific method and the emergence of

Alternative Construction: Contemporary Natural Building Methods, edited by Lynne Elizabeth and Cassandra Adams ISBN 0-471-24951-3 © 2000 John Wiley & Sons, Inc.

professional architects and engineers, the construction industry has developed a detailed experimental understanding of building materials that has led to an increased reliance on laboratory testing over field experience. This has made possible the far larger and more elaborate buildings necessary to shelter, feed, and transport the exploding population, and the now-hallowed concept of the "independent laboratory test" has done a great deal to separate promotional, distorted, or outright false information from fact.

However, laboratory tests are expensive, so materials and systems with no financial backing from affluent individuals, commercial enterprises, trade associations, or government tend to be undertested or remain untested, and therefore dismissed by the modern building community. Because much of what we know about the various alternative materials to be discussed in this chapter *is* empirical or anecdotal knowledge, it seems worthwhile to examine the distinction between empirical field knowledge and "lab results." Anecdotal knowledge is simply "something I heard or saw"; both the quality of the information and reliability of the source can vary enormously. In contrast, laboratory testing is, in the ideal, clearly and thoroughly presented, replicable, and untainted by commercial or promotional intent. As a practicing engineer, I have received a wealth of anecdotal information that met all the stated criteria for laboratory testing, and have, conversely, reviewed lab reports from well-known, accredited labs that were so poorly done as to be useless or even misleading. So the distinction blurs, and the building professional must exercise intelligent discretion in reviewing all information encountered. This point is made not to dismiss the great value of laboratory results nor to assign validity to the many unfounded claims that have been made about alternative materials; rather it is to suggest that anecdotal knowledge about alternative materials should not be dismissed solely on the basis that it has not been laboratory tested.

Engineering design of building structures requires a knowledge of site (soil, topography, and climate), engineering principles in general, and physical properties of the building materials. Applying for and receiving a building permit in a place like California requires that the engineer or architect demonstrate that an alternative method or material will have satisfactory properties of fire and seismic safety, durability, ventilation, and sanitation.

The following discussion of the physical properties of these materials has categorized them, if not quite cleanly, as being "natural" (i.e., unprocessed or minimally processed materials) or "industrial" (i.e., reused or reprocessed industrial waste products, or more intensively processed materials). The discussion is based on personal experience with most of the materials described, interviews with various individuals skilled in alternative building, and a review of available literature. It is neither comprehensive nor exhaustive (additional references for further reading may be found in the notes at the end of this chapter and in Appendix A of the book), but is intended to offer other working professionals a sense of how the materials behave and how they can be designed into building structures.

Natural Materials

STABILIZED EARTH STRUCTURAL DESIGN PRINCIPLES

Following industrial and biological processes far enough back in time, one can see that in a sense *all* building materials are stabilized earth. That pithy insight aside, this discussion looks at a wide range of historic and new methods of stabilizing earth to form building structures of varying strength and durability.

Earth (exclusive of organic topsoils, which are never good construction material) covers the planet in a layer of varying thickness and appears in a virtually infinite array of physical and chemical properties. Forensic geologists can now pinpoint the origin of a soil sample to its source on or near the earth's surface, often within a few meters. Of prime concern to an earth builder is the particle gradation: How much of an earth supply is gravel (larger than 2 or 3 millimeters), sand (from 1 hundredth to 2 or 3 millimeters), and fines (less than 75 thousandths of a millimeter). The fines, in turn, contain silt (essentially very fine sand) and clay (leached minerals whose macroscopic properties change with wetting or drying). In historic adobe, cob, and rammed earth structures, the best building soils typically contained between 20 and 30 percent clay as the stabilizing binder, and relatively little silt. Such structural materials typically attain compressive

A test block of rammed earth is prepared before constructing walls to check compactibility, strength, and color of soil mix. From a workshop demonstration led by David Easton. (Photo by Robert Bolman.)

strengths between 200 and 800 pounds per square inch (psi), and 300 psi is considered by some adobe codes to be a minimum.[2]

In modern times, lime, fly ash, bitumen, and, especially, portland cement are sometimes added as stabilizers, and in sufficient quantity they can turn virtually any soil into buildable soil. With any of these stabilizers used as additives, strengths of 1,000 to 2,000 psi are easily attained, and I have experience with cement-stabilized (8 percent by weight) rammed earth reaching long-term strengths in excess of 6,000 psi. In some cases, such as with the tire houses developed by Michael Reynolds or the earthbag systems developed by Nader Khalili (see Chapter 8, "Modular Contained Earth"), little or no chemical additive is required, as the earth is stabilized by containment in durable surrounds. In other cases, such as with compressed adobe block or rammed earth, strength is developed or greatly increased by physical compaction.

The strength of historic, and most modern, earthen structures also relies on sheer massiveness; for example, the relative weakness of the material is compensated for by the low height/width ratios (e.g., 2-foot-thick rammed earth walls are typically only 9 or 10 feet high). Adobe and cob buildings may incorporate buttressing or curved walls, and sometimes vaulted or domed roofs, that are inherently stronger than straight walls connected at right angles. Lacking other design criteria, builders and engineers often use empirically derived height/width ratios such as those published in the *Uniform Code for Building Conservation* (UCBC). Guidelines established by unreinforced masonry practice are also helpful, such as the limits on size and proportions of piers and openings.

In nonseismic areas (zone 2B or less in the parlance of the *Uniform Building Code*, UBC), the need for stabilization is driven more by durability than strength. Traditional societies in Asia, Arabia, North Africa, and the American Southwest have derived hundreds of useful years of service from adobe buildings by simply replastering every year (the plaster is the same earthen material), usually as part of a cultural event in the spring. In lieu of annual replastering, a modest amount of binder in the wall material, especially with good roof overhangs and a raised foundation, can provide decades of serviceable life. The application of cement plaster to an earth wall will *not* improve the building's service life, because the relatively impermeable cement stucco traps moisture, leading to an unseen and rapid erosion of the earth wall beneath. Any applied plaster must be able to "breathe"; that is, it must have a vapor permeability commensurate with its substrate. For earthen structures not highly stabilized with cement or lime, protection from moisture (especially in freeze/thaw climates) can be the primary design concern.

Structural design in high seismic risk areas is overwhelmingly dominated by earthquake loads, because seismic loads are proportional to mass. In strong seismic events badly designed structures, such as those in poor regions of the world, often fail catastrophically, although many have shown surprising strength and ductility.[3] Durable earthen structures in seismic risk areas tend to rely either on cement binders (typically 5 to 10 percent of the earth mix) with reinforcing of steel or plastic mesh (earth-retaining geotextiles are sometimes used), or on a very stable geometry of thick or curving walls. If reinforcing is used, the working stress design method is conservative and appropriate, as reinforcing steel will bond to cement-stabilized earth with strengths of 400 psi and higher.[4] Walls can be designed to cantilever from wide foundations under out-of-plane loadings and should in all cases be well locked together at the tops with some form of reinforced bond beam. (*Editor's note:* Alternative strategies are presented in Chapter 5, "Adobe," under "Engineering Considerations.")

Centuries of earth building experience and modern research and testing have shown that earthen construction can be safe, even in the most seismically active areas. Sometimes modest amounts of materials such as cement and reinforcing are needed, but always an understanding of the specific soil, of the specific building method, and of engineering principles in general is required to ensure safety and durability.

PLASTERED STRAW-BALE STRUCTURAL DESIGN PRINCIPLES There have been many historic, indigenous buildings of pure straw, such as the reed houses of the marsh cultures of Peru, southern Iraq, and California, but no one has seriously attempted a revival in reed structures. However, particular promise and interest lie in the concept of building with stacked and plastered bales of straw—the same bales that modern farmers need to dispose of. Over the past decade or so, people have been building and experimenting with straw bales as construction materials (either as load-bearing structures or as infill within post-and-beam walls), and there now exists a body of test results and anecdotal knowledge about straw-bale structures that, although modest and inexact, gives some basis for understanding how these buildings work. Specifically, architects and engineers have learned the following:

1. The type of grain from which bales are formed (wheat, rice, barley, etc.) does not appear to matter, although the higher silica content in wet-farmed crops such as rice make these bales slower to rot (but harder on tools).

2. The interior moisture content of the bales should be kept below 20 percent (ideally about 15 percent, as in air-dried wood) before installation and plastering.

3. The denser the bale, the better for building. Seven pounds per cubic foot (dry density) is considered minimal for load-bearing bale walls.

4. Stacked straw bales will settle in the first few weeks, but will settle less if pounded heavily into place. Builders should either allow time for such a wall to settle or precompress the wall with strapping, tying the top of the wall to both sides of the foundation. This strapping also serves to stiffen the entire structure.

5. The stacked bales should be pinned for stability during construction, usually with rebar or bamboo. Extensive pinning (either driven through the bales or strapped in pairs to the outside) is mandated in the first straw-bale building codes, and although useful for job site stability, the contribution of the pins to the completed structure is uncertain and mostly untested.

6. Like earth walls, straw-bale walls are inherently massive and derive much of their stability and strength from simple geometry. I know of no case of a wind loading failure in a straw-bale structure, including two instances in which hurricane-force winds struck unplastered walls. The existing straw-bale building codes provide empirical guidelines for wall geometry, restricting height/width and height/length ratios and establishing minimum thicknesses.

7. Perhaps the most important lesson learned to date is that straw-bale walls need to breathe; that is, they must have vapor-permeable surfaces. Anecdotal evidence and test results indicate that the only places on a bale wall where moisture membranes may be appropriate are at the very bottom, separating the bales from the foundation; on horizontal surfaces (such as the top of the wall and windowsills); on interior walls adjacent to plumbing fixtures; and possibly on the exterior side of the first course or two. Straw-bale walls are vulnerable to moisture-induced decay, but problems and failures to date can be traced to outright leaks of water, not to water vapor within the walls.

Test results have shown that unplastered straw-bale walls, particularly when pinned and/ or compressed, constitute an effective load-bearing system, but they are generally much more elastic than the standard materials used for doors, windows, and cladding. Typical strength values for a single straw bale are as much as 70 psi in compression, with an elastic modulus of about 200 psi. These tests tell us that straw bales by themselves can provide useful strength and ductility, particularly under gravity and in-plane loading; however, the addition of *any* plaster to a wall—regardless of the plaster strength—transforms the wall into a hybrid straw/plaster structural element. The rigid plaster becomes the primary load-carrying element, and the straw serves as a structural, insulating substrate. As discussed in the next section, plaster applied directly into the straw is the primary structure, not just deadweight as some engineers have treated it. This condition can and should be recognized in design.

Scratch-coat exterior plaster is applied in a training workshop to a straw-bale wall prepared with stucco mesh. Notice that higher-gauge mesh over construction paper is used to strengthen and define corners. (Photo by Robert Bolman.)

In its short history, straw-bale design has often been based on the conservative, empirical prescriptions given in the first straw-bale codes and guidelines, but a straw-bale wall can also be designed as a structural sandwich panel in which the plaster skins act as very thin, reinforced concrete walls and the straw-bale substrate acts as a weak but effective shear element. Scratch and brown-coat stucco layers (as defined in UBC), with normal stucco mesh or heavier wire fabric, can be used for gravity and in-plane loads, and the sandwich mechanism clearly functions both for out-of-plane and vertical loading. The UBC allows 180 pounds per linear foot (plf) for $\frac{7}{8}$-inch three-coat stucco under in-plane shear, and engineers in California have adopted this guide for straw-bale construction. Recent tests on stucco-plastered straw-bale walls report failure loads of 1,538 pounds per linear foot in in-plane shear,[5] and up to 6,763 pounds per linear foot in compression.[6] Lateral design in seismic risk areas, given the current incomplete understanding of the straw/stucco assembly, typically calls for a supplemental bracing system except in the smallest and simplest structures.

It is important that engineers and architects not forget the empirical evidence of the last century. There exist many load-bearing straw-bale homes in Nebraska and Wyoming that have peacefully endured nearly a century of snowstorms, high winds, temperature extremes, and human occupancy without ever having had the benefit of rebar pins, precompressing, engineered foundations, or other features that are now deemed crucial. Those houses stand, uncracked, unrotted, unburnt, possibly as much a testament to the value of common sense in construction and maintenance as to the strength of straw bales. They quietly remind us that straw-bale construction can easily be strong and durable. As a methodology for engineered design of plastered straw-bale buildings evolves over the years to come, it should continually reflect back to these impressive examples.

Industrial Materials

POZZOLANS The modern use of pozzolans as a partial replacement for the cement in concrete began many decades ago and is not new or "alternative" to the construction industry, but a trend in recent decades toward greater usage is now redefining acceptable practice. A "pozzolan" is a siliceous or siliceous and aluminous material that in itself possesses little or no cementitious property but that will, in finely divided form and in the presence of moisture, chemically react with calcium hydroxide at ordinary temperature to form compounds possessing cementing properties.[7] Often restricted by building codes to minor fractions of the cementitious material in a concrete mix, pozzolans have held a relatively minor role in the concrete industry, but three issues are now changing that minor role to a much bigger one:

1. *Economy.* Portland cement, the primary "glue" for structural concrete, is expensive and unaffordable to a large portion of the world's population. Some pozzolans, for various reasons, are also expensive, but the most abundant and widely available, fly ash, is not. Blended cements that replace up to 60 percent of the portland cement with fly ash have now been successfully used in structural applications—for example, in a seven-story commercial structure in Nova Scotia.[8] Because portland cement is typically the most expensive constituent of concrete, the implication of increased fly ash utilization is obviously greatly improved concrete affordability. In some traditional cultures (India, Egypt, and other parts of Asia) inexpensive pozzolans are still manufactured from burnt clay soils or reject bricks.

2. *Durability.* A wide variety of environmental circumstances are deleterious to concrete, which include the presence of reactive aggregate,

high-sulfate soils, and freeze–thaw conditions, as well as exposure to salt water, deicing chemicals, and acids. These problems have historically been countered, with partial success, by increasing strength, often by reducing water/cement ratios. There is now a substantial body of laboratory research and field experience demonstrating that a careful use of pozzolans is enormously beneficial in countering all of these problems (and others). Pozzolan is not just a "filler," as some engineers think, but a strength and durability–improving additive (a discussion of the relatively simple pozzolanic reaction by which all this magic happens is readily available in the cited source.)

3. *Environment.* Portland cement requires enormous heat in its manufacture, making it expensive not just to the consumer, but to the atmosphere as well. For every ton of cement produced, roughly ½ ton of CO_2 (greenhouse gas) is released by the fuel consumed and an additional ½ ton is released in the chemical reaction that changes raw material to clinker, making the production of cement responsible for more than 6 percent of all the greenhouse gases released by human activity.[9] High-volume use of pozzolans is (or can be) not just an effective use of "waste" material and not just an economic savings, but can also provide a noticeable reduction in greenhouse gas buildup. From another perspective, high-volume pozzolan utilization in blended cements also presents a way for the cement industry to supply the ever-growing world market without having to build new production facilities.

The following discussion addresses the use of pozzolans in concrete, but it is equally valid in cases where cement is used to stabilize earthen structures. Some pozzolans are manufactured specifically to augment concrete mixes in one way or another, and others are mined directly from the earth (the name "pozzolan" derives from early uses of a cementitious volcanic ash mined near Pozzuoli, Italy). The most commonly used pozzolans, however, are industrial waste products and can be categorized either as cementitious and pozzolanic materials (including high-calcium Class C fly ash and granulated blast furnace slag), as highly active pozzolans (including condensed silica fumes and rice hull ash), or as normal pozzolans (including low-calcium Class F fly ash). These pozzolans are described as follows, in descending order of quality.

1. *High-calcium Class C fly ash (HCFA).* HCFA is the residue collected from the smokestacks of coal-fired power plants generally using lig-

nite and subbituminous coals. Class C fly ashes are in themselves mildly cementitious and have been combined with lime or even calcium carbonate soils to produce moderately strong concretes (of more than 4,000 psi).

2. *Granulated blast furnace slag (GBFS)*. GBFS is the ground residue from iron smelters and is also mildly cementitious in itself, but it becomes strongly pozzolanic in combination with water and cement. Its utilization dates back 200 years, and currently it is widely used in Europe. The 1997 UBC restricts its use to 50 percent of the cementitious material in a concrete mix.

3. *Condensed silica fume (CSF)*. CSF is a superfine powder of almost pure amorphous silica that is a waste product of the silicon metal industry. Although it is difficult (and expensive) to handle, transport, and mix, it has become the chosen favorite for very high-strength concretes such as in high-rise buildings. It is often used in combination with cement and fly ash.

4. *Rice hull (or husk) ash (RHA)*. RHA is the least known of the four pozzolans discussed, but it is enormously promising on a global scale. The world's primary staple crop is rice, the milling of which generates 100 million tons of hulls, or chaff, annually. Like straw, hulls have historically been burned in the fields, but the resulting pollution is increasingly causing health problems. Research in India and the United States has found that if the hulls (or straw) are burned at a controlled low temperature, the ash collected can be ground to produce a pozzolan very similar to (and in some ways superior to) silica fume. Simultaneously, rice farmers have found that their biomass wastes can be burned to produce electrical power, and several such power plants are now operational in the rice-growing regions of the United States (California, Arkansas, and Louisiana); more are planned in the United States, Brazil, the Philippines, and Japan. It is now known to be easy to burn the hulls in such a way as to produce optimal power and generate the desired quality of ash for pozzolanic use. The implication, then, is that the hull/straw disposal problem worldwide can be ameliorated by building small (1 to 5 megawatt) plants in rice-growing regions that cleanly dispose of the crop waste, generate electricity for the area, and provide a high-quality cement substitute.

5. *Low-Calcium Class F fly ash (LCFA)*. LCFA is the residue collected from coal-fired power plants generally fired with anthracite and bituminous

coals. Though generally less effective than Class C fly ash, Class F fly ash is nevertheless an abundant and useful pozzolan. Concrete for the seven-story structure in Nova Scotia (noted earlier) was designed with 60 percent Class F fly ash. In the United States both classes of fly ash are restricted by the 1997 UBC to 25 percent of cementitious material "in special exposure conditions" (Section 1904), but this restriction has been widely interpreted to apply to all concrete. That restriction, however, is antiquated in that it is based on studies of fly ash from older coal-fired plants that burned less cleanly and left carbon residue in the ash. Almost all commercially available fly ash today comes from the more modern power plants, is "cleaner" and therefore more reactive, and need not be so restricted as a concrete ingredient.

Concrete experts are now calling for increased usage and high-volume usage of pozzolans, especially fly ash, and some have proposed that all concrete should contain fly ash unless there is a compelling reason not to.[10] In light of economic conditions and the abundant evidence of their performance, derived both from research and reports from the field, it seems inevitable that regular and high-volume usage of pozzolans will become standard practice within the concrete industry.

PLASTICS There are, of course, already an extraordinary variety of petroleum-based plastic products in the construction industry. This situation is not likely to change much as long as the price of oil is subsidized, but even so, research on the original plant-based plastics is beginning to revive and products are being developed and improved. Linoleum from linseed oil and soy-based binders, for example, are already in the market. Of current interest to structural engineers is the emergence of several kinds of plastic lumber made largely or entirely with recycled content. Available in various colors, some of these lumbers have strength properties comparable to weak softwoods, although they lose stiffness and can deform in elevated temperatures. Being nontoxic and virtually nonbiodegradable (although some degrade under prolonged sun exposure), they are appropriate for a variety of nonstructural or low-structural applications. They are now being widely used for docks, playground structures, sill plates on concrete and masonry, and light framing applications. There is a multitude of new products available. Some of them are nonflammable—in a fire they just melt, releasing fewer toxins than wood.

STRAW PANELS   The first straw panel products appeared in Sweden and England several decades ago, but the number and variety of manufactured straw panel products has burgeoned in the 1990s. Some products, particularly those with higher densities, have the strength and performance characteristics required for structural applications. Low-density panels, of less than 16 pounds per cubic foot (pcf), are essentially flat basketry suitable for use as coverings and partitions, and they have no structural strength. Medium-density panels (16 to 24 pcf) are all based on the original "Stramit" system, in which straw is heated and compressed so that its own lignins act as a binder. These panels are produced in a variety of widths, lengths, and thicknesses and are typically faced with kraft paper. By themselves they have modest structural strength, some sound attenuation and thermal resistance, and swell permanently on exposure to water. When faced with oriented-strand board (OSB), however, they can be used as structural insulating sandwich panels and in this way are now on the market for use in wall and roof structures.

High-density (35 to 45 pcf) panels rely on binders added to the straw (typically polyisocyanurate). They come in many strengths and dimensions, and they are becoming increasingly available. Panels made from very finely chopped straw are comparable in strength, usage, and appearance to non-structural particleboards, and panels made from longer (1 inch +) fibers have strengths comparable to OSB. Although currently utilized only for nonstructural applications, the long-fiber panels can be cut, sawed, drilled, nailed, and screwed like OSB, and they will presumably start appearing in structural applications.

Conclusion

Almost by definition, alternative materials are less well known to the industry, their structural characteristics are less fully researched and documented, and they are generally perceived as "unproven," at least by modern standards. A closer look reveals many examples that start to build that proof: durable thousand-year-old multistory adobes in the Middle East, plastered straw-bale walls passing undamaged through fire, and high-performance concrete that uses radically little portland cement. To be sure, engineering a design with relatively little background information can be unnerving to someone used to the mountains of available literature on steel, concrete, masonry, and wood, but neither is it "shooting in the dark." There now exists a surprisingly large body of research and field evidence on which to draw, and every new engineered project can add to that knowledge.

An inevitable trend in construction, as well as in other industries, is toward sustainability, that is, the conducting of business in a way that can be continued through unlimited future generations. From a practical viewpoint, this already means making increased use of waste materials, using virgin materials and energy more carefully, and restraining or eliminating pollutant releases into the ecosystem. The trend is driven by economics as much as environmental concerns, and the engineering community can and must rise to the challenge of finding ever more creative and efficient ways to utilize the materials and energy readily at hand. With the forests and low-cost energy disappearing, the landfills filling, and the world's population swelling, it is obvious that we must become very skilled at doing more with less. Furthermore, the structural engineering community must commit to supporting those very large numbers of the earth's population who desperately need any kind of useful shelter, but can afford only the earth, plant fiber, and waste materials near at hand. Otherwise, the many mission statements about protecting public safety and preserving life become hollow rhetoric, for we would then be merely servants of the affluent, obdurately blind to the often harmful effects of our work. It is this engineer's profound hope that his peers will bring their enormous intelligence and creativity to bear in establishing effective use of any and all resources available to shelter human beings in the coming millennium.

Notes

1. H. Houben and H. Guillaud, *Earth Construction* (Marseilles: Editions Parentheses, 1989), 6.
2. P. G. McHenry, *Adobe and Rammed Earth Buildings* (Tucson: University of Arizona Press, 1984), 201.
3. E. L. Tolles, F. Webster, A. Crosby, and E. E. Kimbro, *Survey of Damage to Historic Adobe Buildings After the January 1994 Northridge Earthquake* (Los Angeles: The Getty Conservation Institute, 1996).
4. B. King, *Buildings of Earth and Straw* (Sausalito: Ecological Design Press, 1996), 60.
5. N. White and C. Iwanicha, *Lateral Testing of a Stucco-Covered Straw-Bale Wall* (San Luis Obispo: California Polytechnic University, 1997).
6. J. Ruppert and M. Grandsaert, *A Compression Test of Plastered Straw-Bale Walls* (Boulder, 1999).
7. V. M. Malhotra and P. K. Mehta, *Pozzolanic and Cementitious Materials* (Amsterdam: Gordon and Breach Publishers, 1996).
8. W. S. Langley, *Practical Uses for High-Volume Fly Ash Concrete Utilizing a Low Calcium Fly Ash* (Ottawa: CANMET, 1998).
9. *Environmental Building News,* Volume 8, No. 6 (June 1999).
10. J. Scanlon, *Admixtures—What's New on the Market* (Detroit: American Concrete Institute: 1992).

part II

SYSTEMS AND MATERIALS

5
Adobe

Michael Moquin

Used around the world for millennia, adobe masonry walls, like other earthen construction systems, replace the use of scarce, energy-intensive, and polluting construction materials such as kiln-dried, surfaced lumber and cement, which in most poor countries are extremely expensive. Construction with adobe walls can reduce the amount of wood needed for house building by approximately half, and if the roof system is constructed of adobe brick barrel vaults or domes, wood use is further reduced to an absolute minimum. Adobe also possesses optimal thermal-mass storage and heat transmissive properties for winter heating and summer cooling, an integral part of passive solar design. These properties, in turn, dramatically reduce fuel consumption and its associated pollution.

There are other benefits of building with adobe. Because suitable earth for building is locally available throughout most of the inhabited world, little or no transportation is required. Because, usually, only minimal processing of earthen materials is necessary, production can be local and decentralized, which also promotes self-help housing and keeps costs low. Producing adobe blocks takes only 1 percent of the energy required to produce fired bricks or portland cement.[1] Generally considered labor-intensive, adobe construc-

Alternative Construction: Contemporary Natural Building Methods, edited by Lynne Elizabeth and Cassandra Adams ISBN 0-471-24951-3 © 2000 John Wiley & Sons, Inc.

Contemporary home/studio of sculptor Thom Wheeler in Taos, New Mexico. This 32-foot-high adobe was inspired by the earthen walled towers of North Africa. (Photo courtesy of Thom Wheeler.)

tion can be an economic boon to work forces with limited opportunities, as exist in many parts of the less developed world and in the United States as well.

Further benefits include very low sound transmission. The sense of quiet within thick adobe walls evokes the feeling of security and peace. Earth can also balance indoor humidity better than any conventional wall construction material. Because of adobe's absorption and desorption capacity, the relative humidity of a house interior (finished with mud plaster) is a steady 50 percent ± 5 percent.[2]

An added benefit of adobe earth-building materials is that they can be nontoxic. For those who are allergic or sensitive to modern processed materials and finishes, the healthful interior environment of a natural earthen home is a refreshing alternative. Moreover, through passive solar heating, options for fresh air exchange can be almost limitless, because the earth walls store and radiate heat to maintain an even, comfortable interior temperature.

More has been written about building with adobe than about any other alternative construction material or method. This chapter is an attempt to synthesize elements from many of the best reference works.

History

Over the centuries, indigenous, traditional ways of creating adobe shelter have developed throughout the world. Finely tuned to the culture and climate, the sun, and local resources, these traditional regional adaptations have proven to be both practical and effective. These worldwide adobe methods represent a body of well-tested construction knowledge, useful and necessary in solving the planetary human need for adequate shelter today.

Adobe is regarded as the most important "primitive" building material because of its widespread planetary use. It is currently estimated that over one-third[3] to over one-half[4] of the world's population lives in some type of

Adobe residence of architect John Beck built in Baja, Mexico, with adobes produced on site. (Photo by John Beck.)

earthen dwelling. In China, for example, 45 million people live in below-ground earthen shelters, with another 100 million living in adobe and rammed earth homes.[5]

Adobe mud plasters, along with wattle and clay earth daub, may be the first human-made building materials ever developed. Ten thousand years ago mud bricks were hand-formed and used to construct walls, the process later refined by the use of wooden forms for making uniform-sized blocks. A basic understanding of historic adobe traditions worldwide can demonstrate the value and practicality of sustainable earthen construction, today and for the future.

The word *adobe* originates from the Egyptian *thobe* (mud) meaning "brick." In Arabic this became *at-tob,* which in Spanish became *adobe.*[6]

Many different words around the world are used to describe sun- and air-dried mud brick. Some call it "unbaked brick." The English call it "clay lump," or sometimes "clay-earth." The French call it *brique crue,*[7] or "raw earth"; Germans know it as *lehmziegel,*[8] or simply as clay. In Yemen adobe bricks are called *madar (madrah,* singular).

The earliest development of adobe is most often believed to have occurred in deserts, where trees were a limited building resource. Perhaps, too, it developed in areas of higher rainfall and abundant mud. The human ingenuity expressed in coping with a lack of wood by using the most abundant local material—earth—has continued through time.

The earliest permanent dwellings yet discovered are in the ancient Near East, China, and the Indus Valley. Of these, the oldest excavated site is at Jericho, which dates from 8,300 B.C. The houses, made of adobe, were round or oval, averaging 16 feet in diameter, with walls made of loaf-shaped, hand-molded mud bricks,[9] with indentations on the upper surface to create a key for the mud mortar. The level of adobe craftsmanship demonstrated here suggests that it has much earlier origins.

By 4,000 years ago, the Egyptians were building pyramids 250 feet tall × 350 feet (base length), made with a mud brick core and faced with stone.[10] In Babylon at this time, the 160-foot-high adobe ziggurat was faced with fired brick for durability. By the seventh century B.C., the Tower of Babel had been built of adobe, faced with fired brick and asphalt mortar. Ninety meters high, it has been called "mankind's first skyscraper."[11]

Prior to the Spanish arrival in the American Southwest, the early Pueblo people and their ancestors worked with hand-formed mud bricks and coursed adobe walls to create bluff shelter granaries, kivas, and pueblo apartments.[12]

There are many ways to make adobes as there are builders throughout the world. Table 5.1 presents some representative examples of the various weights and dimensions.

Table 5.1. A Sampling of Adobes from Around the World

Location	Structure	Date	Adobe Size	Comments
Acoma Pueblo, NM		ca. 1645	18" × 9" × 3" (31 pounds)	
Santa Fe, NM	San Miguel Chapel[13]	early 1600s	20" × 10" × 4" (50 pounds)	
Bernalillo, NM	Our Lady of Sorrows Church	1854	24" × 13" × 3" (60 pounds)	The huge adobes for this church were provided by a local patron; strong workers were needed to move these monstrous blocks.

Location	Structure	Date	Adobe Size	Comments
Bernalillo, NM	Bibo Hacienda	ca. 1870	16" × 10.5" × 3.5" (37 pounds)	Beautiful, strong adobes with high straw content.
Albuquerque, NM		1998	14" × 10" × 3.5" (32 pounds)	Produced by New Mexico Earth, the sandy, gravely adobe soil contains only 8 percent clay, this excellent soil is a waste product from gravel production.
Albuquerque, NM	Old Town	ca. 1870	13" × 7" × 4"	"Terrone" (dried sod from near river) walls were 2 feet thick.
Abiquiú, NM	Dar al Islam Mosque	1981	10" × 5" × 3" (8 pounds) 10" × 6" × 2"	Used for arches and walls. Egyptian-style adobe for vaults and domes.[14]
Aztec, NM	Below ground kiva[15]	ca. 1250	9" × 5.5" × 3" (9 pounds)	
Albuquerque, NM	Corbelled horno	1998	6" × 4.5" × 1" (1.7 pounds)	Children can hand-form these adobes.
Tucson, AZ	Commercially produced adobes	Present	16" × 12" × 4" (48 pounds)	The Arizona code specifies that a wall be at least 16 inches thick for greater seismic resistance (in contrast to 10 inches in less seismic New Mexico).
Madera, CA	Hans Sumpf Adobe Yard	1936 to present	16" × 12" × 4" (48 pounds)	This manufacturer perfected mechanized adobe production in 1936 with the "adobe laydown machine."
Geneva, NY	Two-story residence	1845	15" × 12" × 6" (65 pounds)	This building reflects an English influence via Canada.[16] The bricks are very thick, which would take longer to dry; most likely made during a hot, dry time in summer.
Presidio, TX	Low-cost housing	1998	18" × 12" × 4" 10" × 7" × 2.5" (11 pounds)	For walls. For dome and vaults.

Location	Structure	Date	Adobe Size	Comments
East Anglia, UK		1800s	18" × 12" × 6" (80 pounds)	"Clay lump" (adobe brick).
Victoria, Australia		1840–1950	15" × 10" × 5" (46 pounds)	"Mudbrick," traditional size.[17]
Yemen	4- to 8-story houses	1000 B.C. to present	18" × 12" × 2.5" (35 pounds)	Top floor walls are of 8" × 5" ("hand span"); the smallest adobes, also called "arch bricks," for vault and arch construction.[18]
Ali Kosh, Iraq		8000 B.C.	10" × 6" × 4" (15 pounds)	Hand-formed adobes tempered with straw; bonding grooves on upper surface.
		ca. 6000 B.C.	16" × 10" × 4" (40 pounds)	Almost identical in size to adobes currently produced in Arizona and California.
Greece and Rome		Ancient times	18" × 10" × 4" (45 pounds)	Reported to us by the Roman engineer Vitruvius (ca. 30 B.C.), this system of adobe brick sizes also provided for molds to make half-bricks.[19]
Fa, Syria	Corbelled adobe dome[20]	1000 B.C. to present	18" × 10" × 3" (34 pounds)	Walls 28" thick, 3–5 m diameter, 4 m interior height.

Contemporary Use of Adobe

One of the singular characteristics of adobe is that it is a modular masonry material of solid predried elements bonded together with a relatively quick drying mud mortar (½ to 1 inch thick). Thus, this method of earth building is relatively fast as compared with the use of cob or rammed earth (with walls 18 to 24 inches and heavy forms). For example, good adobe building practice in the desert Southwest allows a person to lay up to seven courses a day (28 inches high) during summer for a single brick thickness, or, for double-thick walls, three courses a day. However, most adobe structures built today are only a single block thick (10 to 14 inches in New Mexico, 12 to 16 inches in Arizona), which is relatively thin as compared with historic Hispanic adobe structures having residential walls 2 to 3 feet thick and church walls 2 to 6 feet thick.

Provost's house, Hobart-William Smith College in Geneva, New York, one of 20 adobe homes in the area. It was built by master craftsmen in 1843, using an English adobe technique brought through Canada circa 1800. The sun-dried mud bricks used here are 15 × 12 × 6 inches. (Photo by Richard Pieper, courtesy of Jan Hird Pokorny Associates, Inc., New York, New York.)

Some of the world's most sophisticated indigenous adobe architecture is found in Yemen, where dwellings are four to eight stories high. To make the mud brick, silt-rich adobe soil is taken from agricultural fields or palm groves and mixed with chopped straw or chaff. The unit of measure is the *dhira* (meaning forearm), which equals a *cubit* (approximately 19 inches). There are five sizes of bricks, with dimensions varying according to their location in the building as the walls taper upward. The adobes used for the ground floor are 18 to 21 inches long, two-thirds as wide (12 to 14 inches), and 2.5 inches thick, making the external wall at ground level approximately 1 meter thick. To achieve the greatest structural strength, adobes are laid in an alternating header–stretcher bonding pattern.[21]

With each succeeding floor, the adobes are reduced in size. Thus, by the fourth floor each mud brick is 14 × 9 inches. At the sixth floor and above the mud bricks are 8 × 5 inches (a handspan long), making the upper walls about 1 foot thick. These are also known as "arch bricks" for their use in arch and vault construction. Corresponding to all these different-sized adobes, there is also a half-brick, which is cut longitudinally.[22]

In the United States most adobe homes are being built in New Mexico and Arizona, where ancient Pueblo Indian and Hispanic adobe traditions are still alive. Currently there are approximately 25 professional adobe

Thompson residence, Albuquerque, New Mexico—a very large multilevel house, built circa 1995. (Photo by C.E. Laird.)

builders in New Mexico and about 20 in Arizona. Most are building upscale custom homes. Occasionally, these houses employ passive solar techniques, although aesthetics is usually the dominant consideration. Often, a large porch is placed along the south wall, blocking out the sun's rays, thus preventing winter passive solar heating.

There are approximately 75,000 homes built of adobe or rammed earth in New Mexico.[23] Adobe brick is commercially mass produced in New Mexico, Arizona,

Interior of the Goldblum residence, which employs passive solar design. Completed 1989. (Photo by C.E. Laird.)

and California with the aid of machinery. This adobe mud is semistabilized against moisture damage in the field and at the building site through the addition of 3 percent emulsified asphalt. There are also fully stabilized adobes with 5 percent asphalt emulsion (by weight) to be used in the first three courses of wall construction to prevent any water damage. The adobes in New Mexico are usually 14 × 10 × 4 inches, whereas in Arizona they are typically 16 × 12 × 4 inches, because walls are required to be thicker by code for better seismic resistance in a more active area.

In contrast to the expensive custom adobes being built by the professionals, low-cost, self-help housing also often employs the simple technology of adobe construction. Examples can be found in the American Southwest and around the world. For instance, in rural New Mexico, newlyweds traditionally built a small, one-room adobe house, adding more rooms as the family grew.

Critics of adobe sometimes view earthen shelters as hovels fit only for the destitute. However, most of the adobe homes now being built in the United States are custom homes that average 2,500 to 6,000 square feet—clearly not for the poor. It is hoped that adobe homes and communities can be built for the world's millions needing shelter, without the unnecessary stigma of poverty or impermanence.

Design Considerations

A few basic principles are essential for successful adobe construction. These are primarily concerned with control of moisture and erosion. Obviously, a structure should be sited on higher ground, away from areas of standing water and possible flash flooding. Attention to drainage is critical. For example, roof drainage spouts (*canales*) should be placed on the south side of the structure so that melting snows and ice will drain away instead of freezing and blocking the *canales* with ice dams. Likewise, a sloped ground gradient should be provided to remove water from the base of the walls and foundation. In building with unstabilized adobe and mud plasters, for minimal maintenance it is always necessary to provide an adequate roof overhang and a waterproof foundation. Moreover, the first three courses of adobe should be fully stabilized to protect against rising damp and possible flooding.

Adobe brick is not limited in use to regions of low rainfall. Around the world, it has been used in such countries as England, France, Yugoslavia,

Canales become an architectural detail in the adobe home of designer Tom Wuelpern in Tucson, Arizona. Tom built several adobe and rammed earth houses in this historic downtown neighborhood. (Photo by Tom Wuelpern.)

Costa Rica, Tibet, and China. The roofs have wide overhangs, and foundations of raised stone protect the unstabilized adobe walls from moisture damage. A similar example is found in the use of adobe in New York state in the 1830s to the 1880s. Fifteen large custom adobe homes were built in Geneva, and at least 35 others are scattered throughout the state. The dimensions of the typical adobe brick (15 × 12 × 6 inches) point to a connection with an earlier English adobe tradition brought into the area from Toronto, Canada.[24] These earthen houses were well-built two-story homes that reflected a variety of styles. The largest of the adobe houses in New York state is in Oswego. Built in 1851, this 2½ story adobe home occupies almost 6,000 square feet of floor space and is now a Roman Catholic school.[25] The longevity and scale of these New York houses further demonstrate that sun-dried mud brick can be practical and durable, even for areas with high rainfall.

PASSIVE SOLAR
(see Chapter 3, "Natural Conditioning of Buildings")

Adobe structures have been known traditionally for their capacity to moderate diurnal and seasonal temperature swings. Modern research over the last 20 years has linked adobe's optimal thermal massing capabilities to its density (adobe = 106 lb/ft^3), thermal conductivity (adobe = 0.3 Btu hr/ft^2-°F/ft), and specific heat, or ability to store thermal energy (adobe = 0.24 Btu/lb-°F).[26]

Adobe home construction experience in the American Southwest over the last 20 years has shown that insulating the exterior of the walls to R-20,

The warmth of winter sunshine is stored in thermal mass flooring in the living room of architect/builder C. E. Laird's adobe home in Albuquerque, New Mexico. The window shutters are made of local tamarisk. (Photo by C. E. Laird.)

ceilings close to R-40 to 50, and the outer foundation-stem walls to R-7 greatly enhances thermal efficiency and comfort. Thermal mass floors of adobe can play a key role in improving a building's thermal performance by increasing the surface area of the available heat-absorbent material. Locating thermal mass floor areas near south-facing (in the Northern Hemisphere) windows to receive direct gain solar radiation is the most effective way to guarantee abundant stored heat in the winter.

Insulating the exterior of the mass walled building minimizes the rate of heat loss in winter, resulting in reduced heating costs. In this system the optimum adobe wall thickness is 8 to 12 inches, with a 12-inch wall having only a 7-degree daily indoor air temperature fluctuation. Generally, thick-mass walls such as adobe work much better than low-mass walls such as wood frame.

Code Requirements

Building code in the United States addresses only adobe construction (unreinforced masonry) and, unfortunately, does not include rammed earth, cob, or compressed earth block. Adobe has high compressive strength, but is weak in tension. The national code requires vertical steel reinforcement within the adobe walls in all earthquake zone areas with seismicity greater than 1 (most of the western United States). This adobe code was created largely in response to the aftermath of the Long Beach, California, earthquake of 1933, in which a disproportionate number of unreinforced masonry (fired clay brick) buildings were damaged. Unfortunately, the engineers who created this adobe code treated adobe masonry as if it were concrete, ignoring its unique characteristics.

New Mexico has created its own adobe code, which has served as a model for parts of Arizona, Colorado, and other areas where adobe is commonly built. This revised code better acknowledges the inherent properties and performance of adobe by not requiring vertical steel within the earthen walls. The national code has greatly discouraged adobe construction throughout California and other western states. In fact, because of the higher cost and effort necessary to comply with the national code requirements, adobe homes are virtually no longer built in California, a state once noted for its historic adobe homes and missions.

Codes have also discouraged its use in Europe. There, until about 1800, wood provided the chief fuel for household heating and cooking, as well as materials for housing, industry, and shipbuilding. This dependence on wood resulted in widespread deforestation. As early as the late sixteenth century, governmental authorities in Germany were insisting that new buildings be made of earth in order to conserve the remaining trees.[27] Thus, during the eighteenth and nineteenth centuries, tens of thousands of adobe homes were built. After the First World War, thousands of earth-walled buildings were constructed in response to a lack of transported or processed building material. After World War II, another 40,000 German homes were built of earth. In spite of this venerable history, the German government in 1970 prohibited any further construction of buildings with structural earthen walls.[28]

In contrast, the encouraging research of Julio Vargas of the Catholic University in Lima, Peru, on the use of cane and bamboo reinforcement suggests that there may be many more options for safe, sustainable, simple building methods that should be considered for inclusion in building codes.

Soil Composition and Identification

In 1944 a practical manual titled *Handbook for Building Homes of Earth* was compiled by Wolfskill, Dunlop, and Gallaway of the Texas A&M Research

Two **adoberos** *(adobe workers) spread newly mixed adobe mud into the rows of wooden molds at New Mexico Earth in the north valley of Albuquerque, New Mexico. (Photo by Michael Moquin.)*

Foundation, for the Agency for International Development. Most of the discussion of soil that follows is drawn from this fine work.

The basic groups of soils are silt-clay soils, gravel soils, and sand soils.

❖ *Clay* is the most important soil component because it is sticky and bonds the soil particles together when wetted. It tends to disperse in water and is hard when dry. The best adobes and mud plasters are made with soils of 8 to 15 percent clay. Higher clay content (over 30 percent) causes excessive cracking. There are many types of clay. In the presence of water, some types will swell and shrink when drying, causing cracking (e.g., montmorillonite). Most other clays will not swell and crack (e.g., kaolinite and illite). Red clays found in the tropics (laterites) are very stable. Ideally, the type of soil best for earth houses (adobe) are sandy clays (sometimes called clayey sands).

❖ *Sand* consists of fine grains of various rocks, mostly quartz. The particles vary in size from about ¼ inch to about the smallest grain that can be seen with the naked eye. Separate grains too small to see are either silt or clay.

❖ *Silt* is extremely fine rock particles, so fine that individual grains cannot be seen. Silt particles tend to hold together when wet and compressed. Too much water may make the material spongy, but it does not get very sticky. Silts are not strong soils. They lose strength and become soft when wet. In wet, freezing weather they tend to swell and lose their strength. However, they can be stabilized with emulsified asphalt or portland cement.

❖ *Gravel* consists of small coarse pieces of rock (any type) varying in size from ¼ to 3 inches across. These pieces may be round, flat, or angular. A soil that is clayey gravel is good for adobe making if the gravel is less than ¼ inch.

❖ *Organic* soils have a spongy appearance when wet, with a distinct decaying smell and usually a dark color. Any soil with a substantial

Adobes stacked in a construction yard adjacent to a rambling adobe compound being built in Baja, Mexico. (Photo by John Beck.)

amount of organics is to be avoided. If the pH of a soil is low (acid, 5.5 or less), it has too much organic matter.

❖ *Natural adobe amendments* include sand, straw, and grasses. Straw accelerates drying, minimizes cracking (which can lead to surface erosion), and helps wet adobe bricks to hold shape. Straw also moderately increases the tensile strength of the adobe. Cow manure in mud plasters is used in various parts of the world to improve bonding and waterproofing. Mineral amendments include cement (10 to 12 percent), lime, and asphalt emulsion.

In addition to particle size distribution, the mechanical stability of soils (under dry conditions) is influenced by the shape of soil particles; that is, whether they are rounded, angular (sharp grains of sand), platelike (clay), or fibrous.[29] Water-soluble salts can greatly affect the durability of adobe, for they can absorb water, swell, and cause spalling of surface particles. If salt content is high, the rate of surface destruction may be rapid. (The New Mexico adobe code specifies no more than 0.2 percent salt by weight (dry soil).) The percentage is determined by chemical or electrical conductivity methods.

Soil testing in the field can be simply done, yielding relatively accurate estimates. Often the top layer of soil contains too much organic matter for good adobes, so it is necessary to dig down a foot or so to more suitable earth (clay and sand). A simple sedimentation test can show the relative amounts of clay (most important to know), silt, sand, gravel, and humus. A soil sample typical of that used to make adobes is mixed with twice the volume of water in a flat-bottomed, straight-sided glass jar, or (better) a clear cylinder, the length of which is about 10 times the diameter, is used.[30]

The mixture is shaken well until everything is in suspension, and the jar is then allowed to sit. After an hour or two, the materials settle, sized by diameter: gravel and coarse sand first, fine sand, silt, clay, and then humus (which floats). Once the soil testing is accomplished, test adobes should be made to determine their quality. Excessive shrinking or cracking, or lack of dry strength denotes an unsuitable adobe.

Stabilization of Adobe with Asphalt Emulsion

The definitive instruction manual for stabilized earth construction, entitled *Stabilized Earth Construction: An Instruction Manual*, was written by Richard Ferm in 1981 for the International Foundation for Earth Construction. In 1985 this booklet was given to participants in the International Symposium on Earth Construction held in Beijing, China. Its main focus, in addition to soil selection, is on the use of emulsified asphalt as the best method for stabilizing soils for durable adobe construction. Most of the information in this section is adapted from that source (revised to accommodate space limitations).

Asphalt is primarily obtained as a residue from the distillation of crude petroleum or from natural deposits (brea). Asphalt emulsion is a mixture of asphalt solids and water (typically 60:40 by weight), with an emulsifying agent (1 percent soap).

The term *asphalt* is derived from Lake Asphaltites (now called the Dead Sea), where lumps of semisolid petroleum ("pitch") were washed up on its shores from underground seeps. In ancient times at farming settlements along the Euphrates River in Babylonia and Sumer, this material was used as a waterproof mortar and as a casing over weather-vulnerable adobe bricks. At Mohenjo-Daro (dating from 3000 B.C.) in present-day Pakistan, pitch was used as a waterproof rendering over the brick walls of the water reservoir.

Commercial asphalt emulsions designed for road work in the United States evolved in the 1920s. The stabilization process for adobe was researched initially at Oklahoma State University in the early 1930s as a way to recycle petroleum residue. By the mid-1930s, large-scale mechanized production of asphalt stabilized adobes had begun in Fresno, California, at the Hans Sumpf Adobe Yard.

Asphalt emulsion is often used as a exterior waterproofing layer for below-ground cement block foundation stem walls, as well as for rendering walls to be bermed. It also works well as a moisture barrier at the top of concrete stem walls prior to laying the first course of fully stabilized adobes.

Richard Ferm points out that there are many advantages to stabilizing adobe with emulsified asphalt. It upgrades adobe construction by increasing its long-term durability through waterproofing, which greatly reduces swelling and shrinking. Fully stabilized adobe contains approximately 5 percent asphalt emulsion by weight. Resistance to water penetration and erosion when fully stabilized is the most important characteristic of fully stabilized adobe; water absorption is only 3 percent or less after prolonged exposure. An important advantage is that with such stabilization, adobe retains its strength in wet conditions.

Exterior erosion by water and wind is greatly reduced, and, therefore, so is maintenance. Adobe that is stabilized with asphalt emulsion can endure direct and prolonged contact with water and is even used for the lining of water distribution ditches (where cement is expensive). Emulsified asphalt adobe walls are animal-, insect-, and fire-proof.

For greater safety in earthquake-prone areas, the weight of heavy earthen (flat) roofs can be greatly reduced by replacing thick, unstabilized mud roofs with a thin layer (5 centimeters) of asphalt emulsion–stabilized mud mortar.

It is noteworthy that adobe block, fully stabilized with emulsified asphalt, offers the best performance against water absorption, as illustrated in Table 5.2.

ASPHALT EMULSION STABILIZATION: SOIL PREPARATION AND MIXING[32]

Adobe soil should be screened through ¼-inch wire mesh to remove large gravel particles. To get the best distribution throughout the adobe material, the asphalt emulsion should first be mixed with the water used to create the mud.

The speed at which emulsified asphalt sets can be rapid, medium, or slow. For stabilizing adobe soils, only slow-setting (SS, sometimes called DM for "dense mixture") emulsions should be used. Full stabilization of

Table 5.2. Comparison of Water Absorption Rates[31]

Asphalt-stabilized adobe	½–3 percent
Wood	4–8 percent, up to 15 percent
Average concrete	8 percent
Cement stucco	8–11 percent
Burnt clay brick	8–12 percent
Lightweight cement bricks	20–25 percent
Untreated soil brick	25–30 percent

adobe (waterproofing) is achieved by mixing in 4 to 6 percent asphalt emulsion by weight. The minimum amount should be used for economy; too much will weaken the adobe by lubricating the soil particles. The optimum amount of emulsion is easily determined by making small quick-drying test adobe bricks (2 × 2 inches). Asphalt emulsion weighs 8.3 pounds per gallon, and a cubic foot of earth weighs approximately 106 pounds. First, the number of cubic feet of earth the wheelbarrow holds is determined, then emulsion, water, and earth are mixed in correct proportions. Generally, the sandier the adobe mix, the less asphalt emulsion needed. The more clay present in the soil, the more emulsion needed.

Foundations

The New Mexico adobe code requires a continuous concrete footing at least 8 inches thick that extends not less than 2 inches wider than the stem wall on each side. The stem walls are of concrete, of the same thickness as the exterior adobe wall (10 to 16 inches). Rigid insulation is then attached to the exterior of these thermal mass walls, and then plastered with stucco. Many professional adobe home builders believe exterior perimeter insulation made of extruded closed-cell polystyrene (R-4.3 per inch) works best below grade, as it retains its R-value in high-moisture situations.[33]

An alternative, inexpensive, and simple technology for foundations is the gravel-sand foundation, which can employ a cement grade beam at ground level in locales with strict code enforcement. The most ecological solution requires no cement, with the adobe wall built directly on top of the filled and compacted trench. To perform optimally, the first three courses of adobe should be fully stabilized with emulsified asphalt, with fully stabilized mud mortar between the bricks. To use a gravel trench foundation, it is necessary that the soil substrate allow the percolation of water so that the trench always drains in times of heavy rains or flooding, thus avoiding frost heaving. This gravel foundation system was used in the 1940s by Frank Lloyd Wright in the Jacobs house in Middleton, Wisconsin.

Despite its technical simplicity, the gravel trench foundation is one of the most earthquake-safe systems, as it is a form of base-isolation, which minimizes the transmission of ground motion into a structure.

Insulating Exterior Walls

Professional building experience over the last 25 years and government testing in the late 1970s have demonstrated that for optimum efficiency and comfort, the exterior of mass walls such as adobe should be insulated.

Sheets of rigid insulation or, sometimes, spray foams such as urethane, are the most common materials used. Insulation sheets are attached with 4-inch roofing nails or 6-inch pole barn nails driven into the adobe. The insulation is then covered with metal chicken wire, which provides a mechanical key for the typical three-coat cement stucco plaster.

As yet, no one has invented a reliable way to add exterior wall insulation with natural materials such as straw or cellulose. Reed mats may be useful as a means of holding, for instance, a 1-foot thickness of straw against the walls, to be followed with stabilized mud plaster.

EARTH-BASED MUD PLASTERS (see also Chapter 12, "Earth Plasters") Although not specifically endorsed by the New Mexico building code, mud plasters can be highly effective, economical, and natural. Natural stabilization to prevent cracking is achieved with the use of sand, chopped straw, or cow dung (for workability and durability). One of the traditional English mixes for clay–earth daub (mud plaster) begins with a soil that is approximately 10 percent clay, 40 percent silt, and 50 percent sand and calls for 12 parts soil, mixed with 1 part dung, 1½ parts straw (firmly packed), and mixed with ¼ to ½ part water.[34]

Mud plaster should be applied in successive layers, ½ to 1 inch thick, over a prepared surface that has been dusted off and then wetted for better bonding. To achieve a smooth, durable, finished surface, allow drying between applications. The addition of asphalt emulsion to a mud plaster can render it partially stabilized (3 percent by weight) or fully stabilized (5 percent).

In working with mud plasters, it is helpful to have the mud sticky enough, with a higher clay content if necessary, for easy bonding to the walls, whether they are new or damaged walls under repair. It is acceptable for these sticky base coat layers to shrink and crack, because they will be covered with the finish mud layers, which will have smaller and fewer cracks, due to a lower clay content.

The Bond Beam The bond beam (tie, or collar beam) is one of the most important elements of a seismically resistant adobe (or other unreinforced masonry) structure. It spreads the concentrated loads of the roof beams, ties the roof system to the tops of the walls, and prevents the walls from falling outward during strong earthquakes. The New Mexico code requires that the bond beam be 6 inches thick and two-thirds as wide as the top of the wall. Many profes-

Interior of Williams adobe residence under construction in Placitas, New Mexico, shows cement bond beam running above rough-cut wood window lintels. Note the unplastered kiva fireplace in the corner. (Photo by C. E. Laird.)

sional builders use concrete, which requires two steel reinforcing rods. However, wooden bond beams are easier to work with than reinforced concrete and are more sustainable. Building professionals often assemble a wood bond beam in laminated layers. This can be made of low-processed (not kiln-dried) rough-cut lumber.

In response to the need for more seismic-resistant adobe methods, the modern building code in Turkey calls for four continuous tie beams: just above the foundation level, under the windows, above the windows, and at the top of the walls. These timber tie beams of two-by-fours are made with cross ties at 1-meter intervals, and lapped at corners and intersections.[35]

Running bond beams can be seen in this exterior of the Williams residence before plastering. (Photo by C.E. Laird.)

Unfortunately, in some developing countries, where people build their own homes, they rarely include a bond beam in their construction, because of high wood costs and lack of craft. This omission, combined with a lack of proper maintenance, and resultant water damage, has led to widespread failure of homes in seismically active zones. For this reason, adobe has been discredited as an appropriate building material in the eyes of many engineers and housing officials. However, there are good examples of proper adobe methods.

The region of the former Yugoslav Republic has a centuries-old tradition of building two- and three-story adobe homes despite seismic activity in the area. Stone foundations with mud mortar are built 3 feet above ground level to prevent moisture damage. An important feature of this construction is that for every meter of adobe wall height, a horizontal wooden belt, or bond beam, made of two parallel boards, is placed within the earthen wall, providing strength and stability. Regular maintenance has kept these houses in good shape for many generations.[36]

| Earthquake Resistance | In areas of the world where earthquakes are a threat to life, adobe masonry homes must be designed and built to survive violent shaking. In addition, poor maintenance can lead to serious moisture damage. In such a weakened condition, serious structural damage can occur, and if there is no bond beam at the top of the wall, roof collapse becomes a possibility, particularly in earthquake areas. |

Typical damage to poorly built and poorly maintained adobe buildings during an earthquake includes (1) out-of-plane overturning of walls, (2) in-plane shear cracking, (3) separation of adjoining walls, and (4) roof collapse, which may result from a combination of the three other types of failure.[37]

Although adobe is inherently a nonductile masonry material when subjected to shear and tension forces, structural engineer Fred Webster has identified four fundamental characteristics of good seismic design for adobe construction. First, regular floor plan layouts or shapes exhibit better performance than irregular shapes. Second, wall openings should not be unduly large, and they should not be concentrated on just one or two sides of the building. Third, roof beams should be positively attached to bond beams, and the bond beams should be bolted to the tops of the walls to prevent them from slipping out from beneath the roof. Fourth, wood and reinforced cement bond beams should be continuous. They should be stiff and strong in the direction perpendicular to the plane of the wall. They should restrain the out-of-plane deflection of the adobe wall as much as possible. Bond beams should be flexible but strong in the vertical direction, no more than 5 or 6 inches in depth.[38]

ENGINEERING CONSIDERATIONS: WHY STABILITY-BASED ANALYSIS IS A BETTER METHOD

Two engineering analysis approaches are used to characterize the behavior of adobe masonry in an earthquake: the conventional strength-based analysis procedure and a stability-based approach. The latter is preferred because it can "directly utilize adobe construction's unique characteristics and directly address the collapse potential of these [historic adobe] buildings."[39]

In 1985 a shake table (earthquake simulator) test was done at the University of California, Berkeley, to test the feasibility of "confinement" for adobe in which a surface "skin" (chicken wire netting) prevents out-of-plane failure (falling outward) of the wall due to loss of bearing capacity, thus preventing collapse of the roof.[40] Good construction practice was followed by installing a wood bond beam, which was bolted to the top course of the adobe wall. The roof beams were attached to the walls with 1-inch threaded rod drilled and grouted into the top three courses of adobe block. Then 13,000 pounds of weight were put on the beams to simulate a thick, unstabilized earthen roof. The reinforcement skin was wrapped on both the exterior and interior of the adobe walls. It consisted of 14-gauge (.08 inch nominal) welded wire mesh on two of the walls, and hexagonal 18-gauge (.048 inch nominal) stucco wire, both secured with $\frac{7}{8}$-inch staples.

This research demonstrated a successful and simpler way to achieve seismic resistance with adobe in earthquake-prone areas. In contrast, the strength-based engineering solution of vertical steel rod reinforcement (4 feet on center both vertically and horizontally), required by code in California, is an impracticality that has almost ended adobe house construction in that state.

Recent research relevant to new adobe construction, as well as seismic retrofit of historic adobe buildings, has been undertaken by the Getty Foundation. Its investigators have found that stability-based engineering (in contrast to strength-based) greatly improves the seismic performance of adobe buildings through the use of minor restraints. "Stability based retrofit measures can prevent walls from collapsing and broken sections of the wall from developing large offsets."[41] This research testing the effectiveness of

✦ A SUMMARY OF OBSERVATIONS ON ADOBE STRUCTURES

The following observations are the result of expert investigations of the effects of the severe earthquake that occurred in Peru in May 1970, and of methods of adobe construction put into practice successfully in other countries.[42]

General Causes of Structural Failure

✤ Poor quality adobes
✤ Adobes are often too thick (more than 4 inches)
✤ Inferior workmanship in laying adobes
✤ Wrong dimensions of walls—too narrow in width or excessive in length or height
✤ Very wide door and window openings
✤ Insufficient embedding of the lintels
✤ Too many windows (should be no closer than 4 feet from corner)
✤ Lack of a bond beam
✤ Overweight roof
✤ Improper anchoring of roof to wall
✤ Little protection against weakening caused by erosion
✤ Too tall

Walls

✤ Octave (angular) walls are not recommended
✤ Where two perpendicular walls adjoin, the length of one wall should not be more than 10 times its thickness
✤ Walls should not be more than 8 times their thickness in height
✤ Lintels should go at least 20 inches beyond the edges of window or door openings

Buttresses

✤ The use of buttresses, rather than adding complications to the construction process, will instead give major strength and security
✤ Buttresses facilitate future expansion of the structure
✤ Buttresses permit the best solution at the intersection of the walls where the bond beam is placed (the bond beam being of rough lumber, two pieces in parallel, connected with cross pieces, spliced together at the corners)
✤ Buttresses must be built in at time of construction
✤ Buttresses should be placed in long walls, at wall intersections

Bond Beams

✤ A continuous, horizontal, reinforcing bond beam should be placed and, wherever possible, should coincide with the lintels that span doors and windows
✤ A minimum of two more courses of adobe must be laid over the bond beam, particularly on walls that bear roof weight
✤ Everything must be firmly united at the corners, and there should be no gaps

Types of Bond Beams

✤ Rough lumber with diagonally braced corners
✤ Rough lumber in parallel, spliced together at corners
✤ Rectangular welded mesh, overlapped at the corners and tightly fastened
✤ Reinforced concrete

Roof Systems

✤ Whenever possible, the roof should be of light construction (to perform better in earthquakes) and without exaggerated slope (to avoid outward thrust, which can stress the top of the wall or possibly push it out)
✤ Where a viga rests over a window or door opening, the lintel it rests on should be reinforced
✤ A long horizontal wooden lintel placed into the wall will serve as a support for the weight of the roof (in situations where timber is scarce or very expensive)
✤ The use of overhangs or eaves is recommended to protect the adobe walls from rain, along with a mud plaster coating

retrofitting adobe structures with vertical (over the tops of walls and through the wall base) and horizontal (at the attic roof line and below the bottom of windows) vinyl straps has demonstrated that such elements of continuity "inhibit the relative displacements of cracked wall sections and prevent the principal modes of failure."[42] In addition, extra wall thickness (in contrast to thin, single-adobe thickness of 10 to14 inches) provides greater stability, as demonstrated by the performance of historic adobe walls (20 to 30 inches thick).

An important conclusion to be drawn from these earthquake simulations is that cracks will occur during moderate-to-large earthquakes; however, an adobe building is not unstable simply because cracks exist.

> The seismic performance of an adobe building is just beginning when the cracks have been initiated. It is normal for adobes to be cracked from settlement or other natural causes, and it is part of the historical tradition to repair these cracks. . . . The unique character of adobe buildings can be fully utilized if they are allowed to crack during wall movement caused by an earthquake. . . . It is imperative that the theoretical basis for an engineering analysis of retrofit measures include an understanding of the dynamic performance of cracked adobe buildings.[43]

ARCHES, BARREL VAULTS, AND DOMES: ROOFS OF MUD BRICK RATHER THAN WOOD

The invention of masonry arches, vaults, and domes is one of the most amazing architectural feats of the early civilizations. These curved wall-spanning structures are kept strong by the pull of gravity. Working by means of compression and uniformly distributed, well-bonded loads, these masonry roofing systems can be built with sun-dried mud brick (adobe) or cement-stabilized compressed earth block.

One of the earliest arches discovered appeared in village domestic architecture on the northern plain of Iraq at Umm Dabaghiya (5500 B.C.). The mud and straw walls of the houses were made of *tauf*, an Arabic term for coursed adobe (cob, mud walling). Usually, one of the rooms was divided by an arch spanning its width.[44]

Through trial and error, and over time, an efficient arched mud brick roofing system, barrel vault construction, was eventually developed by leaning a series of arches against a thick, buttressing end wall. This construction method does not require wooden forms for centering. The earliest evidence for such a roofing system has been dated to 3200 B.C. at Tell Rimah, also in Iraq.

One of the most photographed early examples of vaulting is the Ramesseum granaries associated with a temple in Luxor, Egypt—a long

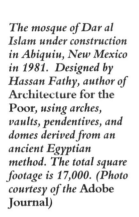

The mosque of Dar al Islam under construction in Abiquiu, New Mexico in 1981. Designed by Hassan Fathy, author of **Architecture for the Poor,** *using arches, vaults, pendentives, and domes derived from an ancient Egyptian method. The total square footage is 17,000. (Photo courtesy of the* **Adobe Journal***)*

parallel grouping of adobe vaults dating from 1400 B.C. These vaults are exceptionally well built, with a four-layer thickness of mud bricks.[45] It is interesting that the Egyptians chose adobe over stone for building these vaults (most likely because stone requires much time-consuming labor to chip and shape the numerous masonry elements).

The present-day Nubian vault design, which is almost identical to the ancient technique, uses the same small, thin adobe bricks (10 × 6 × 2 inches, weighing 8 pounds), with extra straw added for lightness.[46] As was done thousands of years ago, parallel diagonal finger grooves are inscribed on the brick surfaces for better bonding with the sticky mud-and-manure mortar. Again, no wooden forms or centering devices are needed for this type of simplified masonry roof construction.

The largest barrel vault ever built is located near Baghdad, at Ctesiphon, which dates from the fourth century A.D. The large open-ended banquet hall of the palace is covered with a barrel vault 120 feet above the ground. The vault spans the 83-foot width of this huge hall.[47]

CORBELED DOMES In the area of present-day Turkey and northern Syria, an ancient tradition of building corbeled mud-brick domes persists. These beehive-shaped dwellings are good examples of the earliest and simplest method of dome construction, especially for those interested in the reduction of wood usage. Their parabolic shape is achieved by projecting each horizontal course of mud bricks approximately 2 inches over the preceding course.

The resulting walls are 1½ bricks thick (27 inches), each adobe brick being 46 × 25 × 7.5 centimeters (18 × 10 × 3 inches).[48] Each masonry course is well bonded; the construction is kept plumb by eye.

Each domed roof rests on a square (occasionally round) base of mud brick or stone, usually of 3 to 5 meters (10 to 16 feet) interior room diameter. These short walls, 80 centimeters (32 inches) thick, support the corbeled adobe brick dome, which rises to 3.5 to 4.5 meters interior height. When a longer, larger room is needed, two side-by-side units are joined by opening a large shared archway. Repeating units joined together into courtyard houses become a nucleated settlement.[49]

For domes and vaults to be viable in areas other than desert regions, the mud brick must be protected against increased amounts of rain and snow. Fired clay tiles, cement stuccos, and asphalt-stabilized mud plasters are available solutions. In the early 1800s, Wilhelm Tappe of Germany, concerned about the housing problems of the poor, proposed earth-walled dwellings with a circular floor plan for both aesthetic and economic reasons, built in a beehive dome shape for stability and endurance.[50] By laying each circular course of mud brick flat, overhanging the course below, a parabolic dome could be created without the need for wooden centering forms. As protection against moisture damage, Tappe chose fired tile as an exterior surface rendering for his homes.

In areas where wood or cement is difficult or expensive to obtain for roofing, adobe vaults and domes can be an economical, safe, and practical solution. They are especially promising as a way to reduce the need for large wooden beams, which usually come from old growth forests.

New Research

Present-day adobe research in Europe includes work being carried out by Gernot Minke of the Research Laboratory for Experimental Building at the University of Kassel, Germany, as well as CRATerre (International Centre for Earth Construction) located in Grenoble, France. In 1994 the German Technology Center sponsored prototype construction of Nubian vaults and domes, with tapered adobes made by hand, in New Delhi, India.[51] The wall and dome exteriors were rendered with earth (mud) mortar stabilized with cow dung. The finish plaster was then "hydrophobized" with water-repellent siloxane. The lab is also experimenting with casein and lime, water glass, linseed oil, silicone, esters of silicic acid, and bituminous (asphalt) emulsion.[52]

Minke has reported that the shaping of these mud bricks actually reduces the amounts of mortar and labor required. The inverted catenary shape of the vault sections results in better load transfer to walls and foundation. No forms are required for support during construction of the barrel vaulted adobe roofs.[53]

Minke believes that earth is the "best ecological building material that exists" for several reasons: It involves minimal pollution, uses minimum energy in its processing and transport, and is always recyclable.[54] His research focuses on the structural and physical properties of earth mixtures: weather resistance through "hydrophobizing" agents and ways to minimize the clay shrinkage ratio by using additives to change its granularity and increase the binding force. In addition, his research involves the addition of porous mineral aggregates such as expanded clay, glass, and pumice to reduce thermal conductivity by increasing thermal insulation capacity. If these additives are mixed correctly there is no shrinkage. Minke calls the mixture "mineral lightweight loam."

Interestingly, Minke's group has also found that improper stabilization percentages of cement or lime in earth (less than 5 percent for cement and 7 percent for lime) often decrease its binding capability and thus its structural strength, especially in silty loam soils.

Challenges

Earthen construction faces several social, bureaucratic, and political obstacles that prevent its application from being as widespread as it might. First, the developed world, through its institutional decision makers in government, education, banking, and business, has tended to view earthen construction as primitive, unreliable, or inappropriate. These misconceptions are further distorted by the cement manufacturing and wood products lobbies and trade associations, whose powerful influence is widely reflected in national building codes that are biased against earthen construction through their overspecification. In fact, at present, cement associations are advertising heavily that cement should be considered a "sustainable" material.

In contrast, those engaged in building and designing with earthen materials have no organized trade associations to finance the research and documentation of earth-walled construction systems. This situation has resulted

This home/studio of artist Thom Wheeler has 30-inch-thick walls, two and three adobes thick (14 × 10 × 4 inches), with each course bonded in alternating patterns for greatest strength. The building took 38,000 semi-stabilized adobes, which can be seen in this photo taken before plastering. (Photo courtesy of Thom Wheeler.)

in notable gaps in the Uniform Building Code (UBC), such as failure to address fire ratings for earthen walls. Fortunately, in New Mexico and Arizona, state and county building code officials, insurance companies, and banking institutions have recognized the durability of earthen construction and its significance to the culture. They now also have an increasing appreciation of its environmental benefits and economic advantages.

A second and equally imposing obstacle is that the formal education of architects and structural engineers is grossly deficient in its omission of information concerning the structure and performance of adobe and the other earth-walled systems. Incredibly, there is no such academic training available within the United States. Conventional construction as taught in the universities relies on cement, fired brick, steel, and aluminum—all highly processed, costly materials—which must be imported if used in most developing countries. Sadly, at a time when there is a pressing need for more sustainable design and construction, passive solar heating and cooling principles and priorities are nearly ignored.

Another challenging barrier to the acceptance of earthen construction is a pervasive skepticism toward the validity of building alternatives whose sources are outside the boundaries of mainstream, modern, high-tech processes. Many of those who build with earth are not architects or engineers. As a result, they are not accorded the respect and credibility their experience and skills deserve. Great progress toward a more environmentally conscious, sustainable construction can be achieved if those who are actively building with adobe, rammed earth, and compressed earth block are included in the discussion and planning of codes and education. National building code and HUD (U.S. Department of Housing and Urban Development) officials should make an effort to reach out to these practical experts. Given the current situation, there should be an openness to new paradigms outside the narrow focus of "architecture by architects."

Traditions of designing and building with earthen materials have a long history of sustainability throughout the world. As the world population approaches the earth's carrying capacity, ethical and social responsibility demands that we design and build in harmony with the environment so as to guarantee sufficient resources for future generations. Conferences, colloquia, workshops, and earth-building schools can help to bridge the artificial barriers between architects, engineers, and code officials on one side, and earthbuilders and environmentalists on the other. It is hoped that these classical, natural ways of building may be more fully employed to provide the construction of safe, efficient, and harmonious shelter for today and into the future.

Notes

1. Gernot Minke, *Lehmbau-Handbuch* (Ökobuch Verlag,1997).
2. Ibid.
3. Jean Dethier, *Down to Earth: Adobe Architecture—An Old Idea, a New Future* (New York: Facts on File, 1981), p. 7.
4. Edward W. Smith and George S. Austin, *Adobe, Pressed-Earth, and Rammed-Earth Industries in New Mexico.* New Mexico Bureau of Mines and Mineral Resources, Bulletin 127 (1989), p. v.
5. Gideon S. Golany, *Chinese Earth-Sheltered Dwellings: Indigenous Lessons for Modern Urban Design* (University of Hawaii Press, 1992).
6. Mud Village Society, *Building with Earth* (New Delhi, India: Mud Village Society, 1991), p. 106.
7. Bernard Feilden, *Conservation of Historic Buildings.* (London: Butterworths, 1982), p. 415.
8. Ibid, p. 415.
9. Banister Fletcher, *A History of Architecture*, 19th ed. (London: Butterworths, 1987), p. 26.
10. Mark Lehner, *The Complete Pyramids* (London: Thames and Hudson, 1997).
11. Dethier, *Down to Earth: Adobe Architecture—An Old Idea, a New Future*, p. 7.
12. Michael Moquin, "From Bis Sá Ani to Picurís: Early Pueblo Technology of New Mexico and the Southwest," *Adobe Journal* 8 (1992):10–27.
13. F. Boyo, *Popular Arts of Spanish New Mexico.* (Santa Fe: Museum of New Mexico Press, 1974), p. 5.
14. Smith and Austin, *Adobe, Pressed-Earth, and Rammed-Earth Industries in New Mexico,* p. 11.
15. Earl Morris, "Adobe Bricks in a Pre-Spanish Wall Near Aztec, New Mexico," *American Antiquity* 9(4): 436.

16. Richard Pieper, "Earthen Architecture of New York State: Adobe Construction in a Non-Arid Climate," *Proceedings of the 6th International Conference of Earthen Architecture*, 1990, 117–121. Reprinted in *Adobe Journal* no. 10 (1995), p. 118.

17. Ministry of Housing and Construction, *Adobe Guidelines: Mudbrick Building in Victoria* (1989), p. 5.

18. United Nations, Minister of Housing, 1979, "Experimental Dwelling Project," *Adobe News*, nos. 12–15 (1979).

19. L. Wolfskill, W. Dunlop, and B. Gallaway, *Handbook for Building Homes of Earth* (Texas A&M Research Foundation, for the Agency for International Development, 1944). Spanish-language edition, *Manual para la Construccion de Casas de Tierra*, Boletin No. 21, Centro Regional de Ayuda Tecnica, A.I.D., Mexico (1966).

20. Lee Horne, "Rural Habitats and Habitations: A Survey of Dwellings in the Rural Islamic World" in *The Changing Rural Habitat*, The Aga Khan Awards, Beijing (1981).

21. Salma Samar Damluji, *The Valley of Mud Brick Architecture: Shibam, Tarim, and Wadi Hadramut* (Reading, UK: Garnet Publishing, 1992), p. 130.

22. Ibid., p. 130.

23. Edward W. Smith and George S. Austin, *Adobe, Pressed-Earth, and Rammed-Earth Industries in New Mexico.* New Mexico Bureau of Mines and Mineral Resources, Bulletin 159 (1996), p. 11.

24. Pieper, "Earthen Architecture of New York State: Adobe Construction in a Non-Arid Climate," p. 118

25. Ibid., p. 119.

26. Edward Mazria, *The Passive Solar Energy Book* (Emmaus, Pa: Rodale Press, 1979), p. 141.

27. Jochen G. Güntzel, "On the History of Clay Buildings in Germany," *Proceedings of the 6th International Conference on the Conservation of Earthen Architecture* (1990), p. 58.

28. Ibid., p. 68

29. Richard Ferm, *Stabilized Earth Construction: An Instructional Manual.* (Washington, D.C.: The International Foundation for Earth Construction, 1981), p.16.

30. Ibid., p. 12

31. Ibid., p. 33.

32. Adapted from Ferm, *Stabilized Earth Construction.*

33. Ed Paschich and Paula Hendricks, *Timber Reduced Energy Efficient Homes* (Santa Fe: Sunstone Press, 1994), p. 53.

34. John and Nicola Ashurst, "Practical Building Conservation," in the *English Heritage Technical Handbook.* Vol. 2, *Brick, Terracotta and Earth* (Cambridge: Gower Technical Press, 1988), p. 119.

35. Mufit Yorulmaz, "Turkish Standards and Codes on Adobe and Adobe Constructions," in *Proceedings of International Workshop on Earthen Buildings in Seismic Areas*, vol. 2 (University of New Mexico, 1981).

36. Lazar Sumanov, "Traditional Sun-Baked (Adobe) Brick Structures in Macedonia, Yugoslavia," *Proceedings of the 6th International Conference on the Conservation of Earthen Architecture* (1990), p. 133.

37. Leroy Tolles, Frederick Webster, et al., "Recent Developments in Understanding the Seismic Performance of Historic Adobe Buildings," in *The Seismic Retrofit of Historic Buildings Conference Workbook* (National Park Service, 1991), p. 2.

38. Fred Webster, "Earthquake Design for Adobe Structures; A California Perspective," *Traditions Southwest* (later *Adobe Journal*), no. 2 (1990).

39. Tolles et al., "Recent Developments."

40. Charles Scawthorn, "Strengthening of Low Strength Masonry Buildings in Seismic Areas: Dynamic Test of Adobe Building Model," *Solar Earthbuilder*, no. 47 (1986).

41. Leroy Tolles and William Ginell, "Advances in Seismic Retrofitting: The Getty Seismic Adobe Project" (Getty Conservation Institute, 1995), 78.

42. Ibid., p. 67.

43. Tolles et al., "Recent Developments," p. 1.

44. Fletcher, *A History of Architecture*, p. 30.

45. Henri Stierlin, *The Pharaohs' Master-Builders* (Paris: Terrail, 1995), p. 171.

46. Hassan Fathy, *Architecture for the Poor: An Experiment in Rural Egypt*. (Chicago: University of Chicago Press, 1973), p. 9.

47. Fletcher, *A History of Architecture*, p. 94.

48. Horne, "Rural Habitats and Habitations: A Survey of Dwellings in the Rural Islamic World."

49. Ibid.

50. Michael Moquin, "Ancient Solutions for Future Sustainability: Building with Adobe, Rammed Earth, and Mud," *Adobe Journal* 10 (1995):30–35.

51. Arnie Valdez, "Compressed Earth-Block Construction in India: The Applied Research Program of Development Alternatives, Part 2, *Adobe Journal*, no. 11 (1995).

52. Minke, *Lehmbau-Handbuch*.

53. Valdez, "Compressed Earth-Block Construction in India."

54. Minke, *Lehmbau-Handbuch*.

6
Cob

Michael Smith

Cob is probably the oldest earth-building system, and the simplest. It requires no formwork, no ramming, no mechanized equipment, no industrial additives, and only minimal training. Cob, the material, is a combination of sand and clay, both of which can be found in many soils, with straw and water added. After thorough mixing, the stiff mud is piled onto a wall and formed by hand. Building with cob combines the most enjoyable elements of masonry, ceramic sculpture, and cooking.

Cob's long and varied history, from the cold, rainy, gale-swept British Isles to the deserts of Iran, West Africa, and the American Southwest, shows it to be remarkably effective in a wide range of climates. Given an adequate roof and foundation and a protective exterior plaster where necessary, cob buildings have successfully withstood many centuries of snow, frost, and driving rain. Cob's combination of high thermal mass and insulation make it ideal for passive solar buildings. Free of the mortar joints common to fired brick and adobe, cob is more resistant to earthquakes. The curve and taper of modern "Oregon cob" walls make them even stronger and more seismically resistant.

A cob structure is built by hand, without forms, using a malleable material, which makes this the most easily sculptural of all building systems. The

Alternative Construction: Contemporary Natural Building Methods, edited by Lynne Elizabeth and Cassandra Adams ISBN 0-471-24951-3 © 2000 John Wiley & Sons, Inc.

technique is so simple that almost anyone can learn it quickly, allowing owner-builders to take great satisfaction in the creation of their own highly personalized homes. Because certain aspects of the process (particularly manual mixing) go much faster and are more pleasant with a crew, cob construction encourages cooperation and builds community. The work is rhythmic and enjoyable and empowers people usually excluded from conventional building—women, children, elders, and those with disabilities. The silence and safety of cob construction sites are helping to develop a new breed of builder who values meditation and reverence for the process of building.

Because its ingredients are locally available almost everywhere, cob encourages bioregional identity and cooperation, reducing reliance on long-distance transportation. Cob is completely nontoxic and recyclable. It does not depend on manufacturing or mechanization of any kind, giving it an extremely low environmental cost. The economic cost is equally low; a few hundred dollars usually covers the material costs for the walls of a small cob dwelling. Innovative twentieth-century architectural thinkers, such as Hassan Fathy[1] and Ken Kern,[2] have considered simple, inexpensive earthen building techniques like cob to be the ultimate solution to the world's housing crisis.

History

Unbaked earth is one of the oldest building materials on the planet; it was used to construct the first permanent human settlements about 10,000 years ago.[3] Because of its versatility and widespread availability, it has been used on every continent and in every age. Even today it is estimated that between a third and a half of the world's population lives in earthen dwellings.

Earth construction uses a variety of methods, including adobe, rammed earth, straw-clay, and wattle and daub. *Cob* is the English term for mud building, which uses no forms, no bricks, and no wooden structure. Similar kinds of mud building are endemic throughout Northern Europe, the Ukraine, the Middle East and the Arabic peninsula, parts of East Asia, the Sahel and equatorial Africa, and the American Southwest (where it is known as "coursed adobe").

Exactly when and how cob building first arose in England remains uncertain, but it is known that cob houses were being built there by the thirteenth century. A contending theory holds that cob evolved from the mud mixture used almost universally in medieval times to mortar stone

Traditional cob row houses with reed-thatched roofs in Devon, England. Note the exterior lime-sand plaster and subtly buttressed wall at the right. (Photo courtesy of Ianto Evans, Cob Cottage Company.)

walls and to fill the cavities between two stone faces. By the fifteenth century, cob houses had become the norm in many parts of Britain and remained so until industrialization and cheap transportation made brick popular in the mid-1800s. Cob was particularly common in southwestern England and Wales, where the subsoil was a sandy clay and other building materials, like stone and wood, were scarce. English cob was made of clay-based subsoil mixed with straw, water, and, sometimes, sand or crushed shale or flint. The proportion of clay in the mix ranged from 3 to 20 percent, with an average of about 5 or 6 percent.[4] The cob was mixed either by people, shoveling and stomping, or by heavy animals, such as oxen, trampling it.

The stiff mud mixture was usually shoveled with a cob fork onto a stone foundation, then trod into place by workmen on the walls. In a single day a course, or "lift," of cob—between 6 inches and 3 feet high, but usually averaging 18 inches—would be placed on the wall. It would be left to dry as long as two weeks before the next lift was added. Sometimes additional straw was trod into the top of each lift. As they dried, the walls were trimmed back substantially with a paring iron, leaving them straight and plumb, and between 20 and 36 inches thick. With this method, cob walls were built as high as 23 feet, but were usually much lower.[5] Openings for doors and windows were either built in as the wall grew or were framed by setting lintels of stone or wood into the walls at appropriate heights and carving out the openings after the cob had settled and dried.

Many cob cottages were built by poor tenant farmers and laborers, often working cooperatively. A team of a few men, working together one day a week, could complete a house in one season.[6] A cottage begun in spring would receive its thatch roof and interior whitewash in fall, and its inhabitants would move inside before winter. Often they waited until the following year to plaster the outside with lime-sand stucco, so that the walls would have ample time to dry. Cob barns and other outbuildings were sometimes left unplastered.

Yet cob buildings were not reserved solely for humble peasants. Many town houses and large manors, built of cob before fired brick became readily available, survive in perfect condition today. Among them is Hayes Barton, the birthplace of Sir Walter Raleigh, who had such great affection for his childhood home that he offered to buy it from its then owner for "whatsoever in your conscience you shall deme it worth." An estimated 20,000 cob homes and as many outbuildings remain in use in the county of Devon alone, most of them between 200 and 500 years old.[7]

By late last century, cob building in England, considered primitive and backward, was declining in popularity. During the twentieth century, however, public opinion slowly evolved until traditional cob cottages with their thatched roofs are now valued as historical and picturesque. There was virtually no new cob construction in England between World War I and the 1980s, and the traditional builders took much of their specialized knowledge with them to the grave. However, enough information survived to allow a cob building revival in the 1990s, fueled largely by historical interest and the real estate value of ancient cob homes.

Revival and Modernization

THE ENGLISH COB REVIVAL

The English place great value on tradition, and their cob revival has so far been quite conservative in nature. This is hardly surprising, inasmuch as it has been spearheaded by people whose interest in cob arose from their efforts to repair and protect ancient buildings. The first construction project of the English cob revival was a bus shelter built by restorationist Alfred Howard in 1978.[8] Since then there has been an increasing amount of new cob built in England, particularly in Devon. Kevin McCabe received much attention from the press in 1994 for his two-story, four-bedroom cob house, the first new cob residence to be built in England in perhaps 70 years.[9]

The building technique of these revivalists closely resembles that of their ancestors. They mix Devon's sandy clay subsoil with water and straw and fork the mixture onto the wall, treading it in place. Walls are generally 24 inches thick and straight, applied in lifts up to 18 inches high. The machine age has altered the traditional process in only minor ways: McCabe and others use a tractor rather than oxen for mixing cob and often amend the subsoil with sand or "shillet," a fine gravel of crushed shale, to reduce shrinkage and cracking.

In addition to construction and repair, there is a fair amount of research into English cob. Alfred Howard, for example, has built experimental walls using a variety of subsoils. Larry Keefe, a former building conservation officer, has cataloged hundreds of old cob buildings and become an expert on why cob walls fail and why they do not. Larry is the cofounder of a unique program at Plymouth University dedicated to furthering earth architecture, which has sponsored earth-building workshops as well as several international conferences on earth building, called "Out of Earth." One of the best sources of information on English cob, past and present, is the Devon Earth Builders Association (see "International Building Technology Centers" in Appendix B).

The Development of "Oregon Cob"

Concurrent with the renewed interest in cob in England, there has been a parallel revival in the United States, led by the Cob Cottage Company in western Oregon. Its practitioners having less access to (and less dependence on) traditional knowledge, the building system that has arisen here is sufficiently distinct from British cob that it merits a separate name, "Oregon cob."

Construction of this community kitchen and dining room was initiated at one of the first large gatherings of alternative and natural building practitioners organized by the Cob Cottage Company in 1994. Note the arched windows and the living roof. (Photo by Michael Smith.)

By 1989, Cob Cottage Company founders Ianto Evans and Linda Smiley recognized the need for inexpensive, healthful, bioregional housing. They were particularly interested in earthen building because of their experience in developing and promoting Lorena stoves (fuel-efficient cook stoves molded from packed sand and clay). Ianto grew up surrounded by cob in Wales and later witnessed earthen construction in Africa and Latin America. When experimenting with earthen building in rainy western Oregon, Ianto and Linda chose British cob as a model because of its demonstrated durability in a cold, extremely wet climate.

When they started their first cob structure, Ianto and Linda were unable to locate anybody with firsthand experience. They relied entirely on their explorations of existing cob structures in Britain and a very sparse literature on the subject, much of it inaccurate and contradictory. The system they developed involved making loaves of stiff mud, called "cobs." This loaf system had at times been used in Britain (*cob* itself is an Old English word for "loaf"), as well as in Germany, France, North Yemen, and the American Southwest. Its advantages are that the mix can be made at some distance from the wall and easily transported by tossing the cobs from person to person, as in a bucket brigade. As construction progresses, cobs can be thrown to a builder much higher on the wall than a pitchfork can be raised.

Another way in which Oregon cob differs from traditional cob is in the attention given to the quality of ingredients and to the proportions of the mix. Whereas cob builders in previous centuries had to use whatever soil was on hand with little or no amendment, builders today can cheaply import as much sand or clay as necessary to make the hardest, most stable mixture. Furthermore, whereas grain straw was formerly a valuable resource for animal bedding, thatching, and the like, it is now an underutilized waste product available in huge quantities for little cost. Oregon cob is characterized by a high proportion of coarse sand and lots of long, strong straw.

Better ingredients, more precise proportions, and thorough mixing allow the construction of stronger, narrower, and more sculptural walls. In Oregon cob, exterior walls are typically between 12 and 20 inches thick; non-load-bearing partitions taper to as little as 4 inches (but more commonly 8 inches). Most Oregon cob buildings have curved walls, niches and nooks, arched windows and doorways. By adding extra straw in the needed direction, Cob Cottage Company developed a system for corbeling arches, vaults, and projecting shelves beyond the capability of traditional cob.

After inhabiting their first cob cottage for four years and finding it well suited to Pacific Northwest conditions, Ianto and Linda were ready to share their experience with others. I joined them in 1993, when the Cob Cottage Company was formed and the first workshops were taught. Since then, we have taught more than 80 workshops, mostly week-long, throughout the western states and Canada. We have also worked in Australia, New Zealand, Mexico, and Denmark. We have trained at least 1000 people in cob construction, some of whom have gone on to build homes for themselves or teach workshops of their own. As we continue to make alliances with other natural builders, a system of hybrid natural building is emerging that utilizes cob for its best qualities in combination with stone (for foundations and thermal mass), lime putty (for plasters, mortar, and whitewash), straw (in bales or light straw-clay insulation, as well as in plasters), wood (including unmilled roundwood, for roofs and other structural elements), and many other natural materials.

Cob is well suited to a sensitive house design process, based on careful observation of the site and placement with respect to slope, microclimate, and ecology. It offers exceptionally flexible opportunities for passive solar heating/cooling strategies. Rather than the industrial geometry of straight lines and right angles, cob buildings can use organic continuous curves, variable wall widths, and structural buttresses. Ianto Evans has pioneered space-saving designs that reduce building size and cost, including built-in perimeter furniture, personalized spaces to enclose particular activities, and the sculpting of volumes rather than areas. We have observed that curvilinear spaces are perceived as much larger than rectangular ones with the same measured area. The experience of living in these buildings stretches our ideas of what is possible with ecological building.

The cob revival is still in its infancy. Every year we learn more: how to improve our efficiency at mixing and building and how to use a wider range of soil types, new applications, techniques, and designs. For a more detailed how-to-do-it resource for cob building, see *The Cobber's Companion: How to Build Your Own Earthen Home* by Michael Smith (available from Cob Cottage Company, P.O. Box 123, Cottage Grove, OR 97424).

Environmental Benefits

All in all, cob has lower embodied energy than almost any other building system. The energy spent on transportation can be almost zero, inasmuch as the bulk of the material (generally between 40 and 95 percent of the walls' mass) is local subsoil from the building site. In some cases, the only necessary additive is straw, which is lightweight and locally available throughout most of the United States. In most other cases sand or clay is imported from a nearby location. Cob walls can be built without any wood, cement, or manufactured materials whatsoever. In comparison with a conventional stud-framed structure, a cob house of similar size (particularly one with an adobe or other masonry floor) requires as little as 25 percent of the lumber.

Cob can be built with little or no electricity or fossil fuels, making it well suited for use in less-developed countries and on remote sites. Cob is easily mixed by hand and foot, eliminating reliance on machines. Furthermore, as thousands of ancient British cob buildings have shown, the maintenance requirements of cob are minimal; it is common for cob buildings to endure for a century between repairs.[10]

Perhaps best of all, cob creates no waste disposal problems, neither during construction nor when the building is no longer required. The building components of earth, sand, and straw are quickly and easily assimilated into the environment without toxic residues. If natural earth- and lime-based plasters and finishes are used, complex manufactured chemicals can be eliminated completely. For this reason, cob is preferred by many chemically sensitive people. It may be comforting for people to realize that in some future generation their houses may be recycled as gardens.

Design Considerations

MOISTURE ISSUES

The most critical design consideration for cob buildings is protection from prolonged soaking. Cob walls can actually absorb large amounts of water without damage as long as they are able to dry out. If the cob stays wet for a long time, the straw inside will eventually rot, weakening the wall. In extreme cases of soaking, total shear failure can result, usually near the base of the wall where the load is greatest. And rain running down unprotected cob gradually erodes the wall. Traditional British cob builders had a maxim, "Give a cob house a good hat and a good pair of boots, and she'll last forever."

1. *Foundations.* Most surviving British cob structures were built on stone plinths up to a meter high. In less windy or rainy climates, such a high foundation is unnecessary. We generally make our foundations about 18 inches high, using stone, concrete, or other waterproof materials. When the foundation is made of concrete, brick, or

porous stone, a vapor barrier may be advisable, although tests in England have shown that cob's coarse texture prevents moisture rising more than 6 to 10 inches through an unplastered wall.[11] We have also experimented with soil-cement, brick, rammed tires, and rammed-earth bags ("superadobes") as a foundation for cob—each has its advantages and disadvantages. In rainy climates, we particularly like rubble trench foundations because of the extra protection they give against flooding.[12]

2. *Roofs.* In rainy climates, it is important to make sure the roof sheds water away from the cob walls. Long eaves are good, particularly on the windward side of the building. Gutters prevent roof splash from reaching the base of the wall.

3. *Plasters.* Where wind-driven rain occurs, even a high foundation and wide eaves do not always keep the walls dry. This may not be a serious problem, inasmuch as the usual rate of erosion of unplastered walls in England has been estimated at about 1 inch per century.[13] Nonetheless, English cob buildings were frequently rendered with a lime-sand plaster for protection and appearance. Interior and exterior plasters should be made of lime, earth, or other breathable materials so that moisture in the wall can escape easily. Cement-based stucco has been shown to shorten the lives of earthen buildings (both cob in Britain and adobe in the American Southwest) by trapping moisture inside and hiding the resulting damage.[14]

The cob office of the Permaculture Institute of Northern California keeps its walls dry with a unique gutter system. A generous enameled metal roof has been fitted with 4-inch-diameter flexible drainage pipes that follow the roof's curving edges. Holes drilled in this gutter are fitted with heavy chains that direct rainwater to a biological filtering drain and then to a duck pond. (Photo by Robert Bolman.)

STRUCTURAL INTEGRITY Oregon cob, with its careful mixing and high proportion of sand, has enormous compressive strength. Plenty of long fiber (up to 10 percent or more straw by volume) also gives cob better tensile strength than many other masonry materials. Because cob buildings are monolithic, without seams or mortar joints, they are more resistant to earthquakes than adobe. Anecdotal evidence from New Zealand shows that cob buildings can survive earthquakes better than brick or concrete block; a Victorian cob mansion in Nelson, New Zealand, has survived two major earthquakes without noticeable damage, whereas the brick school nearby needed extensive repairs following each occurrence. In seismically active zones, the following strategies should be used to maximize cob's earthquake resistance:

1. *Wall Height, Thickness, and Taper.* Unstabilized, unbaked earth walls can easily reach enormous heights, like the 10-story cob buildings in Yemen. To increase the stability of any wall, however, its center of mass should be kept as near to the ground as possible. This can be done by reducing the total height (a limit of two stories is probably prudent in seismic zones) and tapering the walls. With a taper of 5 percent, a wall loses 6 inches of width for every 10 feet of height. A more severe taper improves stability, but also increases the width of the foundation if the top of the cob wall is not to become unaccept-

This rectilinear walled cob residence in New Zealand survived two significant earthquakes, which tumbled the nearby school. The house also demonstrates cob's versatility—it need not be built only in organic shapes, but can be utilized successfully for conventional Western architecture. (Photo courtesy of Ianto Evans, Cob Cottage Company.)

The best design principles for cob architecture are exemplified in the graceful curving walls and large overhanging roof of this recently constructed residence at Heartwood Institute in Garberville, California. Notice, too, the cob eyebrows over lower-floor windows. (Photo by Michael Smith.)

ably narrow. Non-load-bearing interior partition walls can taper almost to nothing, but exterior load-bearing walls should be a minimum of 10 to 12 inches at the top. Thus, in general, a 10-foot-high wall should be about 18 inches thick at the bottom; a 20-foot-high wall should be at least 24 inches. Thicker walls are more stable and provide better insulation, but also require more material and take longer to build and to dry.

2. *Curved Walls.* Another way to improve the strength and resistance of a cob building is to curve the walls in plan. Curved forms are much more stable than rectangles, which concentrate forces at their weakest points, the corners. Even if a curved wall were to develop a major structural crack from top to bottom, it would not topple easily, unlike a straight wall. The more curved a wall is, therefore, the less thick it must be to have the same stability.

3. *Buttresses.* Buttresses stabilize a building in much the same way as curves. Structural buttresses should be built at potential weak points: where walls end; on either side of major openings like doors; and periodically along long, straight walls. There is no need for the walls of a building to have uniform thickness. In designing with cob, it is easy to make the walls thickest where they will be bearing the most weight or will be subject to shear stresses, and thinner elsewhere.

4. *Openings.* Theoretically, large openings weaken a wall. So do rows of small openings close together with tall, thin pillars of cob separating windows or doors. In New Zealand, however, we have seen ample evidence that cob buildings with large openings nonetheless survive earthquakes well.

THERMAL PERFORMANCE   Cob's combination of thermal mass and insulation make it an excellent material for many passive solar applications. The high quantity of straw and the tiny cavities left by evaporating water and shrinking clay allow cob to insulate better than other masonry materials like stone, concrete, brick, or rammed earth. On the other hand, cob has greater thermal mass than conventional stud-frame construction, straw-bale, or light straw-clay systems.

Cob's R and k values are similar to those of adobe, but vary with the exact mixture. Mixes with more sand result in higher thermal mass; those with less sand and more straw have greater insulation. Other ways to improve insulation include imbedding straw-clay blocks in a cob wall or building a cob cavity wall filled with loose straw, straw-clay, wool, or other natural insulation materials.

When designed to maximize passive solar heating and cooling, cob buildings perform well in many climates. Earthen buildings have proven comfortably warm in winter and cool in summer in desert areas with hot summers and cold, sunny winters, as well as in Mediterranean climates with mild temperatures year-round. Earth buildings are traditional even in places with extremely cold, cloudy winters (cob in Britain; sod in Europe, Canada, and Alaska as far north as the Arctic Circle), but they require a great deal of additional heating. In places with very cold winters, we recommend using cob as part of a hybrid building—for example, as trombe walls to store and release the heat from winter sun or as interior mass walls or floors to regulate indoor temperatures.

AESTHETICS AND PSYCHOLOGY   Having all the fluidity of ceramics, cob is the most sculptural of all wall systems. This sculptability makes curved forms simple to build, which is important both for structural reasons, as noted earlier, and for aesthetic ones.

From the outside, curved buildings blend unobtrusively with rounded landforms. This helps strengthen the idea that the building is a natural part of the landscape, rather than a refuge from it. Exposed earthen walls and gentle, organic shapes seem to answer an innate human need for connection with nature. Cob buildings tend to catalyze their inhabitants toward more

Windows and doors may take any shape imaginable, as in the "Goddess" entrance to Sun Ray Kelley's yoga studio in Sedro Woolley, Washington. (Photo by Michael Smith.)

direct participation in their natural environments and away from dependence on unhealthy and environmentally destructive practices, both in the construction process and in their day-to-day lives.

There is ample evidence that the shapes of buildings and rooms have a profound psychological effect on their inhabitants.[15] Straight walls and right angles, repetitive industrial patterns, and monochrome surfaces are such a deviation from the natural environments in which humans evolved that they may trigger a constant "Something's wrong!" response, contributing significantly to the stress of modern life. In contrast, curved cob walls with natural finishes create interior spaces that seem relaxing and protective. They also appear to be much larger than rectilinear rooms. (An informal survey of rounded cob structures by building and design professionals reveals that they tend to overestimate the floor area by an average of 100 percent!)

Oregon cob structures are works of art that allow the personality and inventiveness of their creators to shine. Windows and other openings can be made any shape desired. Interior and exterior walls can be decorated with bas-relief patterns and sculpted murals, finished with brilliantly colored earthen plasters. Cob buildings often incorporate other natural materials such as attractive stones and sculptural wood. Cob benches and other built-in furniture maximize the use of space and create cozy, personalized spaces.

COB IN HYBRID BUILDINGS Cob has been used in combination with other natural materials to good effect. To optimize thermal performance, for example, we have built several

The bas-relief cob cat over an interior window was sculpted by artisan Kiko Denzer at an experimental hybrid straw-bale and cob structure in Cottage Grove, Oregon. (Photo by Kiko Denzer.)

passive solar houses with cob walls on the south and more insulating straw bales on the north. Another good way to combine cob and straw is to put a few inches of cob against the inside of a straw-bale wall, creating well-insulated thermal mass for passive heating and/or cooling.

Cob has been called "the duct tape of natural building." It adheres well to most other natural materials, including wood and straw. Because it is so plastic, it can be used to fill awkward gaps virtually anywhere in a building. We use it to level out the tops of stone foundations for straw bales or straw-clay, to pack into the seams between straw bales or rammed-earth bags to stabilize them and make a firm plastering surface, and in endless other combinations. Sections of a wall made primarily of other materials can be filled in with cob to create sculptural niches or easy custom windows. For sculptural accents, nothing beats cob.

This cob home was built by Brigitte Miner and her young daughter, Elyse, shown here enjoying their horseshoe-shaped, built-in cob seating area. (Photo by Robert Bolman.)

COB AND CODES   In Britain, cob and building regulations have an interesting history. The first law restricting building materials was the London Building Act of 1667, created in reaction to the great London fire of 1666. Because cob was not used in London at this time because of the unsuitability of the soils, it was not mentioned as a material having satisfactory fire resistance.[16] When the first National Building Act was proposed in 1842, it was based on bylaws of London and Bristol, where cob was unknown. Consequently, cob was no longer legally permissible even in regions where it had been safely used for many centuries.

In 1985, after 135 years of regulatory purgatory, cob was once again permitted by a new clause in the building laws, which states that "any material which can be shown by experience such as a building in use, to be capable of performing the function for which it is intended" is satisfactory. However, lacking definitive information on the proper construction of cob, regulatory authorities are often reluctant to approve its use. Most "new" cob construction in England to date has been either for nonresidential outbuildings or for buildings officially categorized as "restoration" projects. For example, to get local building council approval as recently as 1994, Kevin McCabe set his cob house on the foundation of an old stone farmhouse, with two original walls left standing.

In the United States, where many of the oldest surviving buildings are made of earth (even in heavily seismic zones like central California), there is less awareness of and less respect for our earth building heritage than in Britain and elsewhere. Nowadays, most regulatory change is the result of pressure applied by large, moneyed interest groups. It is significant that increased code acceptance of straw-bale building in California came only after the enormous grain-growing industry of that state was restricted from burning waste straw in the fields. It is extremely difficult to get rich by building with cob, as it requires a great amount of low-skill labor, no specialized equipment, and materials that are almost free. Cob lends itself to low-income, owner–builder, and cooperative community use, rather than industrialized production. It is therefore difficult to see where the financial impetus, that could push cob building into the mainstream, will come from.

The other obstacle to code acceptance of cob is the wide variability of soils from which it is made. Oregon cob requires customized mixes based on a careful analysis of the available soil at a site. The mix proportions depend primarily on the amount of clay in the soil, but also on the stickiness and expansiveness of that particular clay and on the ratio of different particle sizes of silt and sand. These conditions are simple to determine in the field but are hard to quantify and, therefore, hard to regulate.

However, there are no laws prohibiting earthen construction anywhere in the United States, and most building regulations do include provisions for "alternative methods and materials." For example, Section R-108.1 of the Council of American Building Officials (CABO) One and Two Family Dwelling Code states:

> The provisions of this code are not intended to limit the appropriate use of materials . . . or methods of design or construction not specifically prescribed by this code, provided the building official determines that the proposed alternate materials . . . or methods of design or construction are at least equivalent . . . in suitability, quality, strength, effectiveness, fire resistance, durability, dimensional stability, safety, and sanitation.

The interpretation of this section invests a great deal of authority in individual officials. Although several have told Cob Cottage Company that they eagerly await the opportunity to approve cob buildings within their jurisdictions, others are likely to be skeptical of techniques outside of their experience, requiring expensive testing and the approval of a licensed engineer or architect. Architect John Fordice in Berkeley, California, has initiated an effort to develop a rigorous testing program in the hope that a specific cob code can be adopted into the Uniform Building Code, removing the onus of proof from individual builders and reducing permitting costs.

In the meantime, cob builders in different parts of the country are taking various approaches to the code problem. Many owner-builders in rural areas sidestep the issue by building without permits or with less stringent agricultural permits. Most counties have set a minimum square footage (often 120 or 200 square feet) below which no permit is necessary, allowing for the unregulated construction of very small cob buildings. Other builders construct cob within a timber frame so that the cob is not load-bearing. Special owner-builder codes like those in California's Mendocino and Humboldt Counties seem to provide a relatively easy opportunity for permitted cob homes.

One heartening recent development occurred in New Zealand, where it was reported that new codes for cob construction were adopted in 1997. The regulations still require the approval of a licensed engineer for each earth-building project, but allow for substantial variation in mix proportions according to local soil types.

Construction Methods

The materials needed for cob are simple and inexpensive. The base of the mixture is almost any soil easily available (although soils with rich proportions of organic matter, silt, or very fine sand are harder to work with). We usually start with the soil excavated from foundation and drainage trenches. Occasionally, this subsoil already contains sand and clay in the right proportions to make a hard, durable mix that does not shrink or crack. More often, the base soil is amended with clay or coarse sand.

Clay is the basic binder or cement that holds everything else together. It is sticky when wet, hard and brittle when dry. It bonds electrochemically to itself, to water, and to other minerals. It expands when wet and contracts when dry, which can result in severe cracking if there is too much clay in a mix.

Sand is extremely hard and inert. It gives its compressive strength (hardness) to the cob mix. Because sand does not change in size when wet, it stabilizes the clay. As the clay dries and shrinks in the cavities between coarse, jagged sand grains, it pulls tight and locks them together. Oregon cob can be thought of as as building with sand particles mortared together with clay. Sand makes up most of the volume of the mix.

Straw gives cob its tensile strength—the ability to move and bend without breaking and to withstand ground movement and shear forces. Its random dispersal throughout the cob creates a three-dimensional textile, like muscle fiber or the tensile vessels in a tree branch. Straw increases the insulation value of cob by trapping air inside its hollow stems, and its capacity to transport water by capillary action may help to distribute moisture evenly through the cob wall. It is better to use long (8- to 16-inch) strands of a strong-fibered grain straw such as wheat, oat, barley, or rice, rather than hay. The strength of straw varies widely according to variety, planting time, age, and, especially, whether it has been properly stored.

Tools and Equipment   The whole process of cob mixing and building can easily be done without any power tools or equipment. You will need tools for excavating and moving earth and sand: shovels, spades, picks or mattocks, and wheelbarrows. For mixing, you will need buckets for measuring; strong, durable tarps; and a hose and/or storage barrel for water. To apply the mix to the wall you need only a flat-tined garden fork and a "cobber's thumb" (a stick or other object that fits comfortably into your hand and has a rounded end like a thumb). For shaping and trimming, use a machete, hatchet, adz, old hand saw, or a sharp, heavy hoe. Spirit levels are extremely handy for keeping the wall plumb. Other tools to have on hand during the construction process include spray bottles for wetting down walls and plaster; screens of various sizes for sifting materials (especially for plasters); carpentry tools for window and door frames and the like; and masonry tools and trowels for foundations and plastering.

To date, mechanization of the mixing process, using a tractor, Bobcat, or mortar mixer, has been moderately successful. See the section "Mixing Cob" for more details.

Labor Mixing and building cob requires a lot of work, but it is simple, rhythmic, enjoyable work that almost anyone can learn quickly. The only parts that require much training or experience are determining the most suitable cob mix proportions and ensuring that the cob is well integrated with itself and firmly connected to other building elements. The rest of the work (mixing the cob, building and shaping the walls) can be done by almost anyone: low-skill laborers, friends, family, neighbors, children, or elders. This work can be extremely empowering for people with limited building experience, physical strength, or economic means. The relaxed, safe, and quiet atmosphere of a machine-free cob building site encourages the rediscovery of the joy of working together in community.

PRECONSTRUCTION PREPARATION A cob building site should be laid out to allow easy access during mixing and building. Soil, sand, and straw should be piled close to the building perimeter. Near these piles, prepare several flat or slightly dished areas, about 8 feet in diameter, for tarp mixing. You will need water at each mixing area and everywhere along the wall; long garden hoses supplemented with a few 50-gallon drums are ideal. Allow for unobstructed transportation of materials to the mixing areas, and of cob from the mixing areas to the building perimeter.

During site preparation, excavation, and foundation building, always keep drainage in mind. Employ whatever strategies are necessary to prevent flooding of the building in any imaginable circumstance.

Children at a cob building workshop get completely immersed in the medium. (Photo by Michael Smith.)

MIXING COB

Testing Soils and Mixes

To determine the ideal proportions for a cob mix using a given soil, you first need to know something about the composition of that soil. All soils are a combination of up to five basic ingredients: clay, silt, sand, stones, and organic matter. The "shake test" is a simple way to determine which of these ingredients are present and in approximately what proportions. Fill a clear glass jar half full with soil, making sure that any lumps are broken down as finely as possible. Fill the jar with water and shake thoroughly so that the particles become separated and suspended. When you set the jar down, the suspended particles will fall to the bottom in order of weight: first stones and gravel (immediately), then coarse sand (within about three seconds), then fine sand, and then silt, which is simply particles of rock too fine for the unaided eye to distinguish. Clay has chemical properties that cause it to bond with water, so it will stay in suspension and only settle out slowly over hours or days. Organic matter will float to the top of the jar. The shake test should not be used to determine exact proportions for a cob mix, but it reveals the approximate makeup of a soil and, more specifically, whether clay and sand are present, and roughly the proportion of useful coarse sand.

In determining the ideal cob mix for a given soil, the most critical proportionn is the ratio of clay to sand. You need enough clay to make a plastic, cohesive, workable mix, but not so much that the mix shrinks a lot and cracks. Depending on the coarseness of the sand, the quality of clay, and the other components in the soil, the final mix should end up between 5 and 25 percent clay. But mathematics will only get you started; the most important thing is to know how a mix should feel. The best way to learn this is to make cob with someone who already knows. The "right feel" is difficult to describe in words, but the following tests will give you a place to start.

The first critical question to answer in making cob is, "Does this soil have enough clay to stick it together?" The quickest test is this: Take a sample of the soil and moisten it until it becomes plastic. Form it into a ball the size of a golf ball and press it onto the palm of your hand. Then turn your hand over. If the ball sticks as you open and close your hand a few times, it has enough clay to make good cob.

The "crunch test" is a simple way to determine whether there is enough sand in the mix to prevent cracking. Prepare mixtures of soil and sand in different ratios, then add just enough water to make them stick together with about the consistency of pie crust dough. Make a ball of each mixture and squeeze it close to your ear. If the mix has enough sand, you should

Wedging the cob by foot. Experience teaches how to know by feel when the mix of sand, mud, and straw is balanced. (Photo by Robert Bolman.)

hear the sharp grating sound of sand grains rubbing against each other. There is usually a marked difference in sound between the mixes with enough sand and those without.

Ultimately, the only sure way to test the proportions is by making a cob mix, following the directions given in the next section. After you have made enough batches of cob, you will know almost immediately by the feel of the mix whether the proportions are right. In the meanwhile, form a mix into loaves about the size and shape of bricks and wait until they dry all the way through. If they crack noticeably when drying, they contain too much clay. If they crumble easily, they have too little clay. If you can easily break them in half in your hands, they probably need more straw.

Mixing Techniques To be effective, the mixing process must achieve two essential ends. First, the basic ingredients (soil, sand, water, and straw) have to be combined thoroughly and evenly to create a homogenous, consistent mixture. Second, pressure has to be applied to smear the clay molecules across and between all the sand grains. We know of several techniques that meet these basic criteria, detailed in the following paragraphs, and doubtless many others remain to be discovered.

Traditional mixing. The traditional system of mixing cob in Britain started with spreading out a large heap of loose soil on hard, flat ground. Sometimes sand or broken shale or flint was added to the pile, and straw was often spread out beneath it. Several people would tread and turn the pile with special cob picks while others sprinkled water and straw on the soil. The pile would be tread and turned repeatedly until thoroughly mixed. Sometimes oxen were used to tread the mix .[17]

In France, where cob was known as *torchis*, the mix was prepared in the hole from which it was dug, and afterward the hole became the farm pond. Mixing was done by men who linked arms to tread the clay barefoot, dancing and chanting as they did so. Then gravel was added to stabilize the material, short lengths of straw to bind it, and, often, cow dung to make it more adhesive. After this treatment the dance was repeated, and the resulting mixture was left to dry for several days before use.[18]

The tarp method. Our preferred technique for mixing cob is "the tarp method," developed by Becky Bee. The essential equipment is a piece of flexible, tear- and water-resistant fabric (such as woven polypropylene, heavy plastic sheeting, or oiled canvas) between 6 and 8 feet square. Spread the tarp out on a piece of clean and level ground close to the cob ingredients and the building site, and pile soil and sand on the middle of the tarp. If the soil has a high clay content and tends to form hard lumps, it is best to soak it for a day or so before mixing. The amount of the mix can be as large or small as you can comfortably handle.

The first stage of the process is dry mixing, which is easier when done by two people. Grasping opposite sides of the tarp, they rock back and forth together, leaning somewhat backward and allowing their legs to do most of the work. Part of the tarp and most of the weight should remain solidly on the ground as the dry materials roll across the tarp. After a few long rolls, rotate positions 90 degrees and rock in the other direction to make the mixing more thorough. When you can no longer see pockets of unmixed soil and sand, it is time to add water, a little at a time. Then repeat the rocking mixing motions a few times until the water is evenly distributed. The mix should be evenly moist but able to hold its shape, not so loose that it runs off the sides of the tarp.

When the water is thoroughly mixed in, it is time to start dancing! Learn to dance with maximum pressure in order to break up lumps and align and distribute the clay. Once all lumps are broken up and the water is evenly distributed through the mix, begin to add straw (in such a way as to avoid lumps) as you dance. Keep treading until all the straw is dirty and worked into the mud, then turn the mix by pulling one corner or edge of the tarp toward you until the mix folds over on itself. Do this repeatedly as you add more straw, pulling from a different corner each time and making certain to pull the tarp far enough to turn the center of the mix so that you do not end up with an unmixed mass there.

The pit method. Larger, runnier mixes can be mixed in pits lined with tarps. (Unlined pits can also be used.[19]) Make a pit either by digging a shallow hole in the ground or by placing straw bales, with their corners touching, to form a hollow square. The easiest sequence seems to be to mix earth and water first into a smooth, runny mud, then add sand and tread it in thoroughly, and, finally, add straw. Using a pit, a single person can make a very large mix. However, it is more difficult to get the ingredients mixed as evenly as with the tarp method. In a wet mix, both sand and straw tend to sink to the bottom and the corners of the pit accumulate masses of unmixed materials. Make sure to pull the corners of the tarp toward the center of the pit, freeing up these unmixed masses.

Visitors to a Natural Building fair at the Black Range Lodge in Kingston, New Mexico, try their skills at "dancing" in the cob. (Photo by Robert Bolman.)

Mechanical mixing. Several different machines have been used with varying success to mix cob. In both Britain and the United States, people have made large mixes using a tractor or a small Bobcat. Tractors with buckets can transport ingredients, lift and turn the mixture, compress it under their treads, and then carry it to the building. Although they have trouble getting the ingredients thoroughly mixed, tractors provide excellent compression and can produce good cob given enough time and attention.

The other popular cob mixing machine is a mortar mixer, a rotating drum with paddles that move the contents in a figure-eight path. Mortar mixers seem to be capable of making only large, runny mixes with smaller quantities of shorter straw than are generally desirable, or very small, stiffer mixes. Mortar mixers combine the ingredients well, but do not provide very good compression.

Large tractor or mortar mixer mixes have to be quite wet, inasmuch as it may take days to get such a load of cob onto the walls. The greatest danger is that the straw will begin to rot before it has a chance to dry out, unless you have a large building and a large crew ready to apply the cob to the walls. Given wet mixes and low proportions of straw, it may be advisable to rework the cob as you use it, stomping it out and adding additional fresh straw.

Apart from considerations of quality and consistency, mechanical mixing has several disadvantages. First, the noisy and dangerous machines drastically change the calm and welcoming atmosphere of a nonmechanized cob work site. These machines consume fossil fuels and spew noxious pollutants. They have to be driven to the site, requiring vehicle access, and tractors particularly can cause soil compaction and other damage to the site. They can also be expensive.

Specialty Mixes There are endless ways to vary a standard cob mix, and many reasons for doing so. This is a new science and we have barely begun to explore the possibilities, but these are a few specialty mixes we have tried:

For hot, dry weather or slow building, you can afford to make much wetter mixes than usual. These are easier to make, because the extra water makes adding straw much easier. For wet weather or fast building, you should make stiffer, dryer mixes with more sand and less straw. For sculptural details and around windows, we use a finer mix, screening the rocks out of the soil and sand and often chopping the straw. For extra insulation or tensile strength, increase the amount of straw. Another way to increase the insulation value of cob is to substitute ground pumice and pumice chunks for sand. For extra hardness or thermal mass, increase the sand and cut back on straw.

BUILDING TECHNIQUES   There are three basic ways to apply the mixed cob. Each technique is particularly suited to specific circumstances, but it is common to use all three in a single wall.

Pisé (from the French, *pisé de terre,* which is rammed earth) is fast and easy, particularly if you can mix close to the wall where you will be building. It works best for thick walls with few details. We generally use it at the bottoms of walls, where walking the wall is not as scary.

The *Gaab-cob* method (of Oregon cob coinage) is probably the most versatile and efficient technique for general wall building. Because it works best with fairly moist mixes, its main disadvantage is evident when gaining height rapidly is a priority.

Using *cob loaves* takes more preparation time than the other two methods. Its main advantage is the control you have in placing and shaping each cob with your hands. This is particularly important when you get to windows, niches, and sculptural details. The other advantage is the ease of transportation. Especially when a wall grows higher than you can reach, it is convenient to toss cob loaves to a co-worker standing or sitting on its top.

Pisé** and **Gaab-Cob   We use *pisé* to mean the traditional British technique of treading cob by foot onto a wall. Simply transfer large patties of cob onto the wall, either by hand or with a flat-tined garden fork, then walk on them. Your weight will cause the fresh material to stick to the old, and you can go back over it with a cobber's thumb to improve adhesion. Because you cannot consolidate the

Cob tossing is demonstrated at a Natural Building fair. (Photo by Robert Bolman.)

edges of the wall with your feet, you should either smooth them with your hands or use a "persuader," a wooden paddle that allows you to pat and smooth the edges of the wall while standing on top.

Pisé is easiest on wide walls that are fairly low. As a wall becomes narrower and higher, it is more difficult to keep your balance while walking on the wall, and harder to pass the cob directly off the fork. Because you cannot do much delicate molding with your feet, *pisé* works best on walls without windows, niches or other sculptural details.

Gaab-cob combines the speed of *pisé* with the precision of the loaf method. It helps to have a fairly loose, moist mix, so you can bond and mold it easily. Put the cob on the wall in large handfuls or forkfuls, as you would for *pisé*. But instead of walking on the wall to adhere the new cob to the old, work it in by hand as you would for cob loaves (see the following section), using either your fingers or a cobber's thumb.

Cob Loaves Another traditional building technique involves kneading the mix into loaves, or cobs, which can be conveniently transported to the wall and tossed to a builder on top. The kneading provides yet another step in the compression process, producing an even, workable consistency. The size and shape of the loaves is unimportant, as long as they are not too heavy to be easily handled by everyone on the work site, nor so long that they break in two when thrown.

To make the cobs, kneel on the edge of the tarp, lean forward, and grasp a large double handful of mix. Squeeze the material together with your hands, and roll it a few times on the ground, using the weight of your upper body to compress a loaf that will hold together when tossed.

In order to create a monolithic, earthquake-resistant wall, it is important to bond each cob into the mass of the wall. The goal is a three-dimensional textile of interwoven straw, buried inside a strong and durable mass of sand and clay. Use your fingers or a cobber's thumb like a sewing machine, pushing straw through the top layer into the one beneath. When you are finished, it should be difficult to pull any cob off the wall and impossible to tell where any one cob stops and another starts. Leave the top surface as rough as possible to maximize surface area for drying and to improve bonding between courses.

When building with either cob loaves or Gaab-cob, you need some kind of scaffolding to stand on as the wall grows higher. Try to keep your waist level with the top of the wall so that you can use the full weight of your upper body to bond the cobs together.

Drying and Bonding   The factor limiting how quickly you can build height in a wall is the drying time. If you add too much new material before the cob beneath it has had a chance to solidify, the wall will sag and lose its shape, requiring a lot of trimming later. The ideal is to build onto a damp, soft surface, with the wall 6 inches below stable and solid. The rate of drying depends on many factors, including air temperature and humidity, wind, and the proportions and wetness of the mix. Except in very dry conditions, the maximum height practical to build in a day is generally 1 foot.

If the top of a wall dries completely while building is still in progress, you will never achieve a perfect bond when adding new material. When you leave for an extended period of time, cover the wall with wet straw, burlap, carpet, or newspaper and then plastic to retard drying. Should the wall become dry and hard, wet it down before building on it.

Trimming and Shaping   As the walls rise higher, check the plumbness of the wall each time you move to a new section of the building to add more cob. If it is not vertical, trim it back to where it should be, or strap on long, straw-rich cobs to fill it out, before continuing to build on top. Remember that you may choose to deliberately taper your walls for extra strength and earthquake resistance.

It is almost always necessary to trim cob walls. Even if you build slowly and with great precision, a wall generally ends up with bulges, lumps, and divots, which look messy and make plastering harder. It is difficult to trim cob that is either too wet or too dry. Try to trim each section of wall a day or two after it is built, as it reaches a "leather hardness," firm and solid but still moist inside.

The most useful general-purpose trimming tools are a machete and an old hand saw. To trim tight corners and concave spaces, use a shorter-bladed tool like a hatchet, adz, or sharpened hoe.

Corbeling   The key to many kinds of sculptural details is corbeling, which allows the deliberate and controlled widening of a wall as it goes up, in such a way that the weight of the projection and anything that rests on it is carried back into the wall. We have built corbeled cob benches and shelves that project up to two feet from their base yet are strong enough to support heavy objects like books and people. Corbeling from several directions at once produces arches, niches, vaults, and domes.

The most effective corbeling requires special cobs. Corbel cob loaves are longer and flatter and have additional fiber running in the longitudinal direction. To make corbel cobs, first make a standard mix that is slightly sticky. Lay a mat of long straws running parallel on the ground in front of you. Make oblong cobs and roll them in the bed of straw so that they pick up straws in the lengthwise direction. Then knead until these straws are fully incorporated into the cob.

To make a corbeled shelf, first build up a section of wall until it is relatively flat and level. Then apply the first course of corbel cobs so that their front ends project a couple of inches out of the wall. "Sew" these cobs into the wall below, taking advantage of the extra straw to get a really good bond. When the first layer is firm but not dry, apply a second course in the same manner. Be patient. Let each course of cobs project only a couple of inches beyond the one below. Test each corbel to see how solid it is before applying more weight.

Arches, niches, vaults, and domes are constructed in the same manner. To make an arch, for example, build up both sides with regular cob until they are flat and level, then begin corbeling out from both sides at the same rate. The symmetry need not be perfect; it is much easier to trim later than to aim for perfection during the building process. Remember to build up the surrounding area with regular cob as you go, to support the arch. Be extra careful as you near the top. Allow the arch to dry substantially before placing the "key cob" across the top.

Another way to build an arch is to use a rigid mold to support the weight of the cob during its initial drying. This allows more rapid completion of the arch and reduces the need for trimming later. The jury is still out regarding whether this technique produces arches that are as strong as unsupported, corbeled ones.

CONNECTIONS

Roofs and Wooden Elements

Wherever cob meets wooden elements in a building, it is important to ensure a secure bond so that the elements do not shift apart over time. This is particularly true where the cob and wood are both structural, such as, for example, where a load-bearing cob wall meets a structural post or roof structure. Usually, the construction schedule dictates that either the wooden structure will be erected first and the cob built later, or in the opposite order. Both approaches work, but in general it is easier to fit the more fluid cob up to and around the wood.

Wood frame first. Erecting the wooden framing first makes it easier to get it plumb, level, and square. To connect cob to the wood, you need some sort of key. A convenient way to make a connection is to drive nails partially into the surface of the wood to grab the surrounding cob. Another keying system is to affix a two-by-four or similar piece of wood vertically to the wooden surface that will meet the cob. When this key is surrounded by cob on three sides, any sideways shifting between the two materials is prevented.

During its initial drying and shrinking, the cob must be able to slide down the wooden post without getting caught, or it may crack. Interestingly, nails do not seem to interfere with this process. What does cause difficulty is to have a large post completely surrounded by cob. Do not build a load-bearing wooden structure inside a cob wall. Instead, build it adjacent to the cob or as much as halfway buried.

We often erect roof beams and rafters that will eventually bear on the cob walls before the cob walls are built. This allows us to stretch a tarp over the roof for protection from sun and rain. In addition, because many of our buildings are irregularly shaped and the roofs themselves curved, this prevents endless calculations about how high the walls have to go. One way to anchor the roof structure into the cob is to attach an inverted "T" made of two-by-fours or other stout wood under each rafter. Be sure that each piece of wood is firmly attached to those it touches with strong galvanized nails, screws, or bolts and that the horizontal anchor piece is at least a foot below the top of the wall.

Cob wall first. If the cob wall goes up first, you must bury some sort of anchor in the cob wall. Two main anchoring systems are "dead men" and "gringo blocks." A dead man is a chunk of wood, buried in the cob wall, to which the wooden structure can later be attached. This can be done either by leaving part of a wooden dead man exposed and screwing or nailing into it, or by securing a heavy wire or cable around the dead man and using that to connect the post. The cable system works well for rafters, particularly when their exact location has not been determined.

A gringo block, borrowed from Southwestern adobe construction, is a wooden box, made of two-by-sixes or similar heavy boards, with no top or bottom. When buried in the wall with one surface exposed, it can be used for attaching wooden elements (including posts, door and window frames, counters, cabinets, and the like) with screws or nails. Use at least two gringo blocks for each side of a door frame—one near the bottom and one near the top. When attaching a structural post, use a gringo block or dead man for every 3 feet of height.

Straw-Bale Walls In connecting cob walls to non-load-bearing (post-and-beam) straw-bale walls, the objective is to attach the wooden structure to the cob (as explained in the preceding section) and then the bales to the timber frame. This technique makes a lot of sense in damp climates like the Pacific Northwest, where there is reason to expect a cob wall to outlast straw bales by a long time. If, on the other hand, bales as well as cob are to be load bearing, finding an adequate fastening system is more challenging. The following are the techniques we have explored to date.

If the cob wall goes up first, you need a way to thread wires or cable through the bales and then cinch them tightly to the cob. One system is to set a wooden dead man through the cob wall, about 16 inches from the junction with the straw-bale wall, with both ends exposed. A wire is threaded through a bale and then wrapped around screws or fencing staples in the ends of the dead man. Alternatively, lengths of tubing can be set through the cob (also at least 16 inches from the bales) and then a wire or cable is run through the bale, through the tube, and back to itself, tensioning it with a turnbuckle to pull the bale tight to the cob. Either of these systems should be repeated toward the middle of every second course of bales, from the foundation up to the bond beam or ceiling.

If the straw wall goes up first, you can drive thin, strong wooden stakes or short lengths of rebar into the exposed ends of bales for the cob to grab onto. Make the stakes fairly long (8 to 12 inches exposed) and drive them at different angles, so the walls are not able to pull apart.

If the cob wall and the straw wall go up at the same time, you have the greatest number of options for a secure connection. In addition to the aforementioned techniques, you can also create a "running bond" by building cob into the areas that would normally be occupied by half-bales at the end of a straw-bale wall. Then, before the cob is completely dry, pin the junction together by pounding stakes of rebar, stout wood, or bamboo through the ends of the bales, through the cob beneath, and into the straw bale two courses below.

Windows and Doors There are many options for installing windows and doors in cob. For operable windows and doors, the simplest approach is to make a heavy wooden frame and build it into the cob as you go, or secure it in place later using gringo blocks. Fixed windows may be simply cobbed into place. Use a wooden sash window, an insulated double-pane unit, or even a piece of heavy plate glass. Because you can sculpt the cob around the glass to make the window whatever shape you wish, it is unnecessary to cut the glass to shape. However, it is best to avoid burying plate glass too deeply in the cob, or the pressure from wet cob may cause it to crack. If this should happen, replacing a broken pane is easy. Simply carve out the cob around the window until you can remove the broken glass, install a new piece of glass and cob it back in place.

There are two options for spanning above windows: arches and lintels. The process of making a cob arch is described in the section on corbeling. Wooden lintels create a different look and can be very decorative. Particularly when a window is tall, it is wise to wait until the cob on either side dries substantially before arching across the top or placing a lintel. Otherwise, the wet cob beneath may settle and shrink somewhat, causing either the window or the cob to crack.

Plumbing and Wiring Installing plumbing and electrical wiring in a cob wall is simple. Pipes and wires can either be cobbed in place as the wall is built or added after the walls are finished. In the latter case, wet down the cob and use a hatchet, adz, or other sharp tool to carve out a channel in the wall deep enough to cover the pipes and wires with a few inches of cob. There is certainly no fire hazard in burying insulated electrical cable directly in a cob wall. However, for ease of maintenance and modification, you may wish to install a conduit.

The "hairy" exterior of a cob wall that has been plastered only up to the window level. Brigitte Miner residence, Oregon. (Photo by Michael Smith.)

PLASTERS AND FINISHES Because we get little wind-driven rain in inland western Oregon, roof eaves generally keep cob walls dry. We often leave the outside of a cob building unplastered, exposing the rough texture and rich color of the cob. In windier climates, however, and on interior surfaces, it is usually desir-

A contemporary mantel has been fashioned out of cob and finished with a simple earthen plaster to create this handsome Southwest-style hearth. (Photo by Frank Meyer.)

able to plaster the cob to make it more resistant to wear and weather and to alter the color and texture.

Regardless of the intended final finish, we usually start with an earthen plaster to smooth out the wall, fill in divots, and mask protruding straw. The composition of earth plasters is similar to that of cob, with a higher proportion of sand to prevent cracking. Often both earth and sand are screened for a finer texture, and horse or cow manure or finely chopped straw is used for fiber. Further coats of earthen plasters may be applied for a successively finer finish, and they may be smoothed with a steel trowel or polished with a rag or sponge. Earthen plasters alone can be used with beautiful effect either inside or outside, but they require the addition of other compounds like linseed oil, prickly pear cactus juice, or flour paste to make them water-resistant.

The traditional exterior finish for British cob homes is a stucco made of lime putty and sand, sometimes with animal hair or other additives. It can be troweled smooth or "rough cast" by flicking it against the wall with a small shovel. Lime plaster sets slowly, but eventually becomes hard, durable, and weather-resistant. Unlike portland cement, it is relatively "breathable," allowing trapped moisture to escape.

For interior finishes, both fine lime-sand plaster and gypsum plaster work well in areas of high wear or constant moisture, such as in kitchens and bathrooms. Often we prefer to paint thin natural finishes like lime washes or clay slip *aliz* (see Chapter 12) over a fine earthen plaster. Using naturally colored clays and the addition of mica, fine straw, and other texturing agents, you can achieve an almost endless range of beautiful, high-quality finishes.

Repairs and Additions

Our Oregon cob buildings have not been around long enough to teach us much about maintenance and repairs, but the English cobbers are experts on the subject. Most of the people building with cob in England originally got into the field through restoration. Alfred Howard, for example, is currently in the midst of repairing a 900-year-old cob triplex.[20]

The interior of this residence built by Michael Smith and the Cob Cottage Company at Cottage Grove, Oregon, features carved wall niches, natural roundwood stairs, and a stove-heated cob bench. Wall finishes are clay slip aliz and lime whitewash over earthen plaster. The floor is poured adobe. (Photo by Michael Smith.)

One thing is clear, however: Well-built cob buildings can go a very long time without maintenance. As long as the roof is in good repair, very little seems to affect cob walls. Over the centuries, British cob buildings constructed on stone foundations sometimes develop major cracks where ground movements put strain on the building. There are now well-established techniques for bridging and repairing these cracks.[21]

We do know from direct experience that cob buildings are easy to alter and add onto. Windows, niches, and sculptural details can be added at any time by moistening and carving the walls. When connecting new rooms to a cob building, it is essential to thoroughly wet and roughen the joining surfaces. Then a channel is carved into the old wall to create a keyway for the new cob, and some wooden stakes are driven into the area for a better bond.

Conclusion

RESEARCH NEEDS   The cob revival is still in its infancy. During the few years we have been building with cob in the United States, the rate of innovation has hardly diminished, and doubtless many techniques described in this chapter will become obsolete in the future. We have a system that seems to work well under a variety of climatic conditions and with a wide range of soil types.

Cob residence of Mike Carter and Carol Cannon under construction (1998–1999) in the suburbs of Austin, Texas. The house features two large circular rooms and a thatched roof. (Photo courtesy of CobCrew, PMB 367, 815-A Brazos, Austin, TX 78701.)

However, there are questions that remain unanswered, that require time and/or specific research to determine:

1. What is the precise bearing capacity of cob walls of different thicknesses or mix proportions?
2. What are the maximum wall height and minimum wall width and curvature necessary to withstand various seismic conditions?
3. How can we build more earthquake-resistant foundations without increasing the use of portland cement?
4. What simple variations will increase cob's thermal resistance?
5. How can we protect exterior walls (particularly those without roofs) from rain and snow, using only natural materials?

FUTURE TRENDS Of course, it is impossible to predict accurately what direction the cob building revival will take in the years to come, but a few general trends and issues are likely to remain prominent:

First, as more and more cob homes are built in diverse climatic, geographic, and cultural conditions, the technique is likely to become increasingly regionally diverse. For example, strategies will be developed where it is necessary to deal with earthquakes, extremely cold winters, and the unavailability of good quality sand and straw. Just as traditional Welsh cob differed in many specifics from that in Devon and from *torchis* in France, one would expect that eventually cob homes in Texas will both look different and be constructed differently from those in Oregon or Massachusetts.

Second, there will be increasing interest and pressure to bring cob into the building mainstream, both in the United States and elsewhere. This will not be as easy as with straw bale or rammed earth, inasmuch as cob represents a greater departure from conventional industrial building practices. There may be attempts to further mechanize the mixing process, to use forms, or to add unnatural stabilizers like cement or asphalt to the mix. Eventually, some kind of cob building code is likely to be adopted, but whether cob will survive that process in a recognizable form remains to be seen.

Finally, the increasing collaboration between natural builders during the last few years points to the development of an integrated natural building system of which cob is only a part. Expect to see more hybrid buildings incorporating earth, straw, wood, stone, and other natural materials. In this trend, we hope, there will be less focus on individual materials and techniques and greater emphasis on finding the best, most sustainable solution to regional building situations using the natural materials close at hand.

The lessons derived from cob building have been surprising in their abundance and profundity, yet clearly we are only beginning to learn and should record more carefully what we do learn. Demonstration structures, particularly houses, should be monitored and analyzed. Much more can be published, particularly about the psychological and spiritual effects of natural buildings. We should acknowledge that cob building is more than a cheap, environmentally benign way to build. It is a significant tool for the development of a more sane and sustainable culture.

Notes

1. Hassan Fathy, *Architecture for the Poor: An Experiment in Rural Egypt* (Chicago: University of Chicago Press, 1973).
2. Ken Kern, *A House of Clay* (North Fork, Calif.: Owner Builder Publications, 1985).
3. Larry Keefe, personal correspondence, 1998; Jean Dethier, *Down to Earth—Adobe Architecture: An Old Idea, a New Future* (New York: Facts on File, 1981).
4. Clough Williams-Ellis, John Eastwick-Field, and Elizabeth Eastwick-Field, *Building in Cob, Pisé and Stabilized Earth* (London: Country Life Limited, 1919).
5. Devon Historical Buildings Trust, *The Cob Buildings of Devon 1: History, Building Methods and Conservation* (Devon Historical Buildings Trust, 1992).
6. Williams-Ellis, Eastwick-Field, and Eastwick-Field, *Building in Cob, Pisé and Stabilized Earth*.
7. Devon Historical Buildings Trust, *The Cob Buildings of Devon 1*.

8. Jeanie McElwain, "The Three Masters of English Cob," *The CobWeb* 5 (autumn 1997).

9. Clive Fewins, "The Breath of New Life in Old Mortar" *London Times* (12 March 1994).

10. Williams-Ellis, Eastwick-Field, and Eastwick-Field, *Building in Cob, Pisé and Stabilized Earth*.

11. Larry Keefe, *The Cob Buildings of Devon 2: Repair and Maintenance* (Devon Historical Buildings Trust, 1993).

12. See, for example, Elias Velonis, "Rubble-Trench Foundations," *Fine Homebuilding* (April 1996).

13. Keefe, *The Cob Buildings of Devon 2: Repair and Maintenance*.

14. Ibid; Devon Earth Building Association, *Appropriate Plasters, Renders and Finishes for Cob and Random Stone Walls in Devon* (1993); Jean-Louis Bourgeois and Carollee Pelos, "Plaster Problems," *Spectacular Vernacular: The Adobe Tradition* (New York: Aperture, 1989).

15. Christopher Day, *Places of the Soul: Architecture and Environmental Design as a Healing Art* (Glasgow: Collins, 1990).

16. Devon Earth Building Association, *Cob and the Building Regulations* (1996).

17. Williams-Ellis, Eastwick-Field, and Eastwick-Field, *Building in Cob, Pisé and Stabilized Earth*; Devon Historical Buildings Trust, *The Cob Buildings of Devon 1*.

18. Paul Walshe and John Miller, *French Farmhouses and Cottages* (Weidenfeld & Nicolson, Ltd., 1996).

19. Kiko Denzer, "Pit Cob" *The CobWeb* 5 (autumn 1997).

20. McElwain, "The Three Masters of English Cob."

21. Keefe, *The Cob Buildings of Devon 2: Repair and Maintenance*.

7

Rammed Earth

David Easton

Throughout the history of home building, four distinctly different technologies for constructing solid walls of raw earth have evolved along separate tracks. These techniques are wattle and daub, mud bricks, cob, and rammed earth. Originally each technique sprang naturally from the creative minds of vernacular builders and from the necessity of using available resources. Over time, refinements in technology and material selection improved performance and durability.

Rammed earth has experienced a strong revival in recent decades. The primary reason is that the process of building rammed earth utilizes bulk materials, compacted into form work set directly on a foundation, thereby providing greater opportunities for mechanization. Where traditionally materials were moved and compacted by hand, now front-loading tractors can move soil quickly and efficiently and pneumatic ramming tools can speed up the compacting process. Improvements to forming systems have contributed significantly to the speed of wall installation, and, unique to rammed earth, low water contents allow for quick curing and rapid construction sequencing.

The search is clearly on for alternative building products and systems that can respond to the environmental crisis driving the architectural side of

Alternative Construction: Contemporary Natural Building Methods, edited by Lynne Elizabeth and Cassandra Adams ISBN 0-471-24951-3 © 2000 John Wiley & Sons, Inc.

Rammed earth walls after removal of forms, Monterey, California. The level of detail obtainable with this forming system, and with rammed earth in general, is limited only by the skill and imagination of the form builder. (Photo by Cynthia Wright.)

the sustainability movement. Few materials provide as many benefits as does raw earth used in a structural wall system. Earth is regional, unprocessed, low-cost, heat-storing, load-bearing, durable, and recyclable. The only factor currently limiting its widespread acceptance and implementation by the residential construction industry is a misplaced but lingering perception by the home-buying public that earth is not "modern" enough.

As one of a handful of professionals who have been involved in rammed earth construction since the mid-1970s, I have the advantage of witnessing firsthand a slow and steady reawakening of consumers to the enormous benefits of living within structures enclosed by earth walls. At first there is skepticism—questions are invariably raised about cleanliness and durability—but once he or she is inside a house with earth walls, the thermal and acoustic superiority captivates the home owner. The uniquely secure and serene characteristics imparted by mass walls emphasize how flimsy and inferior frame and sheetrock construction is. In the past decade, word has begun to spread that rammed earth houses capture an unsurpassed interior quality. Best of all, as people discover that such structures do not wash away in the rain nor fall down in earthquakes and that they retain high market value, the momentum of the rammed earth renaissance continues to grow. I believe an earth building industry is on the horizon.

History

Biblical references to bricks of mud and straw, coupled with the extant remains of the era of Egyptian pyramid builders, leave little doubt as to the antiquity of adobe brick construction. Less well documented are the origins of rammed earth, although archaeological evidence from numerous sites around the world shows that the technique of packing moist soil into form work was used as early as the seventh millennium B.C. The Great Wall of China, built over a period of several centuries, beginning as early as 2000 B.C., is in part constructed of rammed earth. All of the early civilizations of the Near East were built of mud bricks and rammed earth. The earliest written record of rammed earth dates to the first century A.D., in Roman historian Pliny the Elder's *Natural History:*[1]

> Moreover, are there not in Africa and Spain walls made of earth that are called framed walls, because they are made by packing in a frame enclosed between two boards, one on each side, and so are stuffed in rather than built, and do they not last for ages, undamaged by rain, wind, and fire, and stronger than any quarry-stone? Spain still sees the watchtowers of Hannibal and turrets of earth placed on the mountain ridges.

In central France, using soils washed down from the Alps, **pisé de terre** *(or rammed earth) has for centuries been the wall system of choice. (Photo by Cynthia Wright.)*

Throughout the centuries, rammed earth continued to be built in much the same manner as described by Pliny—"enclosed between two boards, one on each side." Examples of rammed earth structures built by this method that are hundreds of years old survive today in many parts of North Africa, southern Europe, and both the Near and Far East. In fact, in those parts of the world where access to mechanization is economically unrealistic, the traditional method of ramming earth by hand is still practiced. Given Pliny's observations of Roman legions using rammed earth to construct watchtowers at least as early as the first century A.D., even without other written or archaeological evidence it may be safe to assume that the technology survived through the Middle Ages.

In France, during the later part of the eighteenth century, a builder named François Cointeraux became aware of the wide use of the technique of ramming earth, known in French as *pisé de terre*, or simply *pisé*. His fascination with its simplicity and durability led him to publish several manuals relating the principles of the technology, which in turn prompted the publication of numerous articles not only in France, but in England and the United States as well. Subsequent articles published in trade journals indicate that numerous projects in rammed earth were completed during the nineteenth and early twentieth centuries. As always, the attractions to the method were its simplicity and its dependence on readily available low-cost materials.

Of perhaps greatest interest to Western builders is the prolonged use of rammed earth in Europe during the past three centuries. A 1970s survey by graduate students at the College of Architecture at Grenoble (EAG) revealed the astonishing fact that literally thousands of rammed earth houses and barns, some as old as 400 years, survived in the region around the Rhone River. Beneath their plaster facades, entire villages, including public buildings, were found to be constructed of rammed earth. When constructed properly, rammed earth buildings survive far longer than timber frame and, where freeze–thaw cycles can damage mortar joints, monolithic rammed earth walls even outperform stone.

Significant changes have been made to rammed earth form work within the past two decades. In this photo, the plywood and waler system developed by Rammed Earth Works is used to construct a preschool in Leon, Nicaragua. (Photo by Cynthia Wright.)

Traditional methods of building rammed earth structures by hand in small sections of moveable form work are still in use today in many parts of the world, and still immensely practicable. In an economy in which money is in far too short supply to invest in manufactured building materials, rammed earth can make the difference between shelter and homelessness. It is a process involving the investment of labor, not dollars—sweat equity, not outside capital. The form work, once constructed, can be used again and again. The work force, after a brief training period, can be mobilized to build an entire village. Training programs conducted by CRATerre (an organization started by the students at EAG) have resulted in successful construction projects on the village scale in Africa, Asia, and South America.

Along with four volunteers from the EarthWright Institute, I had the opportunity to conduct a training program in traditional rammed earth construction on-site in Leon, Nicaragua, in 1993. Limited funds for the purchase of foundation and roofing materials had been acquired through numerous granting agencies, and the mayor of Leon had donated a vacant lot in the neighborhood. The plan was for the residents of the Barrio de Guadalupe to build a four-classroom preschool using local soil and the investment of their own labor.

Under the supervision and encouragement of Alan Wright, director of the Sister Cities Project, the lot was prepared, underground water and sewer lines were installed, and a block and concrete foundation was built, onto which would be constructed the rammed earth walls. With the arrival of the five-member team from the EarthWright Institute, along with plywood forms (enough for two wall sections) and specifications for hand rammers to be fabricated by a local metalworker, work on the walls began in earnest. For the next 10 days, a 40-member labor force, made up of men, women, and children, hauled by bucket and compacted by hand more than 60,000 pounds of raw earth into load-bearing walls—enough for four classrooms and a small auditorium. Local carpenters put the roof on (also with the help of the neighborhood volunteers), masons applied a plaster veneer to the earth walls, and the preschool Los Carlitos opened to the praise of the mayor and other city dignitaries and with the pride of the Barrio de Guadalupe. Most encouraging of all was that upon completion, a second school was constructed of rammed earth, its builders using the skills and the confidence gained at Los Carlitos.

In 1995 four instructors from the EarthWright Institute traveled to São Paulo, Brazil, to conduct a similar program, this time to build a cottage as part of an orphanage in a village outside the city. Forms and rammers were fabricated to specification, the foundation prepared, and soil stockpiled on-site. The work force was composed of eight strong young men, the four instructors, and the architect along with three members of his staff. In four days the walls for the 500-square-foot cottage were completed, and on day five carpenters began the roof. The successful completion of this simple bungalow drew national attention and led to the construction of a second noteworthy project, as well as to the formation of a South American association of earth architecture.

High-Tech Earth

In the early 1970s, in France, Australia, and the United States, there began a reassessment of the potential for using rammed earth construction in a new, more sophisticated marketplace. Through a series of serendipitous events, as well as an increasing awareness of the need for more "environmental" building solutions, cottage industries in rammed earth sprang up almost simultaneously on three continents. Over the next 10 years, with the practitioners working primarily alone but maintaining communication, the techniques for designing and building in rammed earth evolved at a rapid rate. In France, the CRATerre group managed to gain government support for the construction of a major housing development intended to showcase earthen materials technologies. In California, Arizona, and New Mexico, along with a half-dozen other rammed earth builders, I began to gain an increasingly appreciative clientele and enthusiastic (if still somewhat skeptical) press coverage. Most significant, in Western Australia, where a ravenous termite population precluded the use of wood walls in any form, rammed earth provided so many benefits to home owners that, once launched as a viable alternative to brick wall construction, it quickly captured as much as 20 percent of the market in some areas.

Contributing to the success of contemporary rammed earth in the 1980s and 1990s were improved forming systems, wider use of off-site soils (which resulted in greater control over architectural finishes), and an

The Lenton Brae Winery near Margaret River in Western Australia. (Photo by Cynthia Wright.)

increased awareness of the environmental and thermal benefits of thermal mass walls. As the public continues to gain information on the cost to the environment of manufactured materials and nonrenewable fuel sources, building systems such as rammed earth will increase in attractiveness. Not only do they save on resources and utility expenses, they will retain their resale value better over longer periods of time, thereby representing a far wiser construction investment.

Yet continuing to restrict the use of contemporary rammed earth and affect its ability to capture a greater share of the new home market is that it is still a relatively slow construction process, even with the use of tractors and air compressors. The home building industry today is required to produce finished houses at an incredible rate in order to stay in business. Rammed earth, in which walls for a single-family home may take as long as three weeks to build, cannot compete with the price demands exerted by competition such as frame and stucco, despite the additional value mass walls represent. If rammed earth is to compete in production home building (in other words, if it is going to become "affordable"), either one of two things has to happen: The price of manufactured building materials has to increase in proportion to their cost to the environment (the green tax), or monolithic earth walls have to become faster to construct.

Uniquely aware of this dilemma, I began, in 1989, to develop technologies that could make monolithic earth walls significantly faster to construct. In contemporary rammed earth construction, the slowest component of the system is the assembly and disassembly of the form work, followed closely by the labor of shoveling the soil material into the forms. The challenge was to speed up the forming process and eliminate the need to shovel soil into the forms. Drawing on techniques used to construct swimming pools and tunnel casings, in which material is conveyed in large hoses using high-pressure air shot against the side of the excavation, initial experiments were begun to see whether raw earth building might successfully adapt portions of these technologies. Over a three-year period, numerous test projects were completed. Successes outnumbered failures significantly enough to continue the development process. Early results did indicate that soil materials, if selected properly, could be conveyed through hoses with high-

Pacific Gas and Electric Company, in conjunction with REW, constructed this Energy Showcase home in 1993 to demonstrate the conservation characteristics of the of the PISÉ technology. This view of the exterior kitchen elevation shows the built-in recycling bin storage under the sink. (Photo by Cynthia Wright.)

pressure air and impacted against a one-sided form work. The material thus impacted remained in place and reached design strengths. Over the next three years developmental work continued on mix, designs, placing and finishing techniques, forming systems, and, most critical, door and window block-outs and coordination with the other trades. Today, the PISÉ (Pneumatically Impacted Stabilized Earth) process stands poised to enter the production home building market as a sustainable wall system that can compete with the demands for quick construction turnarounds.

Before PISÉ could be confidently offered to the marketplace, it was necessary to complete a successful and highly visible pilot project. In 1992, my wife and I were contacted by a representative of Pacific Gas and Electric Company (PG&E) regarding the company's willingness to build a demonstration home as part of its EnergyWise Showcase Home program. PG&E had committed a certain portion of its budget to energy conservation and the showcasing of new, green technologies, primarily lighting, glazing, and insulation strategies. A complete new wall construction technology was somewhat of a stretch for this organization, but upon touring the first two test projects and viewing the plans for the proposed model house, the committee members decided in favor of the project.

Working with a major utility company on a house designed for energy efficiency, which utilized a revolutionary wall construction technology, proved to be an ideal collaboration. The PG&E showcase home, 4,000 square feet under three separate roofs, with southern glazing, western shade, radiant slab heat coupled to a high-efficiency natural gas water heater, and fluorescent lighting throughout, opened for public tours in May 1993. For a six-week period, 5,000 visitors marveled at the natural cooling and quiet interior spaces.

As a result of the success of the showcase home and the ensuing publicity, I was contracted to provide design consultation and wall construction services for the new Fetzer Vineyards Administration Building in Hopland, California. Fetzer Vineyards is rare among wine makers in its strong commitment to principles of organic farming and sustainability. The Vineyards' new administration center was intended to manifest that commitment and provide a healthy work environment for its employees. The building, designed by Valley Architects of St. Helena, California, and engineered by Zucco Fagent Associates of Santa Rosa, California, was constructed in 1996 utilizing approximately 500 cubic yards of clay soil from Fetzer Vineyards properties, blended with sand washed down the Russian River during the winter of 1995. Work involved approximately eight separate form setups and wall shootings of 80 cubic yards each. At the time, it was by far the most ambitious use of the PISÉ system.

Environmental Considerations

SUSTAINABILITY

Solid walls of earth reduce the use of construction materials with high embodied energy, reduce construction waste, conserve energy in the operation of the built structure, and offer significant longevity.

Energy conservation during operation is the result of the heat storage capacity of thermal mass. A properly oriented and fenestrated building can gain free heat from the winter sun, store that heat in the wall mass for nighttime reradiation, and thus reduce the need for the input of energy produced by fossil fuel combustion. Conversely, during the summer, assuming that the structure is properly shaded and adequately ventilated at night, mass walls can absorb excess heat from the living spaces and help to maintain cool temperatures, thus reducing the need for artificial cooling (see Chapter

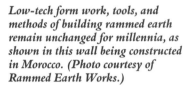

Low-tech form work, tools, and methods of building rammed earth remain unchanged for millennia, as shown in this wall being constructed in Morocco. (Photo courtesy of Rammed Earth Works.)

The forming system developed by French rammed earth builders as a result of lessons learned during the construction of the Village de Terre. (Photo courtesy of CRATerre EAG.)

3, "Natural Conditioning of Buildings"). Of course, the performance of mass walls varies from climate to climate. In some regions, and with optimum design, energy savings can be as high as 80 percent over conventionally built and insulated frame houses.

Walls of earth represent a low embodied energy investment because the basic material, raw earth, has no manufacturing component added to it before it is ready for use. In some cases there are investments for screening at the quarry and transportation to the job site, but in other cases a suitable soil can be found on the site itself, thereby presenting an ideal material from the standpoint of embodied energy. Portland cement is sometimes added to the soil to increase its strength and resistance to weathering, but raw earth can often be used directly to construct durable and structurally adequate wall systems. In a low-tech application, raw earth can be transported by manpower in buckets or wheelbarrows and compacted by hand. This represents the least embodied energy investment of all. At the other end of the technology scale, where rammed earth utilizes heavy machinery powered by diesel fuel, the energy investment can be rather high. In comparing total embodied energy of one type of structure (or one type of wall system) with that of another, the cost of manufacturing and transportation must also be considered, and where raw earth is the major component in the wall system, these costs are usually quite low. There are instances, however, in which the energy invested in the construction of a solid earth wall can be higher than for a conventional lightweight frame wall, such as, for example, where earthquake safety considerations require the use of steel reinforcing and higher wall strengths.

Longevity must be a factor in any equation used to evaluate the return on an investment. In a comparison between one type of construction system and another in regard to embodied energy, value is assessed as a ratio between capital investment and cost of operation. This method of analysis is called *life cycle costing*. In terms of economic value, to invest twice as much money to obtain a product that lasts three times as long, or that requires one-third as much maintenance over its useful life, has to be considered a better long-term investment. Likewise, to build a structure or wall system that can last far longer, require less maintenance, and use less energy to heat and cool would be considered a better overall energy investment, even if initial embodied energy were greater.

Replacing a conventional American wall assembly, consisting of wood studs, fiberglass insulation, Sheetrock, plaster, paint, plywood, Tyvek, lath, stucco, and paint, every 50 or 70 years is not sustainable building. A solid earth wall, even built with high-tech components of trucked-in screened soil, 10 percent portland cement, and diesel-powered equipment, offers a superior return on investment over stick-frame. It provides not only energy savings for the owner in five to seven years, but, with a functional life between 500 and 1,000 years, a rammed earth wall can dramatically enhance global environmental health by saving natural resources that would otherwise have been used for replacement construction.

Construction waste is the fourth consideration that weighs heavily in favor of solid earth walls versus other more traditional wall systems. Rammed earth and PISÉ are basically raw earth, which, on the job, is placed either within or against a reusable form work. Any remaining portion of the earth stockpile can be used on the job site for pathways and exterior paving. The forms are either converted to framing and roof sheathing or moved to the next job site. Construction waste, at least from the wall component of the project, is virtually eliminated. Perhaps best of all, even when time for demolition of the walls finally arrives, the earth from which they were built is completely reusable.

ADAPTABILITY TO CLIMATE Historically, rammed earth houses have been built in an extremely wide range of climate zones, form the snow-covered Himalayan mountains to

Architect Paolo Montoro used both rammed earth and adobes to construct this rural retreat near São Paulo, Brazil. (Photo by Cynthia Wright.)

the scorching deserts of North Africa. Countless old examples survive today in cold, wet regions such as Great Britain and the Caucasus mountains of Georgia. The monolithic properties of rammed earth in contrast to those of the brick and mortar joints of adobe make the walls less susceptible to water damage in wet climates. Even today, especially with the addition of cement stabilizers, rammed earth walls can be constructed in virtually any of the earth's climate zones.

In the past several decades home owners have grown accustomed to greater indoor comfort than their ancestors expected. Maintaining the necessary control over temperature fluctuations requires that in extreme climates, rammed earth walls be coupled with some sort of insulation barrier to mitigate outside temperature fluctuations. Two-inch rigid insulation board, installed in the center of a wall, can improve energy performance significantly while maintaining the natural appearance of both interior and exterior wall surfaces. Alternately, rigid insulation can be installed against the exterior face of the wall and then protected with conventional lath and stucco.

Determining which climates are extreme enough to warrant the additional expense of the insulation requires a careful energy evaluation. Both local weather data and the actual design of the house come into play, as well as the source and projected expense of the supplemental heating (or cooling) to be used in the structure.

Engineering Considerations

Although earth construction has a long and successful history, it nevertheless continues to have many detractors, especially within the engineering community. This is primarily the result of two factors: an inadequate body of technical data on the strength and performance of earth walls and, I believe, an abundance of misinformation and unfounded prejudice.

Prejudice against earth construction is the result of broad generalizations that lump all earthen walled structures into the category of "substandard." A misperception is that earth walls fail in earthquakes and wet weather. The reality is that well designed and constructed earth walls do withstand the lateral forces exerted on them as a result of earthquake loading (in many cases better than stone or wood walls). In addition, walls constructed of properly selected soils, when provided with adequate roof overhangs and foundation drainage, survive for many hundreds of years, much longer than wood walls. Numerous studies conducted by engineers in the United States and Europe have measured and recorded the design and material makeup of historic earth structures that survive both earthquakes and extreme weather conditions, identifying those elements that contribute to durable and successful earth wall design and installation. Likewise, these investigators have studied structures that have failed, identifying the reasons for failures as well. These data are available to architects and practitioners who aspire to build structures with equal or greater expectations for longevity. I anticipate that the prejudice against earth construction will decrease rapidly as misinformation is gradually replaced with the true story of earth wall durability.

The numerous case studies that document surviving earthen walled structures do not provide the type of technical data normally used by professional engineers and building officials as a means to design and evaluate what is safe in creating a structure with load-bearing earth walls. Current engineering practice is to design a structure on the basis of the specific test values of proposed materials, rather than from empirical data. The *Uniform Building Code* (UBC), a design guidebook for engineers and inspectors, defines concrete and masonry building materials and specifies procedures for designing structures that employ these materials. Because "earth" does not technically fall within any of the categories listed in the UBC, engineers frequently find themselves without adequate precedent and the necessary guidance for designing earth walls.

Cast concrete, concrete blocks, and fired brick have all been used in modern construction extensively and, as a result, are supported by vast quantities of both laboratory and in situ test data. Universities offer entire classes on safe design procedures for concrete and masonry structures. National and international institutes exist solely for the purpose of improving the knowledge base and disseminating information on the uses of concrete and masonry.

Humble earth has no such powerful or well-organized advocates. Most of the relatively few architects and builders who have mastered the use of earth in construction gained their knowledge of the material, their expertise in its applications, and confidence in its durability through a long and ardu-

Peeler residence in Reelsville, Indiana. Prior to the post–World War II wood-frame construction boom, rammed earth had a growing following in the United States. (Photo archives of John and Lydia Miller.)

ous apprenticeship. Unfortunately, in nearly every case such "apprenticeships" have been without the benefit of a teacher/artisan who had prior knowledge of the trade. The skills of the earth builder generally ceased to exist at about the middle part of the twentieth century as architects came to rely ever more on "engineered" materials such as cast concrete and concrete block. Earth building became a lost art.

Fortunately, in the past two decades, as awareness of the environmental impact of manufactured building materials increased, there began a resurgence of interest in earth construction. As more architects began to design with earth, more builders began to perfect the special skills necessary to assemble a well-built earth structure. The ensuing publicity resulted in greater exposure to the public and, in turn, an increase in the number of home owners requesting the comfort, durability, beauty, and environmental benefits of earth housing.

The engineering profession was required to keep pace. Guidelines began to take shape for the design of safe earth walled structures, following existing design procedures for other (masonry) wall systems. Test data conforming to current standards began to accumulate. Code agencies began to develop standards for reviewing plan submittals and for inspecting earth buildings under construction. Permits were issued, bank loans were approved, insurance policies were written, and, perhaps most significant of all, earth buildings were reselling at prices higher than conventional frame and stucco. Earth was coming of age.

Today we have hundreds of test results from a wide range of soil types and mix designs. We can predict the effect of various cement stabilization ratios on a soil with a known particle gradation. We can have confidence in the compressive strengths, the shear values, the bonding force to reinforcing steel and anchor bolts. Engineers at last have the knowledge they were lacking in order to design in compliance with the building codes. They can treat earth like any other "modern" material.

How an earth structure is designed varies from region to region. In areas of high seismic risk, steel reinforcing and strong roof-to-wall connections are essential. In low-risk areas, reinforcing can be eliminated, and height-to-thickness ratios, length of wall, and fenestrations become the design determinants. In high rainfall areas, roof overhangs and foundation drainage take precedence.

The point is that current engineering practice is to design on the basis of known values and anticipated conditions. How a building will perform when subjected to an earthquake, hurricane, or typhoon must be predictable. But with the test values on earthen materials beginning to accumulate as a result of studies under way in Germany, France, Australia, Brazil, and the United States, engineers and inspectors are gaining a much needed confidence. Over time, as more earth buildings are constructed and as more data accumulates, we can anticipate an increase in the engineering community's confidence in earth construction and a subsequent reduction in the cost to design and build. These developments will, in turn, contribute to more construction, continuing reductions in cost, and eventually a system that is truly economical. In terms of economics, this is the system of supply and demand at work.

At the time of this writing, structural engineers in earthquake-prone California are designing rammed earth and PISÉ structures using what is known as the *working stress method of analysis.* Assumed compressive strength values ranging between 800 and 1,200 pounds per square inch

A 1,300-square-foot, passive solar, rammed earth house, built by Tom Wuelpern of Rammed Earth Development. This modern classic was one of several new homes fit neatly into small, zero-setback lots adjacent to a historic district of simple working-class adobes in downtown Tucson, Arizona. (Photo by Tom Wuelpern.)

(psi) are used, certain strength/safety reductions are imposed, and steel reinforcing based on these compressive values is specified, following concrete design guidelines. Bond beams with additional reinforcing must be added to the tops of all walls, and anchor bolts are embedded as a means of fastening the roof assembly to the wall system. Each piece of steel and each bolt has a predictable value and effect. Outside earthquake-prone areas, 12-, 18-, or 24-inch-thick earth walls are being designed with and without bond beams. The walls rest on reinforced concrete foundations, and the roof systems are firmly bolted onto the wall tops. In Australia, 10 percent cement stabilization results in wall strengths in the 1,000 psi range. Wall thicknesses there, where weather conditions are temperate, are typically 12 inches. In the southwestern United States, 5 percent stabilization and 24-inch-thick walls are common. Greater wall thicknesses resist the extremes of weather and allow a wall of lower strength to support the same roof loads.

Because we live in a litigation-prone society, we are compelled to comply with the accepted standards for design and construction. No matter how loudly we may wish to complain, the reality is that earth and other alternative materials must be in compliance with current standards. The more that is known about a material, the better the prediction of how it will perform under severe loading conditions and over time. The more accurate the prediction, the lower the imposed reduction for safety need be. The irony is that after all of the testing is complete, the ideal design for an earth building may prove to adhere to those same methods and details that were developed over the previous five centuries.

Rammed earth performed better than concrete in this anchor-bolt test to measure wall strength, much to the surprise of the observing engineers. Under more than 8,000 pounds of pressure, this ¾-inch steel J bolt on the right straightened and pulled out of a rammed earth test wall, without the wall failing. (Photo courtesy of Rammed Earth Works)

Table 7-1. Anchor Bolt Pull-Out Test Results

No.	Anchor Size and Type	Embedment Depth	Load at Initial Movement (lb)	Maximum Load (lb)	Failure
1C	⅝ × 14 J bolt	8½ inches	6,100	8,300	Leg on J bolt straightened and bolt pulled out of the wall
1E	⅝ × 12 J bolt	7¾ inches	5,200	8,000	Leg on J bolt straightened and bolt pulled out of the wall

Table 7-2. Anchor Shear Test Results

No.	Anchor Size and Type	Embedment Depth	Load at ⅛-inch Deflection (lb)	Failure
1G	⅝ × 12 J bolt	8¼ inches	3,200	Bending in bolt and crushing of *pisé*
1H	⅝ × 14 J bolt	9 inches	3,800	Bending in bolt and crushing of *pisé*
1I	⅝ × 12 J bolt	7¾ inches	4,500	Bending in bolt and crushing of *pisé*

Testing data source: Schwein Christiansen Laboratories, Lafayette, CA

Construction Methods The basic procedure for building rammed earth has changed very little over the past two millennia. Suitable moist soil is compacted one layer at a time within a sturdy form that has been erected in place where the wall is to be built. Once the form is full and the top layer fully compacted, the form may be immediately disassembled and moved along the wall line to create another section. Unlike mud bricks, which are cast off-site and must thoroughly dry before being laid in the wall, rammed earth is built in place and does not require a period of drying before other phases of construction may proceed.

Forming systems have evolved significantly since the early days, and several different but very sophisticated systems are currently in use by professional rammed earth builders in Europe, Australia, and the United States. Each of the systems has been shown to be effective in completing

The forming system developed by Rammed Earth Works uses simple components: full sheets of ¾-inch plywood, 2-by-10-inch planks, and cabinet maker's pipe clamps. (Photo by Cynthia Wright.)

Shown here is the forming system currently used throughout Australia. The global revival of rammed earth is nowhere more strongly under way than on the continent of Australia. (Photo by David Easton.)

high-quality walls. The decision as to which forming system to use for constructing a rammed earth project should be based on the design of the walls and fenestrations and the likelihood of using the forms for additional projects. A complete evaluation of the various forming systems is beyond the scope of this chapter.

The critical elements in rammed earth construction are the selection of the soil, the moisture in the soil at time of use, and the degree of compaction of the individual layers.

Ideal soils for building durable rammed earth structures contain a blend of particle sizes: small gravel, coarse and fine sand, and clay. Roughly (and historically) the ratio of clay to sand is 30 to 70 percent. A simple test to determine the ratio of clay to sand is to place a cup of dry soil in a 1-quart mason jar. Fill the jar with water, shake it for a few seconds to put the soil in solution, then allow the soil to settle until the water is clear. The heavier sand particles will settle out first and be on the bottom of the jar. The clay particles will be on top. If the line between the two is distinct, it should be easy to make a visual estimate of the ratio between the two.

However, soil is both highly variable and chemically active. For this reason, even if the aforementioned simple evaluation of the soil indicates that the ratio between clay and sand is optimum, it is not safe to assume that walls constructed from this soil will be strong and durable. There are many different types of clay, and some are much better than others for building rammed earth structures. A few are, in fact, completely unacceptable, and to avoid the risk of proceeding with a project that may result in unsatisfactory walls, additional testing is strongly recommended.

The first round of testing should be the construction of small sample blocks that can be evaluated for their strength and resistance to weathering.

Small samples of compacted soils can be used to determine strength and weathering characteristics. (Photo by Cynthia Wright.)

One simple test is to subject the sample blocks to the full force of water from a garden hose sprayer for an hour. Another method, if time allows, is to leave the sample blocks exposed to the harshest weather season. Be sure to document your mixes. If the results of the tests on the sample blocks indicate questionable strength or water resistance, you can modify the mix design by increasing or decreasing the sand component or, as in the case of most rammed earth construction today, adding a small percentage of portland cement to the mix, typically between 5 and 10 percent by volume.

Once the testing of sample blocks has reached a point of producing encouraging results, the construction of sample walls, a garden element, or a small shed would be a logical and conservative next step. Not only is larger-scale testing important to gaining a better understanding of the soil, it will also provide you with good practice in using the forms and experience with proper moisture, mixing, and ramming. The small mistakes that are inevitable as a person learns a new skill will be much less frustrating on a garden wall or a storage shed than on your new house.

With the confidence-building practice behind you, the real work of constructing the walls can begin. It is important to have a well-thought-out set of plans in which the relationship between the form work and the placement of doors and windows has been considered. The foundation for the walls should be constructed of reinforced concrete and extend far enough above the finished ground level to ensure that the bottoms of the walls do not ever sit in water. A stockpile of the selected soil should be available and readily accessible on the site. An adequate amount of level space should be designated as the mix area, and tarps or plastic should be on hand in the event of rain, which could add an excess of water to the soil stockpile.

To begin, the form work, whether it is the type used to construct just one section at a time or an amount suitable for completing a major portion of the wall, is erected and secured to the top of the foundation stem wall. Conduits and boxes for electrical service are installed in the form work at designated locations, block-outs for doors and windows are constructed and ready for installation, and reinforcing steel, if required, is cut and tied. In earthquake-prone California, we tie all of the vertical reinforcing before beginning wall construction, then lay in the horizontal bars at the appropriate intervals as the wall gains in height.

Once the first section of form work has been set and braced, it is time to prepare the soil mix for ramming. Keep in mind that moisture content is one of the critical factors in the final wall quality. If your walls are to be built without cement as a stabilizer, it is possible to prepare very large batches of soil at optimum moisture for good compaction. If cement is being used, then the soil mix must be prepared in batches that are equal in volume to the form work.

The soil is prepared by combining the various components as determined in the preconstruction testing program. In some extremely fortuitous situations, those components are nothing more than the soil from the site. In others, site soil, sand, and portland cement must be blended in ratios determined through the test program. Once the dry ingredients have been measured and deposited on the mixing area, they are combined either with shovels, a garden Rototiller, or the bucket of a front-loading tractor until they are uniformly blended. At this point water is added to the mix, typically with the use of a hose and spray nozzle, while mixing continues. When optimum moisture is obtained, the soil mix is ready for delivery to the form work and ramming in layers. Achieving this optimum moisture level takes a little practice before it is easily recognizable. Basically, the soil should be wet enough to achieve good compaction, yet not so wet that the wall will shrink and crack as it dries.

All the texts (from basically the beginning of recorded history) that describe rammed earth construction refer to the same test for determining proper moisture: With two hands, mold a small ball of soil. There should be enough moisture that the ball stays intact in the open palm. If the ball crumbles, then add more water to the pile. If it stays together, drop it from waist height onto a solid surface. If the soil breaks into its former loose state, then it is right for ramming. If it is stays together when dropped, there is too much moisture in the soil. Either allow it to dry out for a while, or add some more dry soil to the pile. Once the moisture in the pile is correct, the soil can be carried to the form work and ramming can begin.

The reason proper moisture is so important is that with less than enough water, the soil will not bond securely and the wall will be crumbly and weak. If, on the other hand, there is too much water in the mix, the soil will be spongy as you are trying to compact it, will likely stick to the forms, and will eventually shrink and crack. The problem with cracks in a wall is more serious than just unsightliness. They allow water to penetrate during rainy periods and begin a process of accelerated weathering.

With the forms set and the first batch of mix at optimum moisture, wall building can begin. The prepared soil is transported to the form work with a tractor, wheelbarrows, or buckets. It is dumped into the forms in layers

Ramming 8-inch layers of cement-stabilized earth in Arizona. (Photo by Tom Wuelpern.)

approximately 6 inches thick, then uniformly compacted to approximately half its loose thickness. Compaction can be performed by hand rammers or pneumatic tampers and should continue until the layer is hard and dense. Sound is an indicator of good compaction; loose soil has a dull or hollow sound when struck with a rammer. Well-compacted layers resound with a "ringing" or reverberating sound.

After the first layer is rammed to the proper density, another 6 inches of loose soil is placed in the forms, and the process of compaction is repeated until the "solid" sound is obtained. Continue with one layer after another until the form is full. Then (and this is one of the beauties of the rammed earth process) immediately disassemble the form work and reset it to construct an adjacent panel. Panel after panel goes up, and soon the building begins to take shape. Once all of the walls have been completed, bond beams of concrete or wood are attached to the tops of the walls, which are then ready for the roof of choice.

Shoveling and compacting make rammed earth appear to be an enormously labor-intensive process. Few construction methods can rival it for the hard work it entails. The advantages are that the materials are very inexpensive and the results incredibly durable. In fact, in evaluating the input of labor and materials against the usable life of rammed earth and comparing this ratio to a similar ratio for a wood frame wall or even one of the other

The architectural power of rammed earth is the result of the mass itself. Sonoma, California. (Photo by Cynthia Wright.)

alternate materials, rammed earth emerges as one of the best construction investments over time.

Pneumatically Impacted Stabilized Earth (PISÉ)

PISÉ is a modification to traditional rammed earth that I developed in hope of improving the speed with which monolithic walls could be constructed. The goal was to develop a wall building system that could bring the environmental benefits of earth—namely, resource and energy conservation—into the mainstream housing market. To this end, it was essential to eliminate the labor-intensity component from earth construction and to refine a system that integrates smoothly with existing production home building methodologies.

In the PISÉ process, high-pressure air is used to convey and impact the soil mixture (identical to that used for traditional rammed earth) against a one-sided form. The aspects that contribute to the increased speed are a simplified forming system and the replacement of shoveling and ramming by air delivery.

PISÉ utilizes far more costly and sophisticated equipment than rammed earth. It is not an owner-builder's technology. A basic setup is composed of a front-loading tractor or Bobcat, a mixing machine capable of combining soil and cement at controlled ratios, a gunite "gun" with hose and nozzle, and a high-volume air compressor. A six-to-eight-person crew is required to efficiently operate the equipment and produce the daily output of finished wall.

The process itself is quite similar to that used with dry-mix shotcrete (or gunite), which is typically used to construct the structural linings to

A tractor for loading, a mixing machine, a "gunite" gun, and a large compressor consitiute the PISÉ gear. (Photo by Cynthia Wright.)

The PISÉ technique uses high-pressure air and 2-inch-diameter hoses to deliver the soil material, rather than buckets, shovels, and backfill tampers. (Photo by Cynthia Wright.)

After PISÉ is shot against a one-sided form, excess material is shaved back to a guide wire, making the wall straight and plumb. (Photo by Cynthia Wright.)

The main house of the Easton-Wright residence, Napa, California, showing the front entry and breakfast patio. The front door, which is recycled, is set in a cast-in-place door surround. The window surrounds were precast on-site as were many pavers and interior floor tiles. This photo was taken immediately upon completion in May 1998. (Photo by Cynthia Wright.)

swimming pools and underground tunnels. An experienced gunite crew can be trained to install and finish PISÉ in a matter of days.

The differences between gunite and PISÉ are in the mix itself and how it shoots and finishes. In gunite work, the back side of the wall is seldom if ever exposed, and for this reason, typically, not much attention is paid by the nozzle operator to good consolidation on the back side. In PISÉ, however, with the formed side frequently being the exposed inside surface of the wall, special care must be taken around all electrical and plumbing fixtures as well as along the sides and over the tops of all door and window openings. With skill and a good soil, a PISÉ wall can be left without any additional surface treatment, thereby further reducing construction costs and maintenance over time.

The exciting aspect to the further development of the PISÉ process is that if it can be shown to be truly competitive with other production building systems, to the point where several large-scale builders are using it and getting positive feedback from home buyers, then additional resources will become available for yet further improvements.

Whether production earth building ever becomes a reality, earth itself will remain as it always has been—a free and local resource capable of giving shelter to humankind.

Note

1. The translation for this quote was taken from *The Rammed Earth House* by Anthony F. Merrill, Harper and Brothers (1947), 32.

Modular Contained Earth

Earthbag

Joseph Kennedy and Paulina Wojciechowska

with special acknowledgment of the work of Nader Khalili

The use of soil-filled sacks (earthbags) for construction has received growing interest as a natural alternative building technique, mainly in the United States. Earthbags are textile or plastic casings packed with soil, and sometimes sand or gravel, used to construct foundations, walls, and domes. The technique is essentially a flexible form variation of rammed earth construction.

Earthbag construction is one of the most inexpensive building methods on the planet. It uses locally available site soil and common sacks. The technique requires few skills, is significantly faster than earth building methods such as adobe or cob, and, unlike equipment-intensive modern rammed earth, requires few tools other than a shovel. This system is most valuable in regions with no clay or wood and areas prone to flooding, hurricanes, or wildfires.

The beauty of the earthbag technology lies not just in its low cost, but in its freedom of form. In addition to application in rectilinear walled or sym-

Alternative Construction: Contemporary Natural Building Methods, edited by Lynne Elizabeth and Cassandra Adams ISBN 0-471-24951-3 © 2000 John Wiley & Sons, Inc.

During an apprenticeship course at Cal-Earth Institute in Hesperia, California, students learn earthbag construction. This structure has been built with a long bag system called **Superadobe** *by its developer, architect Nader Khalili. (Photo courtesy of Cal-Earth Institute.)*

metrically domed structures, earthbags can be assembled sculpturally to weave a building together as a potter works with coils of clay. In this way, the building can blend into the existing landscape. Besides being adaptable to a variety of site conditions, earthbags can be used with a wide range of fill materials. When built properly, earthbag walls are extremely strong. They are valuable for remote locations and disaster relief shelter because the only manufactured components, the bags, are lightweight and easy to transport.

History

Most historic evidence of construction with the earthbag method is anecdotal. Filled sacks have been used for many decades by armies to create bunkers and other structures, although it is unknown when they were first applied for this purpose. Trenches built with oil-impregnated sacks during World War I remain to this day.

German architect Frei Otto is said to have experimented with the technique, and more recently Gernot Minke, an architect and professor at the University of Kassel, who was once a student of Otto, conducted extensive experiments with soil-filled sacks and tubes to create a number of structures, including domes. In 1978, Minke built a low-cost single-story structure in Guatemala that utilized cotton hoses filled with local pumice gravel. The hoses were tied together with wire every fifth course, and thin bamboo posts supported the walls on both sides.

Some structures built during the 1930s and 1940s were composed of soil-cement–filled burlap sacks, which were then moistened to set the

Architect Afsaneh Boutorabi and Cal-Earth used Superadobe to construct curvilinear retaining walls for a semipermanent landscaping exhibit at the Los Angeles County fairgrounds. These reinforced earthen forms were plastered and used to contain plants and water features. (Photo courtesy of Cal-Earth Institute.)

cement. A similar technique has been adopted at the Colombian community of Gaviotas to create ponds and retaining walls, called *gabiones*.

Throughout recent history earthbags have been used for varied purposes, mostly in fast-assembly relief work:

❖ Erosion control, when filled with earth or pumped with concrete (e.g., filled with soil-cement for quick construction of embankments after World War II)
❖ Flood control, when filled with sand or pumped with concrete
❖ Structural support of collapsing walls by archaeologists
❖ Military bunkers and air-raid shelters
❖ Landscaping, to create free-form retaining or enclosure walls

Earthbag construction is currently pioneered in the United States by architect Nader Khalili at the California Institute of Earth Art and Architecture (Cal-Earth). Since 1990, Cal-Earth, together with researchers and associates, has been developing earthbag construction further, from straight walls to domed structures. Khalili has introduced a number of innovations in technique and form, including the use of uncut polypropylene bags, a system he has named "Superadobe." He continues to develop the method in association with the city of Hesperia and has worked closely with the International Conference of Building Officials (ICBO) to obtain code approval for such structures.

Cal-Earth has developed a three-vault prototype house, building one such house on-site at Cal-Earth and another for a Yaqui community near Hermosillo, Mexico. In Mexico a 15-man crew assembled the complete structure in three weeks, laying approximately 25 feet of bag per hour, per team. This 650-square-foot prototype was initially designed for a small American family and cost $15,000 to build ($US in Mexico). The building, with three offset rectangular rooms (12 × 18 feet each) with vaulted ceilings, was not recognized by the Yaquis as a residential dwelling, because of its grand nature, but was used as a community center. The three-vault house at Cal-Earth, which cost about $3,000, was built entirely by volunteers.

Other earthbag residences have been built by Dominic Howes in Arizona, by Kaki Hunter and Doni Kiffmeyer in Utah, and by Mara Cranic in Baja, California (see Chapter 16, "Variations on Earthbag").

Construction Materials

BAGS

There are two types of bags available in the market: hessian and polypropylene. Both types can be available in tube form on a roll, or cut and sewn into bags (more expensive).

Hessian (Burlap)

Hessian (burlap) is a natural woven fabric that is biodegradable. Use burlap only if the earth used is not pure sand but contains compressible particles of soil. The bag acts only as a form in which to compress the earth and becomes redundant after it has been filled and covered with plaster. Burlap bags are not as durable as polypropylene and can also be more expensive.

In the United States, as of 1999, 18-inch burlap bags cost approximately 50 to 80 cents per yard plus delivery charge. In England (1999) an 18-inch-wide tube of hessian may be purchased for 30 to 50 pence per meter, plus delivery fee. The bags are delivered in rolls of a few hundred meters and may be of any desired width.

Polypropylene

Polypropylene bags are made of woven threads of plastic. Polypropylene will deteriorate if exposed to ultraviolet (UV) rays; therefore, care should be taken in storing the material to protect it from sunlight. If the building is not complete within two months, all exposed bags should be covered with some type of finish to protect them from UV rays. Whenever possible, UV-resistant bags should be used.

Polypropylene socks are available by the roll. Here a length is pulled from the roll to be cut. (Photo courtesy of Dominic Howes.)

In the United States, as of 1999, most manufacturers will deliver a minimum of 1,000 yards at a cost of approximately 22 cents per yard plus delivery charge. In England (1999), an 18-inch-wide tube costs 17 to 40 pence per meter plus delivery fee. Sacks come in a variety of sizes and in tube form, which is much cheaper.

Reused Bags It is possible to utilize recycled bags, which may be obtained from stores or factories that use bagged products. Recycled seed or feed sacks of polypropylene are among the most commonly found.

Considerations An important consideration in choosing a type of bag is the material used to fill it. A good rule of thumb is the weaker the fill material, the stronger the bag material must be. In some cases, once a strong material has set, the bag skin could be removed from the exposed areas of the structure without any loss of structural integrity. On the other hand, if a material such as dry sand without any stabilizers is used as fill, it is essential that the bags be fabricated of a non-degradable material that can resist (in tension) all the dead and live loads. The bag-to-bag connection is crucial to the structural performance of the structure as a whole, so it needs to be carefully engineered to resist both dead and live loads.

FILL Unless the bags are designed to resist structural loads (in tension), the earth fill must be considered as the structural support, like adobe and rammed earth construction. The consistency of the earth fill is very important, because non-structural bags can only be relied on to act as formwork shaping the earth until it "sets." A standard adobe mix of sand and clay soil would make an ideal mixture. In contrast to rammed earth and cob construction methods, however, the bags make it easier to construct curves and arches.

In preparing the soil, all organic matter should be removed as it could create cavities later on. Generally what has been excavated on site is what has been used. Remove any topsoil and excavate the subsoil from the site or locale. Larger stones, as well as sticks and any degradable organics, must be sifted out with appropriate size screens; gravel can used to fill bags in the lower courses for foundations, to prevent the rise of moisture. The bag fill is then prepared.

Fill can be used either wet or dry. For small site walls, dry material can be used, but for structural purposes, the material should be moistened. Water added to the earth makes it more compressible and ultimately more stable. Test the moisture content of the of the soil by squeezing a handful—

it should hold its shape but not feel wet. Experience and practice will soon allow you to gauge proper moisture levels. If the earth is too dry when dug out, it has to be sprayed with water and left overnight, if possible.

Proper moisture content is key to successful earthbag construction. Here water is sprayed on earth and sand for construction of an earthbag house in Moab, Utah. (Photo courtesy OK OK OK Productions.)

STABILIZATION For extremely strong structures that must carry great loads, for structures that are next to or under water, and for arches and lintels, cement can be used to create soil-cement. Another common stabilizer is lime. Most basic wall applications, however, do not require stabilization. (To understand where stabilization is required, see "Domes" and "Openings" under "Construction Process" later in this chapter.)

OTHER MATERIALS Additional materials include barbed wire, which is used instead of mortar to keep the bags between courses from slipping. For a good grip, use 4-point barbed wire. If the bag is narrow (12 inches wide) only one row is needed. If it is wider than 16 inches, two rows may be required.

Regular wire can be used to weave the bags, as in to basket-making techniques. Old nails are useful to pin bags closed, create new shapes, and keep barbed wire in place.

The first course for a free-standing garden wall includes buttresses. Two rows of barbed wire are temporarily held in place by bricks until the second course of bags is laid. (Photo courtesy of OK OK OK Productions.)

Tools

Tools adapted or developed for earthbag construction are easily available or devised. The most essential tool is a simple shovel—to excavate soil, to fill wheelbarrows, and to fill bags directly. A wheelbarrow is needed to transport materials or to directly pour soil into larger bags. Wheelbarrows can also be used to mix cement/lime into the earth to stabilize it or to mix adobe plaster. Coffee cans are handy; they can be used to fill bags and can be tossed to people higher on the wall. Stands to hold bags open have been made with a variety of materials. Tube sections of cardboard or off-cuts of pipe that fit into the longer tube-shaped bags make filling long bags much easier. Mechanical pumps have been utilized at Cal-Earth with great efficiency to fill tubular bags.

A tamper is essential to compact the bags once they are in place. A strong tamper can be created from a 5-foot-long 1¼-inch piece of metal pipe welded to a 6 × 6 × ¼-inch metal plate. A homemade tamper can be made by filling a plastic yogurt cup with concrete mix and placing a stick in the middle. It should cure two weeks before use. A heavy block or a chunky piece of wood can also be used.

A hoe is good for mixing soil stabilizers, if used. Also needed are a blade/scissors for cutting the bags, a level, a ladder, a tape measure, wire cutters, gloves, a trowel, and a watering hose for leveling the ground. A watering hose or water buckets may also be needed to carry water.

TOOLS FOR DOMES AND VAULTS Earthbags have been used in building domes and vaults. Forms or falsework of wood or metal are used to create vaults, and symmetrical domes are most effectively formed with the use of a simple compass, which acts as a placement guide for the bags. It is critical to employ a compass when the con-

Kaki Hunter and Steve Kemble compact a wall course with tampers that were made on-site in the Bahamas with local casuarina wood and imported cement. (Photo courtesy of Sustainable Systems Support.)

struction of a dome reaches the stage at which it must start to curve (spring line). The compass is then extended after each course. An excellent technique is to attach a lightweight pipe to a caster from which the wheel has been removed. This allows for articulation and rotation, and the caster can be easily attached to a 4 × 4-inch piece of wood set in the ground. A piece of L-shaped metal attached to the free end with a pipe clamp makes a useful guide. To set level courses of bags, a small adjustable level is attached near the end of the pipe where the person placing the bags can easily see it. Special compasses to aid in aligning the bags for catenary-shaped domes have also been developed. Portable guide frames strung with leveling string help in positioning bags for straight walls.

Construction Process

FOUNDATIONS

First, the site is prepared. Subsoil removed from the foundation area can be saved to fill the bags, once organic materials are removed. A rubble trench with proper drainage is generally used as a foundation. If the structure is built in an area not prone to flooding, it is necessary to excavate down to undisturbed ground, below the level of ground movement (i.e., heave), frost, and clay. On this solid ground it is possible to build using gravel in the bags, so as to lift the structure above ground level and prevent capillary moisture movement. If the structure is sited in an area prone to flooding, stabilization in the lower courses is recommended. (*Editor's note:* Bag materials are subject to decay and, so, should not be relied upon to support nonstabilized earthen materials in wet conditions.)

Retaining walls can be built with bags, but it is important to provide proper drainage behind a wall and ensure that the bags are properly secured against slippage. Slight "battering" of the wall can help, as well as compacting the bags at a slight angle toward the earth bank for added stability.

Many people are using bags as low-tech foundations for straw-bale and other construction systems. Bags work well, as they provide the required foundation width easily and cheaply. In most situations, bags should be filled with gravel at least a foot above grade, with the final course relatively level to receive the next building material (straw bales, adobe, wood, etc.) Rebar can be pounded into earthbags, while they are still curing, to receive the straw bales. Tubes of polyvinyl chloride (PVC) or another material (¾ to 1 inch) can be placed in the earthbag stem wall to provide a chase for the strapping used later to compact the bales and to tie the bond beam to the foundation. It is probably desirable to create a waterproof membrane between the tops of the bags and the bottoms of the bales. A thin layer of concrete with proper reinforcing can also be poured on top of the bags to create a grade beam and level surface.

FILLING THE BAGS Long bags on rolls should be cut to desired length with scissors or a sharp blade. A good method of building domes is to use one continuous casing per course, unless the wall is more than 25 feet in circumference (it is difficult to fill a bag of greater length). Always stagger the joints. In the first or foundation courses, it is best to minimize bag joints. Higher on the wall, this is not as critical, and shorter bags may be easier to fit. When measuring the required length of bag to be used, allow a foot extra (6 inches per end) for closing the bag.

Bags can be filled in several ways. A common method is to use a stand or to have a couple of people holding the bag while others shovel in the fill. Large bags can be filled in place with the use of a wheelbarrow or bucket. Bags higher on the wall can be filled in place with shovels or cans. Different techniques are used for different-sized bags. It is preferable to fill the bags in place to avoid having to pick them up and move them, although they can be partially filled and then moved to a designated location, where they are further filled.

The bags filled by different people will vary in thickness because of the difference in people's strength and technique. It is therefore advisable for one person or team to build a whole course to minimize changes in thickness. The courses may vary in thickness as long as each course is of a consistent thickness.

As bags are filled, the bottom corners are invaginated, a process dubbed "diddling" by builders Kaki Hunter and Doni Kiffmeyer. Elimination of protruding corners produces a more uniform wall surface and minimizes the use of plaster. The bags are gently laid in place before being tamped. The first bag in a row is usually pinned, and subsequent bags are laid with their open ends folded shut

Throw in two cans of soil, then "diddle" the corners. A metal stand holds the bag open while it is filled in place. (Photo courtesy of OK OK OK Productions.)

next to the sewn end of the previously laid bag. This is much easier than sewing each open end closed. To get a secure fold, it is important not to over-fill a bag. Once the bags in a course are laid, they are checked as to whether they are plumb, or properly placed in the ring of a dome, adjusted if need be, then pressed with feet into place.

Long bags are filled by inserting soil from the ends and picking up the ends, thus causing the soil to move to the middle of the tube. Another technique is to gather the long bag around a 1-to-2-foot length of tube, which holds the bag open and allows it to be let out gradually as it is filled. A mechanical pump can make this process fast and easy.

A piece of pipe, approximately 18 inches long, is used hold open the end of a long bag for filling in place. (Photo courtesy of Dominic Howes.)

LAYING COURSES It is important to lay bags in staggered courses (running bond) so that the joints of one row are covered by the bags of the next row. In earthquake-prone areas, a layer of barbed wire (4-point barbed wire or branches of a thorny plant that will cause friction) is laid between courses to keep the bags from slipping. If the bags are wider than 350 millimeters (14 inches), two rows of wire may be necessary. Wires can be kept in place with nails. Another method to keep bags from slipping is to interlock them by pounding a dimple into each bag, which is then filled by a bag in the row above; however, it is harder to stagger courses with this method.

Removing a "slider" after the bag is in place. The slider allows a bag to be positioned above a lower course without snagging on the barbed wire. End bags have been side-tamped and had their corners diddled and "locked" to prevent slumping on exposed wall ends. The first bag was locked (i.e., pinned closed with a nail). The second bag is held shut by the first bag. (Photo courtesy of OK OK OK Productions)

Tamping must wait until a full course is laid, which will minimize unevenness. Compaction is done gradually in order to keep the level consistent (though earthbags are relatively forgiving). Continue tamping until a "ringing" tone is heard and the bag no longer gives as it is tamped. A large stick (or a two-by-two) can be used to flatten the sides—an evenly shaped wall will require less plaster.

Sticks can be placed between each row as a handy attachment for plaster. "Gringo blocks" similar to those used in adobe construction can be used at door and window openings to simplify door and window framing. An alternative is to set the rough buck for the doors and window as you go, with the bags then laid to it. Because of the flexibility of the bags, this is often the preferred method.

OPENINGS Windows and doors can be incorporated by rough framing ahead of time. If an arch form is used, no lintels are necessary. The "pointed arch" form has been found to be much more stable than a round arch; however, where resources are scarce, any tubular object such as a bucket or barrel will suffice as a temporary form for small apertures. Small openings of 1 foot or less can be created by spanning the opening with wood or metal.

Stabilized fill should be used in bags around arch openings. In creating arches it is important to fill the bags in place and to tamp them into wedge shapes that will not slip when the form is removed. After the first three rows of bags are placed in position, buttressing must be added. The keystone bags, which consist of the top center three bags, are filled in place simultaneously. Alternatively, the keystone bags may be positioned with their ends folded down. Then the bags are closed with a nail or stitch using a piece of wire. Before the form is removed, the rest of the wall should be completed to buttress the arch and at least three courses should be built above the form.

Tamping of "fan bags" around a 5-foot Egyptian arch. Rum Cay, Bahamas. (Photo courtesy of OK OK OK Productions.)

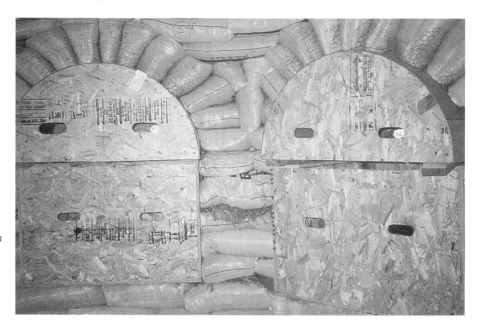

Interior view of arch window forms. The undersides of the fan bags have been cradled in chicken wire to make plaster adhere. (Photo courtesy of OK OK OK Productions.)

To build a form for arches out of wood, cut out the two end arches first. Attach them to each other with 50-by-50-millimeter pieces of timber. Make the form at least 50 millimeters (approximately 2 inches) wider on each side to allow for plaster. Cover them with plywood or any flexible sheet material. Place the form on top of wedges or blocks so that it can be removed easily after completion of the arch.

Openings can also be cut out from a finished unplastered wall as long as the bag fill is unstabilized earth. In vaults, the main openings should be placed only at the ends where the wall has no structural value (see the later section "Testing and Demonstration Structures"). Small vaults as door openings have been built with earthbags, but larger spans are probably not practical.

DOMES A special use for the earthbag technique is to make domed structures. The previously mentioned compass makes these domes easy to build. The only really practical dome is one built on a round base (sufficiently buttressed), with the dome being of a pointed or catenary shape. It is important to lay the bags in corbels (each ring is level) in order to avoid slippage in the upper rings. Hemispherical domes are impossible without additional form work and arc not advised.

For small spans, the bags can be easily used to create "free-form" domes and arch shapes. Kaki Hunter and Doni Kiffmeyer have built a dome with

bags of gradually diminishing sizes. They covered the resulting dome with cob, which was then planted with grass to make a living roof (see Chapter 16, "The Honey House").

A direction worth pursuing is to make domes using bags filled with a straw-clay mixture. An experimental dome at Cal-Earth was built in this fashion, in which predried straw-clay bags were placed over a sacrificial form to create a hemispherical dome.

In dome structures there are two areas that need careful attention: (1) A tension ring around the base of the dome is necessary if there will be little or no buttressing. It is good to have at least one continuous stabilized and reinforced ring in areas with ground movement. (2) A compression ring is needed around any openings created at the top. This must be a continuous ring that stops the forces acting inward. If made of concrete, it must contain several continuous reinforcement rings. For more complex structures, consult an engineer.

Interior of coiled Superadobe dome under construction at Cal-Earth Institute, Hesperia, California. (Photo courtesy of Cal-Earth Institute.)

PLASTERING It is important to plaster a structure built with polypropylene bags as soon as possible, as the bags will deteriorate in sunlight. Bags can be plastered with a scratch coat as they are built. As mentioned previously, invaginating the bag corners creates a more uniform wall surface that requires less plaster. As the bags often do not provide an ideal adhesion surface, an "adhesion coat" of boiled flour paste or glue, manure, and sand is recommended to prime the surface. The sticks laid between bags can also be useful.

To date, most renderings used on earthbag structures have been mud plasters. For lime/sand or cement plasters, a lath may be desirable. Strings laid between courses can offer convenient attachment points for such lath. Domes should be plastered with a cement- or lime-based plaster except in the driest areas, where mud plaster can be practical. Effective waterproofing is an area that needs additional research.

Testing and Demonstration Structures

Experimental earthbag structures at Cal-Earth have passed structural tests (ICBO-approved testing), leading to a building permit for the Hesperia Museum and Nature Center in California, which is now under construction. This is the first permit to be granted under the California building code for earthbag construction. The project features earth-tube lake bank retainers, a museum complex, windcatchers, and passive solar design. Many of the structures are being filled with a cement-stabilized mix for added strength.

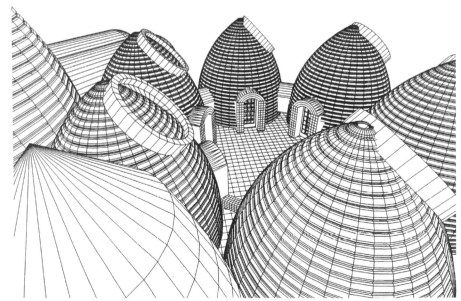

Computer perspective of the Hesperia Museum and Nature Center, Hesperia, California, designed by architect Nader Khalili. Computer rendering by Scott Kuhlmann. (Photo courtesy of Cal-Earth Institute.)

RepTile *exterior finish of Earth One house, constructed by students under the direction of Nader Khalili, is made of cement-stabilized earth applied in balls. (Photo courtesy of Cal-Earth Institute.)*

There are currently more than five structures built at the Hesperia site, and more are planned. Among the interesting features of the original dome project is that it is built partially underground. During construction a torrential rain overwhelmed the drainage system, leaving 18 inches of water in the structure. Because the bags kept the earthen material in place, there was no failure other than in the mud plaster, which was easily replaced.

Other projects at Cal-Earth include a three-vault house, the base of which is 3 feet underground. All materials from the excavation were used in the walls. The vaulted roof is made of a soil-cement mixture applied over forms. This structure features an exterior plastering system dubbed "Rep-Tile," in which balls of soil-cement are applied to the surface and the joints between are grouted. Other small student-built domes are plastered using the same technique.

In addition to his several years in earthbag technology development at Cal-Earth, Joe Kennedy built a sand and burlap structure with Nobi Nagasawa in Prague. Innovations there were the use of nails as pins and use of a metal plate to place bags on successive courses without snagging the barbed wire prematurely. Here, too, custom-shaped bags were used for the first time.

Several structures in South Africa were built by Joe Kennedy using this technique, including experiments with high-clay fills, concrete bond beams on bags, bag additions to existing structures, and bag landscape walls and seats. Additional research in Nova Scotia led to the use of gravel in wet-climate foundations. This technique was applied to the foundation for a straw-bale vault at Shenoa Learning Center in Philo, California. The foundation for a small house project built during the 1997 Natural Building Colloquium Southwest resulted in the dimpling and stick techniques, as well as innovative methods for foundations in straw-bale construction.

Nader Khalili's students and others have been spreading the earthbag concept. Earthbag structures have been built in Arizona, Utah, and Wisconsin and in other countries, including Mexico, Canada, the Bahamas, and Mongolia.

There is still much to be researched until this technology can be considered mature. As natural builders continue to experiment with earthbags, many innovations are sure to develop.

Earth-Rammed Tires

Michael Reynolds

Used automobile tires pounded with earth create the structural building blocks of an Earthship home and provide thermal mass that passively heats and cools the dwelling. This earth-tire brick and the resulting bearing walls it forms are extremely durable. The 3-foot-thick massive walls and the method of incorporating them into the landscape create living spaces that can maintain a relatively even temperature. Thus, with natural ventilation systems integral to the design, an Earthship home can heat itself in winter and cool itself in summer.

The landscape of the United States is cluttered with an estimated 2 billion discarded tires. Of the more than 250 million scrap tires generated each year, nearly half end up stockpiled or in landfills, according to the Scrap Tire Management Council. Used tires are available for more productive

The Earthship was designed in northern New Mexico, where the temperature in winter gets as low as –30 degrees Fahrenheit and in summer as high as 100 degrees. By working with the sun's energy and the earth's insulation, these houses maintain an interior range of 65 to 75 degrees Fahrenheit. (Photo courtesy of Solar Survival Architecture.)

use. The benefits of using scrap tires are particularly enhanced if they replace virgin construction materials made from nonrenewable resources.

Small Earthships may use as few as 200 scrap tires, and larger structures, may use considerably more, such as actor Dennis Weaver's Earth Yacht, which used 7,000 scrap tires in construction.

Design

The design concept of the Earthship is based on a U-shaped module with mass on three sides, glass on the fourth, and a skylight in the ceiling. Earth can be bermed on the three mass walls to provide even more mass. The walls of the Earthship are 2 feet, 8 inches thick and are wide enough to evenly distribute load. Therefore, the bearing walls of the Earthship are also the foundations. Based on observations and reported tests, it appears that tire walls are not as brittle as adobe or rammed earth walls of comparable thickness. This means that a tire wall can deform further than an adobe wall without losing strength. Lateral force tests were conducted and indicated a high coefficient of friction between the running bond of tires, as would be expected. The presence of the tires should allow much more in-plane and out-of-plane deformation in the walls than in an equivalent adobe wall (an extensive engineering evaluation for earth-filled tire construction is available from Solar Survival Architecture, listed in Appendix B).

It is not recommended to create U modules larger than 18 feet wide and 26 feet deep, although they can be smaller. Larger structures are made by multiplying modules. (Photo courtesy of Solar Survival Architecture.)

Design Variations

The basic U module is one of the most efficient structures for earth-rammed tire construction; however, other designs have been built. Because of the extreme heat and humidity in Texas, a tire hacienda was designed with no direct solar gain. The tire walls were built without curves. These straight walls were reinforced with concrete or tire buttresses and columns. Because of their massive nature, rammed earth tire walls should not exceed 10 feet in height.

Any passive solar building needs both insulation and thermal mass. Standard Earthships have circumvented the need for wall insulation, because the earth-rammed tire walls are buried with earth. (Editor recommends insulation even for earth-bermed applications, as heat will continually migrate into earthen walls without some kind of thermal barrier.) Earthships that are not buried require the tire walls to be insulated. Straw-bale "veneer" is a most economical and efficient insulator.

Construction Method

Tires can be used as found, without any modification. Tire walls are made by laying tires in staggered courses, like masonry, and filling them individually with compacted earth. A filled tire weighs more than 300 pounds; therefore, all tires are pounded in place with a sledgehammer and only minor movements can be made. Each tire holds about three or four wheelbarrows of soil when compacted. The tires are pounded level in all directions. Scrap cardboard is used to cover the holes in the tires in tire courses other than the first course.

Tire pounding is done in teams of two—a shoveler and a pounder. Slightly damp soil is easiest to compact. Tests have shown that hand packing of tires is more effective than mechanical methods. (Photo courtesy of Solar Survival Architecture.)

When the tire U has been constructed, two continuous plates are anchored to the tire walls with 3-foot-long rebar stakes. This tie plate receives the trusses for the roof. The greenhouse of the Earthship is framed, and the roof is installed. Voids between the tires are typically filled with recycled cans and bottles and packed with mud or concrete, which creates a relatively smooth surface for the final plaster coat. The roof of this thermal mass building should have a minimum of R-60 insulation to ensure minimal heat loss. Metal roofing is used to provide clean catchwater for household use. Two layers of 6 mil (1 mil = 0.001 inch) plastic "skirt" the building a minimum of 12 feet for bermed structures.

Interior walls, with recycled cans filling gaps, will be further plugged with mud or cement to even out the wall surfaces and reduce the amount of plaster needed. (Photo courtesy of Solar Survival Architecture.)

Fire Safety

People think that because tire piles in dumps burn easily and the fires are so hard to put out, an earth-rammed tire wall would act similarly. On the contrary, an earth-rammed tire wall that is plastered with mud or cement is extremely difficult to burn because of the density of the materials and lack of air circulation. The devastating Hondo fire that hit New Mexico in the summer of 1996 devoured many conventional homes. An Earthship caught in the middle of the fire lost its roof, front face, and all the wood on the inside, but the earth-rammed tire walls and the aluminum can and cement walls remained intact. The structure of the building was left usable, and a new roof and front face were installed.

Code Acceptance

The construction of earth-rammed tire walls has been a code-accepted building method for 20 years, and building permits for Earthships have been granted in many counties across the United States. Earthships also exist in Bolivia, Australia, Mexico, Japan, Canada, and South Africa. Based on a series of tests on tire-constructed walls at the home of Dennis Weaver in Ridgway, Colorado, engineer Tom Griepentrog has stated, "It is my opinion that the [characteristics of this] construction method [are] equivalent to or better than the general quality, strength, effectiveness, fire resistance, durability and safety that is required by the Uniform Building Code."

Light Clay

Robert Laporte and Frank Andresen
with special acknowledgment of the work of Franz Volhard

The name "light clay" derives from the German *Leichtlehm*, which is translated literally as "light loam." This is a composite building material, primarily constituted of natural fiber, commonly straw or wood chip, that has been coated with clay slip and packed into forms to shape it into walls, blocks, or panels. It is generally utilized as an infill product, valued for its insulating capabilities, but it also has some structural capacity that increases proportionally with density and thickness.

Light clay is composed of more fiber than clay, but can be made with varying ratios of fiber to clay. At its densest, however, it is still significantly lighter than cob or adobe. As documented by German earth construction specialist Franz Volhard in his book *Leichtlehmbau* (*lehmbau* means "earth building"), it ranges in density from 300 to 1,200 kilograms per cubic meter, with the common middle density at 600 to 800 kilograms per cubic meter. Somewhat confusing is the fact that light clay is widely known as "straw-clay" in the United States, but the German *Strohlehm* (literally, "straw loam" or "straw clay") is composed of earth with only a small amount of straw

Alternative Construction: Contemporary Natural Building Methods, edited by Lynne Elizabeth and Cassandra Adams ISBN 0-471-24951-3 © 2000 John Wiley & Sons, Inc.

Recently completed residence of timber-framed light clay shows clean lines and solid craftsmanship. (Photo by Robert Laporte.)

Exterior light-clay walls on concrete block footing before plastering. (Photo by Robert Laporte.)

(usually chopped). In other words, what is known as straw clay in Germany is more like an adobe that ranges in density from 1,400 to 1,600 kilograms per cubic meter (*Massivlehm* measures 1,800 to 2,000 kilograms per cubic meter, equivalent in density to rammed earth or pressed earth block.)

The clay serves as a binder and preservative and adds fire protection to the straw. It also deters insects and rodents. Light clay is made by coating each straw with liquid clay slip, or "slurry," and packing the mix firmly into forms. (Light clay is also known by some in the United States as "rammed straw.") It can be packed between temporary forms or "shuttered" in lifts to produce monolithic walls or formed into individual blocks or bricks. Occasionally it is placed free-form by hand, similarly to cob, but this method runs the risk of producing insufficient density (densities below 500 to 600 kilograms per cubic meter are more prone to the growth of mold) and is typically used only for "chinking" or to fill odd-shaped spaces or corners.

Light-clay blocks, drying here on pallets, can be made of clay slurry mixed with wood chips, straw, or other natural fibers and laid in running bond courses. (Photo by Frank Andresen.)

Other natural fibers can be substituted for straw to make light clay. Since the early 1990s, the use of wood chips as an aggregate has increased in popularity in Germany, to the extent that clay-coated wood chips, ready to be dampened and packed in forms, are available there in building supply stores. Additional aggregates include cork, sawdust, coir, sisal, hemp, flax, and similar natural fibers.

Numerous variations of earth and natural fiber construction are used in Germany. *Lehmwickelstaken,* oak stakes wrapped with a mixture of straw fibers and clay paste, are used for ceiling insulation or walls. *Lehmplatten* are clay panels made of mud, reed, and straw, also utilized for interior walls and ceilings. These panels can be made to varying thicknesses and dimensions and, depending on the orientation of the interior reeds, can be formed flat or curved. *Lehmschlag,* a mixture of fatty clay and long straw, is used for floors.

Light clay reduces the use of wood in construction. All infill construction above the foundations can be made with variations of light clay—the outer and interior walls, ceilings, and roof insulation. It is recommended that exterior walls be made 10 to 13 inches (25 to 30 centimeters) thick. Inside walls can be thinner, ranging from 4 to 10 inches (10 to 25 centimeters). Unlike cob, which will slump if built up too rapidly, straw-clay can be constructed layer by layer to its full height without any waiting time. Because of the form work, it lends itself to rectilinear architecture and is traditionally used in conjunction with timber frame construction.

Although monolithic wall forming is the most common method of assembly, precast modular blocks are sometimes preferred if the building season is too short to allow sufficient drying of solid walls. (To abet efficient drying, it helps to construct those walls located in shaded areas first, then

A completed "Eco-Nest"—timber-frame and light-clay housing developed by Robert Laporte's Natural House Building Center. (Photo by Robert Laporte.)

move to walls that get the most sun.) A large structure in a temperate climate can take a few months to dry, whereas mortaring of blocks can be done any time of year as long as the mortar will not freeze. Modular blocks, which use only small molds, may also be preferable in regions where wood for forms is scarce.

The feathery textured surface of light clay is ideal for plastering directly, and it bonds equally well with lime-, clay-, or gypsum-based renderings. It is critical, however, that the walls be thoroughly dry before any rendering is applied. Light clay is surprisingly durable and has a long history in harsh and wet climates, but on walls exposed to driving rains, a covering of clapboard or shingle over a protecting layer of clay plaster may be desirable. Impervious moisture barriers are not recommended, as they do not allow the walls to "breathe" and can trap interior moisture. Light clay, in general, will regulate dampness; the combination of straw and clay balances humidity through even diffusion, helping to maintain the material's insulative qualities.

History

Although the history of earth construction is nearly as long as the history of humankind itself, and many cultures have built with straw and clay, the kind of light-clay construction described in this chapter was fully developed in Germany only during the twentieth century. Wattle and daub, a process of applying mud to woven sticks or wicker, is one of the most ancient methods and has persisted to a small extent in indigenous cultures in Africa, Asia (notably Japan), and northern and central Europe. In Europe, many half-

Shown here is exposed wattle made of flexible sticks, such as willow, woven like a wicker basket over oak stakes. In traditional buildings this wattle is then covered with a "daub," or sticky mud, that is often mixed with sand, rocks, or straw. (Photo courtesy of Frank Andresen.)

timber buildings of three to five centuries ago, that are infilled with wicker-work and *Strohlehm,* are still inhabited and maintained. Techniques of wrapping clay-coated straw around these skeletal sticks, either woven or in rows of spindles, were precursors of the light-clay process.

The technique of *Strohlehmstanderbau,* or packing straw-clay into shuttered forms like those used to build rammed-earth walls, was published in Berlin in 1933. Illustrated methods for mixing and forming a straw-based *Leichtlehm* were first documented in a 1948 book by W. Fauth entitled *Der Praktische Lehmbau.*

Until the 1960s clay and straw construction was still used sporadically in new construction in Europe, even though the rise of modern construction methods had long since displaced earth as an everyday building material. An interregional understanding of the importance of global coherence and an extended consciousness of the effects of particular construction materials on ecology and human health did not evolve until the end of the 1970s. The growing environmental consciousness of the 1980s led to a revival of traditional construction systems throughout Europe, and earth building techniques were studied as viable construction alternatives. In France and Germany, especially, many technical advances began to evolve, including refinement of the lighter-weight straw-clay.

Through the cooperation of craftsmen, owner-builders, architects, and preservationists (restoration specialists and architectural historians) various interesting projects have sprung up in recent years. Many residences have been built by well-established earth construction contractors and owner-builders, using a synthesis of modern and historic methods with common tools and equipment. The State of New Mexico, where light clay has taken a foothold, has recently published standards for non-load-bearing light-clay construction through its Construction Industries Division. (This publication, *Clay Straw Guidelines,* describes the assembly as straw that is thoroughly and evenly coated with clay slip and compacted to form a monolithic insulating wall. It specifies that it be built a minimum 12 inches thick and that the bottom of the light-clay wall be at least 6 inches above final exterior grade.)

Light-Clay Materials Preparation

Materials include water (urine, particularly of horses, has been used instead of water in traditional cultures to strengthen the clay and increase its plasticity), clay, straw, and lumber for tampers and forms (2 × 4 and 2 × 2 stock, and ¾-inch plywood).

MATERIALS

TOOLS   Tools include shovels, hoes, pitchforks, tampers (fashioned from 2 × 4 and 2 × 2 lumber), stepladders, scaffolding, 5-gallon buckets, 55-gallon barrels, and a basic carpenter's kit for building the forms (hammer, saw, drill, drill bits, screwdrivers, wrenches, etc., and fasteners, including drywall screws and lag screws). Pumps for mixing or spraying the clay slip are optional.

CLAY TESTING   A preliminary sedimentation test is performed to determine soil composition. Fill a pint-sized jar one-third full with water and one-third full with *Sedimentation Test* the soil (clay) sample. Cap and shake the jar vigorously until the solids are in suspension. Allow the mix to settle overnight. The solids will layer as clay, silt, sand, and, possibly, small stones. The top (clay) layer should occupy at least 50 percent of the volume of the solids.

How can you know whether you have a clay layer? Clay is smooth, slippery, and somewhat sticky when it is wet. It consists of very fine particles, and when moist it can be easily molded, shaped, and formed.

When there is a suitable clay content in dry soil, a characteristic cracking appears from shrinkage. Exposed clay becomes weathered to a fine powder.

The Ribbon Test   The ribbon test is a test for plasticity. Plasticity roughly means stickiness. Binding or sticking the straw together is the main structural purpose of the clay in straw-clay construction. Rolling a moist sample of clay between the fingers to the diameter of a pencil, and then bending it at a right angle, will display the relative plasticity of the clay. If the test sample does not crack at the bend, it is suitable.

Prototype Test   Before committing the clay candidate to a project, test it in a small prototype wall, following the procedures for mixing, forming, and tamping described in a following paragraph.

ECOLOGICAL CLAY HARVESTING   Exposing a clay layer to the action of weathering (freezing-thawing and wetting-drying) a year or more in advance of harvesting will allow it to weather into a fine powder. Let nature do the work. Using local, natural material and allowing it to mature to optimal readiness in its own time reflects the spirit of the ecological approach to harvesting clay. In its dry state, soil is much easier to handle; so, again, work with nature. Remove all organic material from the surface (grasses, brush, sticks, leaves, etc.). The finer the clay taken from the surface, the more easily it will mix into water. Powder is best. The scraping action of a hoe works well to loosen the clay. Then a flat shovel is used to collect the clay and place it into buckets and barrels. This is the simplest harvesting technique.

MAKING CLAY SLIP After harvest, the clay is strained through ¼-inch hardware screen to remove any stones or organic materials. Fill a 30- to 35-gallon barrel one-third full with water. Add the strained clay. One person can sprinkle the clay into the barrel while another uses a stirring paddle, giant whisk, or hoe to churn the water. Keep the water moving as you add the clay powder, to keep it from clumping. It becomes more difficult to prevent clumping as the mix thickens; the stirring action is similar to churning butter. The desirable consistency of the mix is that of thick cream. To test thickness, dip your finger into the clay slip, withdraw it, and when the slip layer masks the fingerprint, it is the right consistency. The amount of clay required to achieve this desired thickness varies with the type, quality, and moisture content of the clay. Periodically check the bottom of the barrel for lumps of clay. Continue to churn until there are no lumps and all material is in suspension.

MIXING STRAW AND CLAY For mixing the straw and clay, prepare an 8-by-12-foot working platform with plywood on level ground in proximity to the walls to be constructed. This portable workstation is moved as necessary. Break down one-half bale of straw on the platform. Next, take a bucket of slip and broadcast it uniformly over the straw. (Franz Volhard recommends using a plaster pump to spray the slip over the straw.) With several people working, use a bouncing motion of the fork and toss the straw (as in tossing a salad). The

Silky clay slip is broadcast over loose straw. (Photo courtesy of Robert Laporte.)

The straw is then tossed, with pitchforks, like a salad. (Photo courtesy of Robert Laporte.)

desired goal is to have a light, even coating of clay on the surfaces of the straw. Do not saturate the straw with wet clay; just a light coating will do.

The straw-to-clay ratio will vary from wall to wall, depending on the thermal requirements of the particular wall. For example, a north-facing wall (in the Northern Hemisphere) has a higher insulation requirement and, hence, would contain more straw. On the other hand, a south-facing wall will need more mass and, hence, a higher proportion of clay. The dry weight per cubic foot should vary from 25 to 50 pounds for walls and 15 to 25 pounds for a straw roof. In a roof, only enough clay to retard fire and mildew is necessary.

Light-Clay Construction Method

FORMING TIER I

Final preparation before assembling the wall forms includes measuring for and drilling the horizontal holes (that will take the stabilizing bars) in the studs at each tier of the straw-clay wall. Typically, these holes are drilled at 24-inch vertical intervals. Usually, an interior corner wall section is formed up first. Fasten ¾-inch plywood sheathing to the studs (vertical supports) and to the timber posts with removable wooden strips called "walers."

Twin outside forms (24-inches high) are alternately leapfrogged up the wall. With forms secured to both sides of the wall, you are ready to load the straw-clay into the wall cavity.

Two tiers of assembled form work. (Photo courtesy of Robert Laporte.)

Right half of photo shows monolithic straw-clay wall immediately after removal of the forms. Form work is still in place on the left. (Photo courtesy of Robert Laporte.)

LOADING AND TAMPING   Attention to the details of uniform loading and even tamping is required to achieve strong and stable walls, free from flaws.

Prepared straw-clay is rolled into cigar-shaped bundles appropriate to fit the wall cavity, about the size of a loaf of French bread. The cylinder of straw-clay is then placed snugly at the bottom of the cavity, with care taken not to soil exposed wood-work. For protection, the exposed wood surfaces can be temporarily covered with paper. The "cigar" is carefully tucked into all corners.

Begin tamping with your feet and follow with wooden tampers, evenly compressing the straw-clay

Straw-clay is brought from the mixing platform to the forms in wheelbarrows, buckets, or by hand in cigar-shaped bundles. (Photo courtesy of Robert Laporte.)

into every corner and space. Tamping tools can be fashioned from two-by-fours and two-by-twos.

If your posts are keyed (with a vertical "V-groove") for wall stabilization, then it is important that the straw-clay be tucked and tamped carefully at those points. Building up the wall is a simple repetition of the preceding sequence.

INSTALLING STABILIZING BARS   To strengthen and stabilize the wall, horizontal rods are embedded at vertical intervals corresponding to the forms. Locally available straight hardwood saplings of uniform diameter make suitable stabilizing bars (e.g., willow, oak, bamboo).

When the first tier has been completely filled and tamped, install the first stabilizing bar by inserting the rod through the predrilled holes in the studs and posts. It is important for the bar to have a firm base supporting it so that it will not break under subsequent loading and tamping.

Scaffolding must be used to reach upper-tier forms. (Photo courtesy of Robert Laporte.)

INSTALLING TIER II For a clean-looking wall, any stubborn "whiskers" must be tucked into the wall cavity prior to installing the next form. Install the second of the twin outside forms in a position above the first, and adjust the height of the working platform (scaffolding) for safe and efficient loading and tamping. Continue loading and tamping straw-clay until this second tier is filled, tamped, and stabilized with the second horizontal rod.

INSTALLING TIER III Remove the outside form of Tier I by carefully sliding it down before pulling it away from the wall. Otherwise, straw fiber stuck to the form could be pulled out, leaving a shaggy wall surface.

The first outside (bottom) form now becomes the form for Tier III. Install as before, and continue leapfrogging the outside forms in this way until the wall is complete.

When a plate obstructs vertical tamping, the last tier is halved. The bottom half of this last tier is constructed as usual, and the top half is horizontally stuffed and tamped without an outside form.

REMOVING FORMS Once the wall is complete, the form work should be immediately removed, because the forms do not breathe and will support the growth of mildew.

When removing forms, always remove the lag screws first to prevent accidental tear-out caused when a forgotten lag screw claws the wall surface as the form is slid downward. If the forms are not to be used immediately, they should be cleaned and stored. *Cardinal rule:* Always slip the form down first to screed the surface, and then move the form out and away from the wall.

BLIND FRAMING Blind framing is used when a timber frame is completely exposed to the interior. The purpose of blind framing is to provide a nonbearing matrix, or

Straw-clay walls must dry before plastering. (Photo courtesy of Robert Laporte.)

web, against which the straw-clay mixture is stabilized. The timber frame provides structural support. With the use of studs smaller than the wall thickness, an uninterrupted layer of insulation is achieved, thus significantly reducing infiltration and increasing the thermal performance of the wall.

Positioning the studs this way simplifies interior plastering by eliminating the need for bridging over the studs (because the studs are either buried in the wall or are exposed only on the external wall). Eliminating interior studs reduces the likelihood of the plaster cracking as a result of the wood shifting.

Bridging is the process of "floating" the plaster over an unstable substrate, such as wooden studs, the movement of which exceeds the elastic limits of plaster. When bridging is necessary, fabric, such as burlap or fiberglass mesh, is applied over the stud, extending 2 to 3 inches on each side of it. The fabric can be stapled to the stud or imbedded in a thin layer of plaster, preceding the rough coat.

A wall surface of this nature, without studded interruptions, has a finished quality that may be left unplastered.

INFILLING Infilling is the traditional technique used in conjunction with timber framing. In Europe it is common to see the framework exposed to the exterior, hence the term "half-timbered." In Japan and other Asian countries, the framework is usually exposed on both sides.

An infill is a nonstructural wall component utilizing a secondary frame, or lattice, anchored to the structural framework. This matrix, often woven of thin branches (traditionally willow) and referred to as "wattle," holds the straw-clay mixture in place—a variation of wattle and daub ("daub" refers to the application of sticky mud).

Light-Clay Roof Construction

PREPARATION Typically, the dry weight of the straw-clay mixture for roof construction varies from 15 to 20 pcf (pounds per cubic foot) and is prepared in the same way as the mixture for wall construction.

This lighter mixture provides greater insulation and contains sufficient clay to protect the straw from fire and mildew.

INSTALLATION With roof framing and ceiling decking in place, the rafter cavities are ready for loading. For maximum thermal performance it is important that the entire rafter cavity be filled with straw-clay, except for a ventilation channel. To maintain the structural integrity of the roof, it is absolutely vital to vent moisture out of the roof cavity.

Light Clay with Wood Chips

One of the most recent innovations in the development of modern German clay-building techniques is the use of wood chips bound together with clay as an insulation infill for exterior and interior walls. Since its introduction in the latter part of the 1980s, contractors are able to offer another natural monolithic clay infill system. Wood-chip light clay is comparable to straw-light-clay in terms of its physical qualities, but its production and manufacturing process is easier and quicker and it requires less drying time, settling, and tamping. It also provides a more efficient and less labor-intensive construction method.

Using an ordinary cement mixer, wood chips in different sizes are mixed together with a clay slip (other machinery may be required to produce the clay slip). The size of the chips varies between rough sawdust and chunks up to 2 inches in diameter, depending on the chipper. The chips can be dry or green, but they should be bark-free, especially when poured into

Wood-chip clay infill system for a timber-frame house, using laths to cage the 12-inch-thick wall. (Photo courtesy of Frank Andresen.)

thick wall sections. Generally, the mixture is three to four buckets of wood chips to one bucket of clay, depending on the required strength and weight, the quality of the clay, and the size of the chips. The mixture should be blended in the mixer for a minute or two, until all the chips are coated with slip; then it is ready to pour into the wall forms. The mix is spread in place with a board or stick to fill all voids—there is hardly any tamping involved.

Depending on the system you are using, the density ranges from 1,000 to 1,500 pounds per cubic yard of dry mix. The insulative value for a 12-inch exterior wall with plaster can be as high as R-25, depending on the quality of the wood chips, the clay, and the pressure with which the mixture is packed. Vapor barriers are unnecessary because of the clay's inherent capacity for moisture absorption and dispersal.

A variety of forming systems and materials can be employed. When reed mats are used as an infrastructure, studs should be no wider than 12 to 16 inches on center, depending on the quality of the reeds. With the use of wooden laths or bamboo as a light framework to cage the infill, these distances can be nearly doubled. Because the infill is anchored with laths or matting, the whole system is stiffer and stronger and shrinkage is virtually nil. Reed mats can be applied quickly with a staple gun by connecting the wiring to the framework. Wooden laths take more time to install; however, they may be more available in certain areas and, in the end, they work just as well. Slip forms can also be used to form walls that require no mats or lathing to act as a cage. In this system, vertical reinforcing, such as saplings or 2× material, should be placed in the middle of the wall to add lateral strength. Many types of plaster can then be applied easily to any of these systems. Exposed beams should be covered.

A 12-inch-thick exterior wall requires up to eight weeks of drying time during a normal temperate summer. Thicker walls can be used if there is a sufficient drying season and the additional insulation value is required. It is possible to make bricks, blocks, or panels with clay and fiber that can be used in a predried state. With

Reed lath is applied with a pneumatic staple gun. (Photo courtesy of Frank Andresen.)

the use of these dry materials, the range of construction techniques is even greater and the building season can be extended. Hollow wall systems can be laid up and filled with cellulose fiber, and unfired bricks can be used on the interior to create thermal mass. When you are using a wet system, everything must be dry by the first frost, so do not start too late or too early—work with the seasons.

Home sweet eco-home—EcoNest interior.
(Photo courtesy of Robert Laporte.)

10

Straw-Bale

Kelly Lerner, Bob Theis, and Dan Smith

Plastered straw-bale construction—which was born out of desperation as a method of building on the Nebraska plains in the nineteenth century—has evolved into an elegant alternative building solution for the twentieth and twenty-first centuries.

Straw-bale has exploded in popular interest in the last decade. Although but a few dozen straw-bale buildings stood in 1990, only 10 years later there are several thousand, including speculative homes and multifamily housing. Straw-bale construction is now poised to go mainstream and may be used by production builders within the next decade. This phenomenal growth has taken place absent government subsidy and with only grass roots marketing efforts.

Historical Use	Straw has had a central place in construction for most of human history. Roofs thatched with straw were traditional across northern Europe, Russia, and the northern portions of eastern Asia, including Japan. The *tatami* floor

Alternative Construction: Contemporary Natural Building Methods, edited by Lynne Elizabeth and Cassandra Adams ISBN 0-471-24951-3 © 2000 John Wiley & Sons, Inc.

The one-bedroom, one-bath Goodwin-Zeik residence has its own photovoltaic electric system and hydronically heated floors. This post-and-beam house was designed to look like the conventional pitched roof homes of northern New Mexico. (Photo courtesy of the architect, Tara Teilman-Way.)

mats of Japan are flat straw bundles with woven grass faces and cloth edges. In cold weather the tipis of North America were insulated with loose straw between the inner lining and the outer cover.

Adobe and cob construction use straw as tensile reinforcement for earth mixes, and at the other end of the spectrum, light-clay uses straw as the principal element with clay as a coating and binder. In addition, there is a whole range of earth plasters that incorporate chopped straw as an element. Cow dung, for instance, a common plaster in much of the world, is composed essentially of straw that has been processed by the cow.

The natural building movement, then, is rediscovering a building material that has been around forever. It was just forgotten about for a few generations during industrialization.

EARLY BALING During the mechanization of American agriculture, the need to simplify the storage and transport of hay and straw led to the development of the mechanical baler in the mid-1800s. The potential for building with the result-

Historic photo, circa 1908, of Simonton house under construction near Purdum, Nebraska. (Photo courtesy of Matts Myhrman.)

Hipped-roof straw-bale house built in Nebraska in 1936 and still occupied. (Photo by David Bainbridge.)

ing blocks of straw was quickly realized when the Sand Hills of western Nebraska were opened to homesteading in the mid-nineteenth century. Importing lumber to the area was very expensive, and unlike most of the ground surface of the Great Plains, the thin sod that overlaid these sand dunes was too fragile for building sod cabins.

Thus, between 1896 and 1945 some 70 bale buildings were constructed in this region, of which 13 were known to exist in 1993. Most of these remaining structures seem to be in sound condition, and the interior straw appears, upon examination, indistinguishable from recently harvested straw.

CURRENT REVIVAL The seed of the current straw-bale revival was a one-page description by Roger Welsh of historic Nebraska straw-bale construction, published in the popular 1973 book *Shelter*. When John Hammond's straw-bale studio was published in an article in *Fine Homebuilding* in 1983, another wave of interest arose.

Through the 1980s and into the 1990s, more modern pioneers learned of (or reinvented) the method and began to spread the word by teaching and writing articles. In 1993, Matts Myhrman and Judy Knox launched a newsletter on straw-bale construction, *The Last Straw*. By April 1994 the network of individuals pursuing straw-bale construction was spreading at an incredible rate.

When our architectural office designed its first straw-bale house in that same year, use of this building genre was largely unknown. The permit we obtained was only the second one issued in the State of California for straw-bale construction.

Jonathan Hammond's straw-bale residence in Winters, California, as it looks today. The original studio is on the left; the right wing and garden walls have been added since the project was first published in 1983 in **Fine Homebuilding.** *(Photo by Jonathan Hammond.)*

Within four years there were two very popular books on the subject, quite a few videos, a California Straw Building Association with 100 members, and similar organizations in Texas, New Mexico, and Arizona. At the close of 1999, there are an estimated 200 straw-bale buildings in the state of California. Our office has designed more than 40 straw-bale projects in that time, and this kind of building has become the predominant focus of our work.

Structural Approaches to Straw-Bale

Bale structures are typically categorized as either "load-bearing" or "post and beam," based on whether the vertical loads are carried principally on the bales themselves or on a frame. However there are also several newer "strengthened bale" approaches, incorporating composite designs with stressed skins, exterior ribbing, constrained bale design, and alternate geometries such as vaulting.

LOAD-BEARING The classic Sand Hills, Nebraska, style of load-bearing construction is simple, direct, and intuitive. Although generally appropriate only for single-story structures, higher loads can be accommodated through significant precompression of the bales. This style of load-bearing construction is especially popular in Arizona, Colorado, and Canada.

Today many systems are used to precompress the bales between the footing and the box beam or bearing plate at the top of the wall. A system of full-height threaded rods every 6 feet is direct and effective, but awkward to install, especially in ultradense, rice-straw-bales. More common is strapping with polyester, wire, or wire mesh, often combined with some method of compression such as course-by-course "stomping" by foot or pressing on the wall with a backhoe.

Whereas the advantage of load-bearing structures is the simplicity of material, the disadvantage is the complexity of the structural action. In seismic areas, for two-story structures, or in high snow load areas where building department engi-

The walls and roof of this load-bearing straw-bale house in Guadalupe, Mexico, were assembled in one working day. (Photo courtesy of Development Center for Appropriate Technology.)

neers want a calculable system, a load-bearing system makes for a more complex approval process.

It should be pointed out that even in load-bearing construction, the plastered skins of the walls (typically cement plaster) play a significant role in the structural character of the walls. The ensemble really functions as a sandwich panel or a weak stressed-skin panel—which is much stiffer than either the bales or stucco alone.

POST AND BEAM   The post and beam constitutes a larger catchall category of methods, combining a vertical load-bearing frame with the bale wall. In general, adding a frame greatly simplifies the engineering and approval process, especially for two-story buildings or those in earthquake-prone areas.

The character of the walls (as well as the process) is affected by whether the bales are used as *infill* panels between supports, the bale walls are a continuous and intact *fabric* with added supports, or the bales are a *wrap* around an existing wall.

"Infill"   *Bale Wall Infill between Full-Width Supports.* When bales are used as infill between supports, the posts are usually full-width vertical supports such as concrete block, concrete, short 2 × 4 fin walls, or full-width wood I beams. The top beam is then a concrete tie beam, a glu-lam beam, or a box-beam. The post-and-beam approach is the most structurally straightforward in

The TJI box-beam header and roof are supported by steel posts in this 1176-square-foot, three bedroom straw-bale built by Habitat for Humanity near Santa Fe, New Mexico. (Photo by the architect, Tara Teilman-Way.)

terms of dealing with the unusual aspects of bale walls, and the most conservative.

Fabric *Post and Beam with Continuous Bale Wall Alongside.* The post-and-beam system with a continuous bale wall is often typified by an exposed heavy timber frame with the bale wall running alongside. The general advantages of this system are that the frame and roof can go up first and protect the bale raising, the frame can be engineered for a greater variety of loads, and the walls can have more architectural freedom of movement. The disadvantage is that a heavy, exposed braced frame adds a significant additional layer of cost and labor.

Bale Wall with Light Notched-in Posts. Many walls are now incorporating a light post-and-beam frame notched into a continuous bale wall, so that the frame is not exposed. With the frame flush to the face of the bales, it is easy to add strap or cable X-bracing—significant in seismic California. We have found that this approach, using 4 × 4 posts, interfaces very well with conventional wood stud framing.

Wrap *Bale Wall Wrapped Around an Existing Shell.* A common approach to insulating an existing house, barn, or steel industrial building is to wrap a bale wall around the structure. The bales are typically wrapped outside the existing skin of the building and then tied to it. This method has been used by Kelly Lerner to retrofit existing buildings in Mongolia with insulation.

STRENGTHENED BALE "Strengthened-bale" approaches combine bales with another material to form a stronger composite structure. Typical strategies include using enhanced stucco stressed-skin panels, lighter infill posts, exterior ribbing-stiffened baskets, and vaults.

Stressed Skin Panel with Stronger Concrete Skins Increasing the strength of the concrete surfaces of straw-bale walls enhances their function as stressed-skin panels. Several people have substituted gunite for stucco. A system designed by Gary Black, University of California, Berkeley, professor of architecture, for high or multistory walls uses 3-inch gunite skins and innovative cross wall "spars" between the bales to tie the two skins together.

Strongly Enmeshed Bale Wall Strongly enmeshed bale walls are sometimes used to enhance the inherent seismic resistance of the flexible bales by treating them as if they were shock

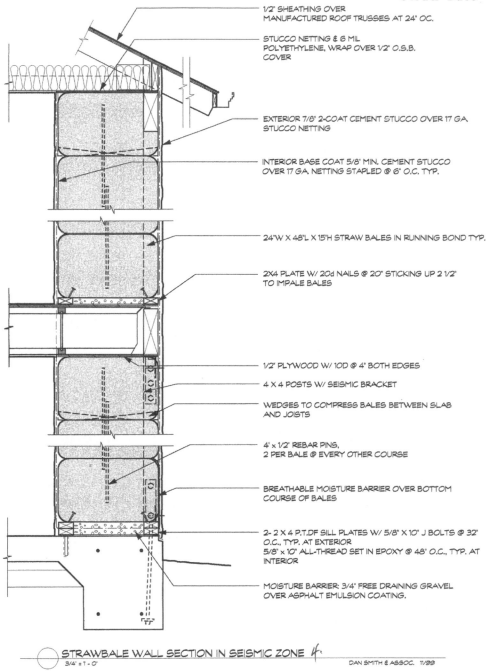

1/2" SHEATHING OVER
MANUFACTURED ROOF TRUSSES AT 24" OC.

STUCCO NETTING & 6 MIL
POLYETHYLENE, WRAP OVER 1/2" O.S.B.
COVER

EXTERIOR 7/8" 2-COAT CEMENT STUCCO OVER 17 GA.
STUCCO NETTING

INTERIOR BASE COAT 5/8" MIN. CEMENT STUCCO
OVER 17 GA. NETTING STAPLED @ 6" O.C. TYP.

24"W X 48"L X 15"H STRAW BALES IN RUNNING BOND TYP.

2X4 PLATE W/ 20d NAILS @ 20" STICKING UP 2 1/2"
TO IMPALE BALES

1/2" PLYWOOD W/ 10D @ 4" BOTH EDGES

4 X 4 POSTS W/ SEISMIC BRACKET

WEDGES TO COMPRESS BALES BETWEEN SLAB
AND JOISTS

4' x 1/2" REBAR PINS,
2 PER BALE @ EVERY OTHER COURSE

BREATHABLE MOISTURE BARRIER OVER BOTTOM
COURSE OF BALES

2- 2 X 4 P.T.DF SILL PLATES W/ 5/8" X 10" J BOLTS @ 32"
O.C., TYP. AT EXTERIOR
5/8" x 10" ALL-THREAD SET IN EPOXY @ 48" O.C., TYP. AT
INTERIOR

MOISTURE BARRIER: 3/4" FREE DRAINING GRAVEL
OVER ASPHALT EMULSION COATING.

STRAWBALE WALL SECTION IN SEISMIC ZONE 4
3/4" = 1'-0"   DAN SMITH & ASSOC. 11/99

Straw-bale wall section in seismic zone 4. This two-story bale wall section shows enhanced strength mesh stucco shear walls for seismic zone 4. It is assembled with the addition of steel strap X-bracing on the ground floor. (Illustration by Dan Smith and Associates, www.dsaarch.com.)

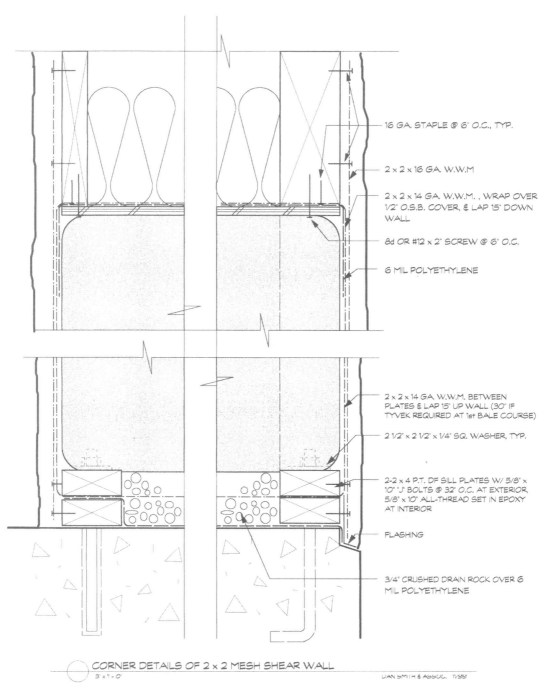

16 GA. STAPLE @ 6" O.C., TYP.

2 x 2 x 16 GA. W.W.M

2 x 2 x 14 GA. W.W.M. , WRAP OVER 1/2" O.S.B. COVER, & LAP 15" DOWN WALL

8d OR #12 x 2" SCREW @ 6" O.C.

6 MIL POLYETHYLENE

2 x 2 x 14 GA. W.W.M. BETWEEN PLATES & LAP 15" UP WALL (30" IF TYVEK REQUIRED AT 1st BALE COURSE)

2 1/2" x 2 1/2" x 1/4" SQ. WASHER, TYP.

2-2 x 4 P.T. DF SILL PLATES W/ 5/8" x 10" "J" BOLTS @ 32" O.C. AT EXTERIOR, 5/8" x 10" ALL-THREAD SET IN EPOXY AT INTERIOR

FLASHING

3/4" CRUSHED DRAIN ROCK OVER 6 MIL POLYETHYLENE

CORNER DETAILS OF 2 x 2 MESH SHEAR WALL

3" = 1' - 0"

DAN SMITH & ASSOC. 11/99

Corner details of 2 × 2-inch mesh shear wall. This section shows the details of the mesh wrapping. Note: recent testing has led to a preference for 4 × 4 upper sill plate. See testing at the EBN web site: http://www.ecobuildnetwork.org/. (Illustration by Dan Smith and Associates.)

absorbers, with the mesh in tension constraining the bales in compression. This system was tested by structural engineer David Mar and builder John Swearingen on a vault-shaped roof structure designed by Dan Smith and Associates. The assembly was found to resist seismic loads up to 1.1 G. David Mar and Kelly Lerner have collaborated in using this system with mesh and wire ties on buildings in Mongolia and China. We have designed several permitted projects now in California Seismic Zone 4 areas using enhanced mesh walls as the primary system for one-story loads.

Exterior Bale Ribbing   Another approach is to gradually stiffen a bale wall with exterior bracing or ribbing. We have been using a system of exterior rebar or bamboo ribs to stiffen bale vault structures. The rebar ribs are spaced 2 feet on center, or two per bale, and are tied through the bales, creating a "basket" effect. The same approach holds promise for stiffening tall walls, replacing the interior pinning while improving the stressed skin panel action.

Similar systems include the walls of Bob Cook's straw-bale studio in Tucson, Arizona, that were stiffened with 1 × 2-inch fir (in *The Straw Bale House Book*), Bill and Athena Steen's reed-stiffened walls in Mexico (see Chapter 14, "Straw, Clay, and Carrizo: Obregon Project, Northern Mexico"), and Kelly Lerner's use of interior and exterior wood ribs in Mongolia.

Vaults   Using the analogy of masonry vaults and domes, bales seem to be a natural. However, their elasticity makes the curved structures act more like baskets, with gradual stiffening achieved with external ribbing. Vaults, like load-bearing structures, depend on compression, and they require well-compacted

This vaulted straw-bale pavilion was built during the 1996 Natural Building Colloquium at Shenoa Retreat Center in northern California. It was the first bale vault with compound curvature; the walls were shallow arcs in plan, but flattened as they rose to interlock at a horizontal ridge. (Photo by Kelly Lerner.)

bales and wedges between the bales. The vault shows promise of minimizing the use of wood framing while maximizing the effect of the superinsulating envelope.

Basic Structural Specifications

For the best review of the structural aspect of straw-bale construction, refer to Bruce King's *Buildings of Earth and Straw*. The following paragraphs here give an overview of how bale walls perform in the testing done to date.

The allowable compressive stress for straw-bale walls in the current codes is 360 to 400 pounds per square foot (psf), or 2.5 pounds per square inch (psi). Tests on plastered walls have shown an ultimate of 3,000 to 6,000 plf. However, tests have shown a maximum "failure" stress of 70 to 84 psi with unplastered bales laid flat. In this instance, bales were compressed to half height, from which they fully rebounded. With bales on edge, failure was at 17 to 21 psi, when the strings broke.

The weight of a bale wall is more in the plaster, at 12 psf per face, than in the bales, 16 psf.

IN-PLANE LATERAL LOADS

Tests have shown that unplastered bale walls have significant, but elastic, resistance, resisting up to 179 plf with a lateral displacement of 4 inches at the top of an 8-foot-high wall.

Recent tests on plastered bale walls at California Polytechnic-San Luis Obispo show values for "stucco" greater than required by code: 1,500 plf ultimate for a two-sided plastered wall. With a 4× safety factor, this is a 375 plf design strength. Failure occurred at the stucco netting attachment to the sill plates and beam, with no cracking in the field of the plastered bale wall. These results compare favorably with the 2 × 180 plf code allowance for stucco. Because failure occurs first at the stucco netting attachment, we recommend a 4-inch nailing pattern at sill plate, posts, and beam.

Clearly significant improvement can be achieved with heavier reinforcement and stronger attachment for a super-stucco approach. Enhanced mesh and attachments should raise the pounds per lineal foot per face to 240 to 360. The ultimate strength, using the constrained bales approach, could be over 2,000 plf.

In resisting lateral stress from wind, the inherent strength of the plastered bales is generally adequate for one-story buildings. However, in seismic areas, such as California's Zone 4, X-bracing often becomes necessary, as well as mandatory structural analysis by an architect or engineer. Methods of X-bracing include flat 3-inch (12 gauge) steel straps, cables, and threaded rods.

OUT-OF-PLANE LATERAL LOADS The width of the bale walls is an advantage with out-of-plane loads, with an aspect ratio for a 10-foot-high, 2-foot-thick wall of 5:1, as compared with 20:1 for a 6-inch-thick wall. In tests performed in New Mexico, a plastered bale wall handled a load of as much as 50 psf with "no apparent sign of structural distress or failure."[1] Even an unplastered bale wall performed well, with a maximum 1-inch deflection at 26 psf, simulating a 100-mile-per-hour (mph) wind.

Engineer David Mar and contractor John Swearingen of Skillful Means Builders checked their calculations for compression struts by testing this barrel vault assembly of straw-bale, wire netting, rebar, and cement stucco. This new structural approach provides strength and safety that substantially exceeds code seismic requirements. (Photo courtesy of Daniel Smith and Associates.)

Here is where the bale wall's inherent nature, performing as a stressed skin panel (even imperfectly) helps. As Bruce King has described, the typical plastered bale wall works quite inefficiently as compared with a perfect sandwich panel, but nevertheless is 20 times stronger than an unplastered bale wall.

The Berkeley vault test provides a dramatic example of extreme out-of-plane stress.

GUIDELINES FOR LOAD-BEARING WALLS For unbraced bale walls, the Pima County codes give height/width guidelines of 5.6:1 for height (10 feet, 6 inches for a 23-inch wall), and length/width limits of 13:1 (25 feet for a 23-inch wall). Wall openings should be less than 50 percent of the wall length, and at least 4 feet from corners. California prescribes load-bearing limits, allowing 400 plf in bearing, but it leaves seismic engineering up to the professional, and the vast majority of California bale buildings are non-load-bearing.

Environmental Benefits

Patterns of development are as critical to ecological health as the building materials used. Each year in the United States an area twice the size of New York is paved over (168,000 square miles). To the extent that straw-bale construction fuels a desire for virgin land, it has a negative overall environmental impact, in spite of its many benefits. The average size of a house in the United States has been growing steadily, from 1,392 square feet in 1968 to 2,120 square feet in 1996, and, again, building a larger house will offset the ecological benefits of using environmentally superior materials.

Moreover, straw makes up but a small portion of the total material and energy required to construct a building. The environmental impact of straw-bale buildings varies greatly, depending on the size of the building and the detailing of the wall, roof, and foundation systems.

POLLUTION REDUCTION— STRAW AS A WASTE PRODUCT

Because of its high silica content and long stems, straw does not easily decompose in soil and poses a disposal problem in many areas. California farmers had adopted the practice of burning straw, emitting 56,000 tons of carbon monoxide and 3,000 tons of particulate annually from the combustion of rice straw (1.1 million tons) alone.[2] Thus, straw burning practices in California generated more than twice as much carbon monoxide as all the state's power plants combined.[3] Field burning also emits carbon dioxide, methane, and methyl bromide, greenhouse gases responsible for global warming.

Legislation in California will completely ban field burning in the coming years. Even composting straw in fields has its drawbacks. Rice straw decaying in flooded rice paddies contributes 10 percent of all atmospheric methane.

Instead of producing pollution and headaches for farmers, waste straw could be diverted into construction. Using a conservative figure of 450 three-string bales per 1,500-square-foot house (walls and ceiling), the 200 million tons of waste straw annually produced in United States could build more than 10 million super-insulated houses every year, six times the U.S. housing starts in 1997.

STRAW DISPLACING WOOD

With thoughtful design, straw can displace significant amounts of wood used in buildings. Construction of the average 1987 single-family house consumed 15,824 board-feet of framing lumber and more than 7,000 square feet of structural and nonstructural panels.[4] Replacing stud walls with straw-bale walls and substituting a wood floor system with the slab-on-grade foundation typical of straw-bale projects can reduce lumber use up to 50 percent.

Very simply shaped (rectangle or square) single-story load-bearing buildings with rigid roof-bearing assemblies and small-dimension window bucks offer the greatest savings of wood. With more complex shapes, two-story designs, or heavy lateral loading, small-dimension post-and-beam systems offer more possibilities for wood conservation.

The construction industry consumes 46 percent of the total U.S. timber harvest. In the unlikely event that all 1.5 million new homes in the United States were built of straw, with a modest wood savings of 33 percent per house, 7.8 billion board feet of lumber and 3.5 billion square feet of panels could be saved. Simply building 5 percent of all new houses in the United States with straw bales each year (75,000) could save 380 million board feet of lumber and 170 million square feet of structural panels.

SUPERINSULATION The most striking benefits of straw-bale construction stem from its inherent insulation value. Calculated by Joe McCabe at the University of Arizona, the insulation value for a three-string bale wall is R-45 to R-57, and for a two-string bale wall, R-42 to R-43. Oakridge National Laboratories in Tennesee rated a three-string straw-bale wall at R-33. The State of California has a conservative default value of R30. In comparison, current U.S. stud wall construction standards require only R-11 or R-19.

Straw by itself is no more insulative per inch than many other materials (fiberglass, cellulose, or mineral wool), but a straw-bale wall assembly with widths of 18 to 24 inches yields an impressive superinsulated wall at a very low cost with minimal embodied energy. Straw-bale construction is a long awaited solution for sustainable, low-tech, low-cost superinsulated walls.

The final measure of any environmental control system is its ability to provide both human comfort and delight. A space with warm, radiant surfaces (thermal mass) often feels comfortable with cooler-than-normal air temperatures, and, likewise, sitting by a cool wall can take the edge off a very hot day. Plastered straw-bale walls provide both superinsulation and sufficient thermal mass, optimally distributed, yielding a stable radiant environment (see "Natural Conditioning of Buildings," Chapter 3).

ENERGY ACCOUNTING— FROM CREATION TO DESTRUCTION An average U.S. house in 1996 used 103.6 million British thermal units (Btu) of energy. Analysis suggests that sustainably produced, minimally processed materials[5] with long life spans have the fewest negative environmental impacts. Straw performs well as compared with conventional wall system materials. Straw-bale's primary processing (baling) takes place at its production site—the field—with low energy inputs, no water required, and minimal pollution. Ideally, straw can be grown locally and costs of transportation to the building site can be minimal. Straw-bale requires little installation energy or energy-intensive companion systems. It saves heating and cooling energy over the life of a building and releases no toxic materials.

⬛ ADDRESSING THE BASIC CONCERNS

Fortunately, many historic experiments with baled straw and hay—in climates as diverse as those of Arizona, Nebraska, and Alabama—were sturdy enough to survive into the late twentieth century for us to examine. These structures, along with field and laboratory testing, answer many basic questions about straw-bale building.

What makes a good bale for building?

Good building bales should be dry and well compacted with no discoloration from rot or mold. The weight of dry bales (moisture content below 20%) is 7 to 9 pounds per cubic foot, with an average of 60 pounds for a two-string (18 × 36 × 14 inches) bale and 75 to 80 pounds for a three-string (24 × 46–48 × 15–16 inches) bale. A building-quality bale should be dense enough that it will not deform when two people stand on it. The strings should be tight enough that when a bale is lifted by the strings, they allow space for no more than one finger.

What are the best ways to use bales?

Laying the bales flat is most common method, for several reasons. Having the strings of the bales in the flat orientation allows up to 6 inches of notching and curving and *nichos* (wall niches). The bales laid flat are more stable as the wall rises, and less dependent on extra stiffeners.

Setting the bales on edge may save space, but with the strings thus exposed on the surface, notching is not possible, the strings are more likely to be cut (or burned), and pinning is more difficult. Bales on edge have far less load-bearing capacity and are often used in non-load-bearing, post-and-beam infill projects.

In curvature, bales can be "bent" to create radii in plan of as little as 12 feet for two-string and 15 feet for three-string bales. Bales curve only when laid flat.

Doesn't straw burn?

Unlike loose straw, straw-bale wall systems are very fire-resistant. A plastered bale wall does not contain enough air to keep a fire going. Fire safety tests sponsored by the National Research Council of Canada showed that plastered straw-bale walls withstood temperatures up to 1,850°F for two hours before a crack appeared in the stucco. In tests conducted at the Richmond, California, Field Station in 1997 by Cassandra Adams of the University of California, Berkeley, a plastered straw-bale wall passed the American Society for Testing and Materials (ASTM) E-119 test for a one-hour wall. Similar testing in New Mexico showed only 2 inches of charring into the straw where the plaster had cracked and a temperature rise of only 12°F after two hours on the opposite side of the wall panel. Even unplastered bale wall panels are relatively fire-resistant. In New Mexico testing, burning through a panel took 34.5 minutes with temperatures over 1,550°F.

Loose or exposed straw, however, can combust quickly, so builders should take precautions to maintain a clean site, avoid heat- or spark-producing activities, protect the site against arson, and provide a water supply adequate to quench a fire should it occur. Some builders have taken the extra precaution of using a commercial fire retardant or a simple spray-on solution of borax and boric acid.[7]

It is also important to meticulously protect from water infiltration any stacked bales stored on a construction site. Not only will bales be ruined for construction use if they get wet and begin to decompose, but large masses of decaying straw also pose the hazard of spontaneous combustion.

Doesn't straw rot?

Long-term or repeated exposure to water is the greatest danger to straw-bale walls. Fortunately, the moisture content of straw must be more than 20 percent (by weight) to support fungal growth and decomposition.[8]

Other than its possible influence regarding the success and survival of historic structures in Nebraska and Huntsville, Alabama, we know little about the actual long-term effect of moisture on straw-bales.

Testing in widely diverse sites thus far shows rot problems only in sections of walls with leaks or in direct contact with foundations or soils.

Wise use of materials, good architectural detailing and building practices, and regular maintenance generally keep the moisture content of straw below the decay threshold. A watertight roof, large overhanging eaves, elevated foundations, separation from moisture-wicking concrete, adequate door and window flashing, and a skin of good weathering plaster or siding can work together to protect bales from rain and groundwater. Special care should be taken to protect horizontal bale surfaces (windowsills and the tops of walls). Drip edges on windows and walls should lead water away from the walls below. As in conventional construction, moist air from kitchens and bathrooms should be exhausted to prevent excessive interior moisture.

Moisture barriers in straw-bale walls can cause as many problems as they prevent, by trapping moisture inside walls. Most practicing straw-bale professionals advise against using sheet barriers on walls, other than draped over the bottom two courses as a splash guard. A breathable barrier is best, tucked into the joint between bales at the top and extended over the edge of the foundation for several inches. The bottom edge should never be tucked beneath the bale—this creates a "bathtub."

Moisture testing in plastered straw-bale walls suggests that maintaining the breathability of straw-bale walls may be the best insurance against rot.[9]

Isn't straw attractive to vermin?

Infestations by insects and rodents can be avoided by denying access. A coating of plaster on all exposed straw is usually adequate to keep animals out, and should they gain entrance, densely packed straw makes it difficult for them to move around. Termites in North America are adapted almost exclusively to a diet of wood and will not bother a straw wall, but they will prey on wooden window frames and structural members.

Can I get a permit to build a straw-bale building?

The acceptance of straw-bale construction by building departments is growing. Initially, all permitted straw-bale projects had to be approved under the Uniform Building Code (UBC)—Section 104, Alternate Material, wholly at the discretion of local building officials. This is still the case in most locations. Pioneers of this building method have spent countless hours educating building officials, contractors, banks, and insurers about straw-bale construction, and their job is slowly getting easier.

In 1997, Pima County, Arizona, adopted the first (and so far only) prescriptive code for load-bearing straw-bale structures. As long as a project falls within the stated prescriptions, a submission does not require the signature of an architect or engineer.

The California State Assembly passed a model code in 1995, based on the Pima County, Arizona, working code,[10] which has been adopted by many California jurisdictions. More than half of the counties in California have adopted this code or have issued permits under UBC Section 104.

In 1997, New Mexico adopted a statewide code governing the construction of post-and-beam straw-bale projects. Likewise, Austin, Texas, Boulder, Colorado, Cortez, Colorado, and Guadalupe, Arizona, have adopted various straw-bale codes for builders in their jurisdictions.

In areas without straw-bale building codes, permits can be issued at the discretion of the local building official. Meeting with local officials and educating them about straw-bale construction early in the design process can avoid many problems later (see the discussion of building codes in Chapter 2).[11]

Can I get financing and insurance for straw-bale construction?

If a project can receive a permit, it can usually get insurance and financing as well, but be prepared to educate your mortgage broker and insurance agent. Local community-based banks are often the most open to financing straw-bale homes, and some will even give a special rate for energy-efficient homes.

The generous roof overhang on this residence will provide ideal protection for its bale walls. (Photo courtesy of David Eisenberg.)

And, at the end of a building's long life, straw can be composted back into the earth.

Yet, straw-bale construction is not a panacea. As architect Ann Edminster has calculated, in the quantifiable measures of embodied energy, water use, and waste, a straw-bale house designed with no consideration of adjacent materials and systems can use 60 percent more embodied energy and 3.5 times more water than a conventional frame house.[6] But on a more optimistic note, she also found that a low-impact straw-bale house complete with a stabilized rammed earth foundation, straw-insulated thatch roof, earth-based plasters, and *liechtlehm* (light clay) interior partitions used only 7.6 percent of the embodied energy of a conventional frame house and only 21 percent as much water.

The Economics of Straw-Bale Construction

UP-FRONT COSTS: INEXPENSIVE MATERIALS, MUCH LABOR

Many are drawn to straw-bale construction by stories of incredible cost savings. Indeed, many owner-builders have built low-cost straw-bale houses ($10 to $20 per square foot) using salvaged and recycled materials and strictly their own labor. More recently, hopes of lower costs have evaporated as many straw-bale projects increasingly incorporate mainstream building practices, with building permits, contractors, and conventional mortgage financing. Straw-bale construction expenses vary widely, depending on a number of factors: geographic region, complexity of design, refinement of finishing, structural system, owner and contractor labor contributions, and permit requirements.

In areas with average labor costs, the square footage costs for contractor-built straw-bale construction are about equal to those for stud-frame construction. In areas with high labor costs, contractor-built straw-bale can

This wall raising in Guadalupe, Mexico, was a team effort. Here the box-beam header is tied down to the floor plate with metal straps that cinch-in the load-bearing bales. (Photo courtesy of David Eisenberg.)

cost 10 to 15 percent more than stud-frame. Difficulty in permitting, designing for high structural loads (seismic or snow loads), and dealing with designers and contractors unfamiliar with straw-bale construction can drive up costs even more.

As compared with conventional stud construction, straw-bale requires inexpensive materials and low-skilled labor. In wood frame construction, the material/labor cost ratio is 60/40 percent, as compared with the straw-bale ratio of 40/60 percent. Because straw-bale construction is relatively low-tech, it encourages owner involvement in building. Walls account for approximately 15 percent of the total building construction budget, so hiring a contractor for everything else, but self-building and finishing the walls, can reduce building costs by approximately 8 to 9 percent.

By any measure, a straw-bale wall is a superior value for the money. It provides the least expensive method to achieve superinsulation, thermal mass, and thickness, as compared with adobe, rammed earth, *pisé*, double-insulated masonry walls, or superinsulated double-stud construction (see Table 10.1). As more restrictive energy conservation codes are adopted and lumber becomes more expensive, straw-bale construction will become even more financially attractive.

Table 10.1

	Conventional 2 × 6 Stud Frame Wall System	Three-String Plastered Straw-Bale Wall System
Thickness	6¾ inches	26 inches
Insulation value	R-19	R-40–R-50
Thermal mass	0.5-inch gypsum board	1–2-inch plaster
Fire resistance	Unrated	2 hours
Sound transmission	Poor to average	Good to excellent
Embodied energy	913,237 kBtu	873,990 kBtu

TABLE 10.2

	Conventional contractor-built, stud-frame with R-19 insulation	Straw-bale, contractor-built	Straw-bale, partially owner-built (50% of labor)	Straw-bale, partially owner-built (75% of labor), passive solar design
Construction Cost	134,000[a]	134,000	93,000	73,700
Down payment (20 %)	26,800	26,800	18,760	14,740
Finance: 30-year fixed rate at 8%	175,969	175,969	123,179	96,783
Energy use at 1993 energy costs	30,000	18,000	18,000	4,500
Total life cycle costs	366,769	354,769	362,079	264,375
Savings over conventional construction		12,000	113,830	177,046

[a]All figures in $US dollars

LIFE CYCLE COSTS Life cycle costs measure construction, financing, and energy costs over the entire life of a building. Table 10.2 compares life cycle costs for similar 2,000-square-foot houses in a moderate climate over a 100-year building life span.

Opportunities for cost savings, through sweat equity and reduced energy costs, apply most directly to owner-builders and owner-occupants. Because these savings accrue incrementally over many years, there is currently little financial incentive for speculators to invest in straw-bale. The widespread use of straw-bale construction has not been promoted by large companies or manufacturers because there is no proprietary product to market. The primary profits do not come from building, buying, or selling a straw-bale house, but from living in it.

Design Properties Lack of hands-on experience with straw-bales can result in designs that waste wood and metal and ignore the material's design potentials. Often the bales are treated only as environmentally friendly insulation.

The rethinking demanded of us by this material is in its infancy. Many aspects of standard construction interface poorly with bale walls, but are used anyway because of familiarity rather than suitability.

Windows and doors, as currently manufactured, have low tolerances for dimensional variance and flexure along the plane of the wall. Therefore, we currently surround each unit with small (6- to 9-inch) sections of stud wall to give an adaptive buffer between the 2- or 3-inch dimensional variance of a bale wall and the ¼- to ⅛-inch variance of windows. This is not an ideal marriage, but until a high-tolerance, high-flexure window configuration is developed, we are doing what we can.

Early in the straw-bale revival there was much talk about what to use for lintels to carry the bale courses over doors and windows. We have since learned that, provided the floor or roof structure above can span the opening (which is generally the case), the weight in these few bales is so minor that a temporary lintel or prop will suffice until the bales can be fastened to the structure above with baling twine or wire; they are permanently held there when the opening is meshed and plastered.

ROUGHLY MODULAR The masonry metaphor is useful during the design of a building, because it is typical to set plan dimensions according to the unit module and wall heights according to unit coursing. This is generally done with straw-bales, but it is important to remember that balers generally tie off a bale at a signal that a certain length has been exceeded. Thus, although width and height stay relatively uniform, the bale lengths will vary 4, or even 6, inches. For this reason, it is best to keep the layout module at the long end of the variation, because taking up slack by filling gaps between bales is much simpler than shortening bales a few inches.

Because a two-string bale is theoretically 36 inches long, and laid in a running bond, an 18-inch-square module is used for layout, with courses 13 inches high. Three-string bales are theoretically 48 inches long, so a 24-inch square is used with courses generally of 15 inches. Dimensions should be verified for each batch of three-string bales, because the standard height was switched from 16 to 15 inches a few years ago and planning for the wrong height can cause grief.

The variation in length can be vexing at times, especially when the last bale in a course does not fit. Maintain your bale module when possible, but you need not be rigid. When design issues require, a few custom-length bales do not take much work.

In coursing heights, however, there is much less flexibility, because customizing bale heights is hard to do—you don't have the cleavage planes of the flakes in the bale working in your favor. When in doubt, oversize the opening and fur it in with the rough bucks.

SIMPLE WALL MASSES Bale walls behave like basketry; they are hard to break, but they flex easily. As a result, it is useful to lay out the walls with enough mass that they brace themselves.

The cottage at Shenoa (shown on page 217) was designed with a non-load-bearing 2 × 2-foot straw column between a door and a bay window. Until it was stiffened with plaster its flexibility caused the construction crew endless grief, and they decided in retrospect it would have been simpler just to box-in that wall with studs. It is best to keep a full bale between openings—a bale and a half whenever possible. This principle applies especially at corners.

From a visual aspect, this practice results in a solid massing like that seen in fieldstone or adobe walls. It is also more appropriate to the material's nature to distribute stress across a field than to concentrate it on slender elements. Keep openings small enough in relation to their surrounding walls that they are well embedded.

In some situations, such as in a passive solar face where more glass than wall is sought, it generally makes more sense to gather the openings together and create a "window wall" in wood or metal, than to pare the straw-bale fabric down to elements that are visually unstable.

THICK WALL DESIGN Remember that at 18 or 24 inches thick, a wall is no longer a membrane, but a structure of its own. The presence created with such thick walls is a major draw of straw-bale construction.

A window in a bale wall is not a hole in a panel, but a place between this and that side. These smaller places—windows—need design attention because their daylight makes them foci of the room. If you do not bevel back the side walls of a window at about 60 degrees from the interior face of the wall, for example, you create windows with the "San Quentin" (prison) look. Sunny windows with a nice view that have not received the benefit of the small extra effort to create a window seat actually frustrate people, because they are *almost* window seats.

In addition, it is important to realize the extent to which straw-bale walls separate indoors from outdoors. The close connection of interior spaces to the outdoors for which most Americans express a preference must be consciously created, or the spaces may be more insular than desired.

This sense of removal is not only visual. Bale walls are highly effective sound dampers. One of our clients reported that she had no idea a brush fire was being fought by fire engines, helicopters, and water bombers a few hundred yards from her straw-bale house, until the smoke drifted past her windows.

SOFT OUTSIDE/ DENSE INSIDE Probably the primary reason that building with bales can seem implausible at first glance is the appearance of the loose straw on the bales' exterior faces. How can something you can bend with a pass of your hand hold anything up? Indeed, loose straw does not. If you cut away 4 inches from a bale's face, you will find the tightly compressed core that bears the weight when bales are stacked. Probably the most serious mistake people have made in this work is attempting to build with bales that do not have sufficiently dense cores.

Window seat of a small straw-bale hermitage at Santa Sabina Retreat Center, north of San Francisco. Construction was largely completed during a one-week workshop sponsored by Skillful Means Builders and the architects. (Photo courtesy of the architects, Daniel Smith and Associates.)

The core of the bale is also what has to be penetrated when items are fastened to the face of the wall. Landscape pins are often used to secure stucco netting to bale walls, and 12-inch-long pins work much better than 8-inch pins. Junction boxes for electrical work are typically screwed to a 12-inch-long stake, cut from a two-by-four, driven flush to the back of a niche carved in the wall.

We have a saying : "You can nail to a bale with a nail the right scale." The higher the load you want to carry, the bigger or more frequent your "nails." Wall cabinets are generally screwed to ledgers fastened to a row of stakes in the wall. Some people barb the stakes with nails for extra security in heavier loading. For extra-heavy loads, or supersecurity, threaded rod can be run through the wall and a plywood plate secured behind a nut and washer, but we have rarely gone to this extreme.

The looser straw on the bales' surface typically gets a "haircut" with a weed whacker or hedge clipper prior to plastering. In doing this, you soon discover that you can actually carve into this wall. There is no single task that makes the difference between bale and conventional construction as

Niches can be carved into straw-bale walls and plastered. Natural light falls gently into this meditation room, which is part of a residence on the northern California coast. (Photo courtesy of the contractor, John Swearingen of Skillful Means Builders.)

tangible as carving out niches at focal points in the interior walls. Beyond fashioning simple recesses, however, lie creative possibilities for wall sculpting and bas-relief ornament prior to plastering.

UNEVEN/HIGH-TOLERANCE CONSTRUCTION

Straw bales, as mentioned before, are only roughly modular. As a consequence, the resulting walls are only roughly planar. Even if a wall is plumb overall, it is not unusual for it to bulge out and/or hollow in as much as 2 or 3 inches to either side of its theoretical face. In general, people who are drawn to straw-bale construction expect and enjoy the informality of the walls.

For the designer accustomed to the close tolerances possible in typical interiors, a language of detailing has to be created that permits greater dimensional tolerances. Baseboards cannot just be nailed flush on most straw-bale walls. There are two basic directions to take in adapting to high tolerances:

Ceramicist Sue Mullen exhibited an artist's understanding of the flowing character of straw-bale walls in building her own studio in Gila, New Mexico. (Photo courtesy of Matts Myhrman.)

1. Insert intermediary elements that true up the surface, to which low-tolerance elements will be fastened. Window and door bucks have been mentioned in this regard, and kitchen walls often have ledgers set in the walls at appropriate heights from which cabinets can be hung.

2. Select, or invent, details with the gentle irregularities of straw-bale walls in mind. For example, it is possible to straighten the surfaces around a door sufficiently for the installation of wood molding, but it is generally easier to design the frame so that the plaster relaxes into it.

Wet trades (e.g., plastering, mortaring) acquire a definite advantage with this approach, because their material is so adaptive. For example, to construct window stools 12 to 20 inches deep, which must accommodate the splay of the jambs, it is generally simpler to cast or trowel the stool in place than to make one out of wood. For a carpentry-oriented builder or designer, this approach often seems like extra work. In reality. the only difficulty is reorienting to new tasks; the work itself is actually simpler.

Anyone can work to a tolerance of an inch or two. We have seen home owners with no construction experience who are drawn into the finishing work of their homes as it becomes apparent that they will not "do it wrong." Habitat for Humanity affiliates have recognized the opportunities created by this user-friendly approach. Production builders can also appreciate the advantages of higher-tolerance finishing.

PLASTERING Although the use of earth, lime, and cement–lime plasters has increased in recent years, cement stucco has been the covering of choice for most projects. Plasters easily follow the irregularities of bales and key in well with rough bale surfaces. By its very nature, plaster leaves signs of its application and demands attention to surface. The moldable, impressionable, hand-wrought finish of plasters (even machine-applied plasters are finished by

Plaster on the meditation hermitage at Santa Sabina has an appropriate relaxed quality. Reused terra-cotta roof tiles were set on a mortar bed over a bituminous waterproofing material to form the exterior sills. (Photo courtesy of the architects, Daniel Smith and Associates.)

hand), invites us to leave our mark in texture, pattern, and sculpture. It is best to concentrate on edges and intersections. Crown molding, baseboards, and window and door surrounds can all be built up with plaster, and patterns can be pressed or tooled into still-wet plaster.

Empowering, Community Building

Because of the rise of centralized, industrialized production, most people are entirely divorced from creative handwork. For many, the only mediums of hand production left are cooking, gardening, or basic home repair. Very few people have the skills or resources to build their own homes. Those excluded from the industrialized processes of home building—most notably women and children—intuitively understand straw-bale construction, often better than building professionals. Because of its inherent variability and flexibility, building with bales does not demand specialized skills. The work also provides the camaraderie of cooperation and encourages the creativity and handworking skills of novices. Although a desire to enjoy the benefits of building a supportive community is not the primary reason people come to alternative building, this is a documented by-product and a reason people stay involved long after their houses are up. When straw-bale home construction starts with friends coming together to raise the walls, the builders leave with the primordial bond that results from creating a shelter with others.

These newly empowered builders often go on to complete their own buildings, eagerly searching out natural finishes—adobe floors, mud plasters, naturally tinted paints, fiber floor coverings. Spurred on by clients and their own curiosity, professional builders and designers also enjoy the excitement and challenge of creating solutions with a new material.

Raising the header is easy and joyful with so many working together. Guadalupe, Mexico. (Photo courtesy of Development Center for Appropriate Technology.)

Conclusion

Technically, straw-bale construction is in its infancy. It will take some years to perfect and systematize the construction process, not to mention all the structural and performance testing to be done. As designers and builders learn the language of this new material, its strengths and limitations will become apparent. Building costs will decrease and codes will be written. In the meantime, it is amazing to watch straw-bale's "grass-roots" growth and development.

Supported only by a small word-of-mouth network, straw-bale construction is increasing because it responds to the concerns and values of our times. Our world is quickly being forced into an examination of the true costs of materials and processes. From this new viewpoint, straw-bale construction is one of the "cheapest" solutions available.

Straw-bale buildings can divert straw from burning, reduce wood use, increase energy and water savings, and support local economies. Equally important, its friendly building process is inclusive, empowering, and decentralized, which helps build community. Overall, there is a sense that this construction represents a better direction, a path toward a more natural way of life.

Designing with straw-bale feels good, too. The material is flexible and can be used in many architectural configurations. Its casual form encourages creative handwork. Thick walls give designers opportunities to play with light and shadow, texture and form.

Straw-bale construction is certainly not a complete answer to the social, environmental, or design ills of our built environment, but it offers us a step in the right direction.

Notes

1. Tests conducted by SHB AGRA, Inc., Albuquerque, New Mexico, 1993.
2. *California Agriculture* 45, no. 4 (July-August 1991).
3. Ibid.
4. David B. McKeever and Robert G. Anderson, "Timber Products Used to Build U.S. Single-Family Houses in 1988," *Forest Products Journal* 42, 11–18, as mentioned in Cassandra Adams, "The Impact of Architects and Builders on Forest Sustainability," *Earthword Journal,* Issue 5 (1994), 61.
5. Ann Edminster, *Investigation of Environmental Impacts: Straw Bale Construction* (self-published, 1995), 71.
6. Ibid., 73.
7. Matts Myhrman and S. O. MacDonald, "Build It with Bales" (Tucson: Out On Bale, 1997).

8. John Daglish, *The Last Straw: The Grassroots Journal of Straw-Bale and Natural Building*, Issue 7 (1994) (online at www.strawhomes.com).

9. Kim's Canadian Study, "Practical Experience in Alberta," *The Last Straw: The Grassroots Journal of Straw-Bale and Natural Building*, Issue 8 (fall 1994) (online at www.strawhomes.com).

10. Assembly Bill 1314, passed September 1995. Now known as Health and Safety Code 18944.

11. Available through the Development Center for Appropriate Technology (see Appendix B).

Bamboo

Darrel DeBoer and Karl Bareis

Although extensive testing and construction took place with both native and imported bamboo species at Clemson University in South Carolina during the early 1950s,[1] it has only been in the last several years that the potential for bamboo construction has captured the imagination of architects and builders in the western United States. In California, Oregon, Washington, and most notably Hawaii, where bamboo is familiar to the culture and landscape, interest has accelerated because of conferences, workshops, and demonstration projects.

Bamboo is significantly faster growing than trees, especially as compared with a premium framing timber such as Douglas fir, which takes more than a hundred years to reach maturity. This giant grass can reach heights of 60 to 150 feet in a few months, and individual culms reach maturity in 3 to 6 years, although it generally takes about 10 years for cultivation of a mature grove with culms of maximum size and strength.

Because of its high strength-to-weight ratio and its flexibility, bamboo has frequently been used for structural roofing applications. It is particularly appealing, in seismic areas, as a complement to relatively nonrigid wall systems such as straw-bale.

Alternative Construction: Contemporary Natural Building Methods, edited by Lynne Elizabeth and Cassandra Adams ISBN 0-471-24951-3 © 2000 John Wiley & Sons, Inc.

The bold symmetry of a bamboo roof by architect Simón Vélez of Bogotá, Colombia, seen here from the interior looking up. (Photo courtesy of Simón Vélez.)

Otherwise, bamboo is used in a broad range of residential and commercial building applications—in wall systems, floor systems, and numerous finishes. Some bamboo buildings are quite grand, others very humble. In Colombia, South America, architect Simón Vélez has developed a modern bolt and mortar joinery system for the local 6-inch-diameter bamboo, *Guadua angustifolia*, with which he builds elegant, massive structures, such as deluxe residences, ranch buildings, and a country clubhouse. In contrast, in Costa Rica, very simple low-cost housing has been built with the use of smaller-diameter bamboo as a lath and cement as a plaster.

Country club with roof of structural bamboo in Colombia, South America. Experiments to gain the spans achieved by Simón Vélez have taken place far from any inspection. Now that these 60-foot spans and 30-foot cantilevers exist, they stand as proof of what works and as models, which may enable us to attempt only one-quarter the span and still find it adequate for most of our needs. (Photo by Simón Vélez.)

Bamboo has been the construction material of choice, if not necessity, in many Asian countries for millennia. Joinery systems in these countries include an intricate pin system that uses no materials other than the bamboo itself, and lashing systems that use locally growing fibers. These simple joinery methods, based on pegging and tying, have evolved to take advantage of the strong outside fibers of the hollow bamboo. Because bamboo culms split lengthwise very easily, they are not generally joined by nailing. In the United States there has been some isolated experimentation with plastics, glues, and metal space-frame connectors. The recent innovation of the bolt and cement method by Simón Vélez has allowed 26-foot cantilevers and 60-foot spans. This is the future for bamboo joinery.

Workshop participants in Port Townsend, Washington, learn the basics of the bolted bamboo joinery used by architect Simón Vélez. (Photo by Darrel DeBoer.)

Several varieties of bamboo suitable for construction can be grown even in temperate regions of North America. The reality, however, is that at the beginning of the twenty-first century only several thousand poles of construction-grade timber bamboo are grown each year in North America. Unless bamboo is cultivated in ample quantity, bamboo construction of any significant scale in the United States will require the unsustainable importation of culms from Asia or South America. Imported timber bamboo poles in the United States consist primarily of culled, older, ungraded, highly blemished and fumigated (upon import) material from China. What we have to work with does not at all resemble product of the Kyoto system (see the discussions on harvesting and curing in later sections of this chapter), which we may hope to emulate. As a consequence, those most passionate about bamboo building are educating themselves about its propagation and, in some cases, investing in bamboo groves.

Traditional Uses

Bamboo tools, utensils, and buildings strongly influence the lives of half the world's population. This giant grass has been classified as having more than

1,500 different uses; in the area of construction this includes fences, gates, trellises, and almost every part of a building. Bamboo housing has been traced to 3500 B.C., although it must have been the material of choice far longer.

SOUTH AMERICA In South America, structures of the native timber bamboo *Guadua* were popular with both rich and poor until several catastrophic fires in Colombian cities at the turn of this century relegated the use of bamboo exclusively to the poor.[1] Unfortunately, most of those living in these structures consider them a temporary step on the way to a more "permanent" concrete house. It is estimated that today more than 800,000 people live in bamboo structures in Guayaquil, Ecuador, alone. Similar

Simón Vélez has designed several massive bamboo shelters such as this one for horses and livestock on large ranches outside Bogotá, Colombia. (Photo courtesy of Simón Vélez.)

statistics can be found for other parts of Latin America. The work of a few—especially Simón Vélez and Marcelo Villegas in South America and Bobby Mañoso in the Philippines—is making bamboo structures quite acceptable among the wealthy once again, which is critical for ultimate acceptance by all.

In Andean Colombia, timber bamboo is used as structural posts or split open and flattened for lath, finish wall surface, ceiling surface, and flooring. Flattened poles are cheaper than round. In Colombia, only 5 to 10 percent of the poles in the bamboo yards are old enough to be structural. Most poles are sold for scaffolding and concrete form work. A system for reinforcing concrete, with *Guadua* splits woven into cables, has been developed by Oscar Hidalgo.

Population pressures have significantly reduced the supply of bamboo accessible to the public. The National Bamboo Project of Colombia has planted 2,000 hectares in the last several years in hope of bringing back the plant so critical to the establishment of human settlement.

ASIA In rural Thailand, parts of Indonesia, and many Pacific islands, bamboo is used in every aspect of life. In most of the rest of Asia—especially Japan—bamboo is more often used decoratively for walls, flooring, and the occasional roof frame.

It is worth noting that the cultures of Japan and China are attracted to bamboo's natural green color and understand the value of perfectly unblemished poles. Harvesting bamboo on the day of use to make, for exam-

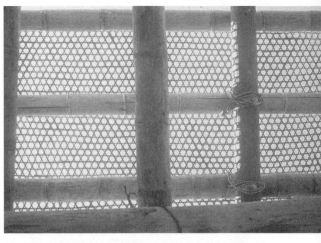

Bamboo dominates the crafts in the many parts of Asia where it is native. We have only begun to explore the possible methods we could use to construct with bamboo. Imagine being able to weave an entire building like a basket! (Photo by Simón Vélez.)

ple, sake cups and a dispenser, demonstrates great respect for a guest. The "new" green color reflects an aesthetic of renewal, and the older waxy green color, short-lived, expresses strength and special awareness of a season or annual festival. Structures of green bamboo always exude functionality, such as a booth display for bonsai plants. The roof material is also of a temporary nature; the most traditional covering—seldom used since the advent of blue polyethylene tarps—is a *Sasa palmata* bamboo leaf thatching woven onto a light bamboo frame over *Phyllostachys nigra f. henon* rafters. Larger structures can also incorporate *P. bambusoides* posts and beams with simple lash and mortise joinery. This temporary structure is light and easy to assemble and disassemble with local, replenishable materials.

Permanent structures in Japan usually rely on heavier structural materials such as wood posts and pine log beams, with bamboo incorporated as rafters and purlins for the ceiling, over which some form of thatch is applied. The model for this type of framework comes from the thatched roofs of farmhouses. Until this century, *Miscanthus japonica* thatch was the most common rural roofing material. It was easily available, and the periodic rethatching offered a chance to replace weaker rafters. What allowed this type of roof to last more than 10 years or so was the smoke from the flueless fireplaces of the traditional farmhouses. (See the later discussion on the traditional Japanese curing methods.)

The aesthetics of the Japanese country farmhouse carry over to the elegantly simple bamboo elements used in the Japanese tea ceremony. The refined utensils and architectural elements all derive from the rustic *wabi sabi*. This mood-evoking term refers to the understated design of the traditional tea room.

Much of the way a region makes use of bamboo is rooted in a tradition of isolation. In some areas, truly inventive ideas used for centuries would be rejected in other places as entirely impractical. For example, in southern China, flattened bamboo commonly spans 5 feet for flooring, and the slightly convex surface feels comfortable to the feet. Unique traditions of specially detailed doors, window lattice for both ventilation and personal expression, and the use of bamboo for inexpensive household furnishings are identifiable by region rather than species of bamboo.

Bamboo Building

CHARACTERISTICS

"To compare materials, one can look into their strength and stiffness," says Dr. Janssen, "It is even more interesting to also look into their mass per volume, and ask, 'How much strength and stiffness can I buy with 1 kilogram?' "

Bamboo is an extremely strong fiber—twice the compressive strength of concrete and roughly the same strength to weight ratio of steel in tension. The fiber strength is rendered 1.9 times stronger by virtue of the hollow, cylindrical shape. Tables 11.1, 11.2, and 11.3 compare a number of characteristics of bamboo with those of other building materials. "In the first table, bamboo ranks second," says Jules Janssen. "In the second table, it ranks first."[2]

Bamboo has roughly twice the compressive strength of steel. This bamboo roof system by Simón Vélez uses no sheet goods, such as plywood. (Photo courtesy of Simón Vélez.)

Table 11.1 Strength

Material	Stress	Mass per Volume	Ratio
Concrete	8 N/mm^2	2400	0.003
Steel	160 N/mm^2	7800	0.020
Wood	7.5 N/mm^2	600	0.013
Bamboo	10 N/mm^2	600	0.017

Table 11.2. Stiffness

Material	E-modulus[a]	Mass per Volume	Ratio
Concrete	25,000 MN/m^2	2400	10
Steel	210,000 MN/m^2	7800	27
Wood	11,000 MN/m^2	600	18
Bamboo	20,000 MN/m^2	600	33

[a]E-modulus = Young's modulus.

Source: Jules Janssen, 11/29/96

Moso is quoted as having a "Tensile strength parallel to the grain" of 1,960.6 kg/cm^2. Its average density (specific gravity, SG) is quoted as 0.61. This converts to a tensile strength of 192 Newtons per square millimeter (N/mm^2).[*] By comparison, *Dendrocalamus asper* is quoted (Surjokusumoa) as having a MOE of between 227 and 307.27 N/mm^2, between 16 and 57% stronger than *Moso*. The average density (SG) of *D. asper* is quoted at 0.7, which is 11.5% heavier than *Moso*.[**]

[*]Zou, F., in 1981 and Zou, H. M., in 1985 translated from Chinese by Lou Yiping, an associate professor working under Fu Maoyi at the Fuang Research Institute.

[**]Surjokusumoa of Bogor University, Subyakto Indonesian Institute of Sciences (LIPI), with further references by Widjaja and others.

ENGINEERING Lashed and pegged joinery has been used successfully for millennia. It allows for movement, and if natural fibers like jute, hemp, rattan, or split bamboo are used while still green, they will have a tendency to tighten around a joint. Unfortunately, in most of the world seasonal moisture changes cause bamboo to expand and contract by as much as 6 percent across its diameter,[3] causing a slackening of the joint, and not all joints remain accessible for tightening. The joint of preference has become that one developed by Simón Vélez in Colombia. He relies on a bolted connection, and because the bolt alone concentrates too much force on the wall of the bamboo, the void between nodes is filled with a solidifying mortar. This

Table 11.3. Various Characteristics: A Comparison of Bamboo with Other Building Materials

	Bamboo	Spruce	Concrete	Steel
Tensile strength	35–300 N/mm²	90 kp/mm² = 9 N/mm²	$\frac{1}{10}$ of compressive strength	250–350 N/mm²
Compressive strength	64–110 N/mm²	43 kp/mm² = 4.3 N/mm²	12.6–126 N/mm²	250–350 N/mm²
Shape characteristics	1.9 times stronger than solid because of hollow cylindrical shape. Efficient in compression and bending		Because it's solid, more weight to get same load-carrying capacity with reinforcing in tension zone	Most efficient in tension. Capable of most work in smallest cross section
Fire resistance (seconds until ignition)[4]	Acts as firebreak when growing (61.2 untreated).	(19.1)	Will not burn, but prone to spalling as steel expands	Loses elasticity, risk of quick failure
Embodied energy Btu/ft³	Minimal, unless imported	Transportation can be high.★	42–96,000[5]	91,618[6]
★★Regenerative capacity/ year[7]	80–300% (28,000–50,000 lb/acre)	3–6 % (16,000 lb./acre—pine)	None	None
Time to maturity	7–9 years	60–80 years		
Subsequent maturity after initial harvest	1 year	60–80 years		
★★★Conforms to Natural Step's four system conditions	Yes	Yes	No, fails first three	No, fails all four

★From forest to mill to lumber yard to construction site.

★★Although this is the only U.S. study that compared the yields of bamboo and pine, W. H. Hodge of the U.S. Department of Agriculture (USDA) estimated that bamboo annually produces six times as much celluosic material per acre as southern pine.

★★★1. Substances from the earth's crust must not systematically increase in the biosphere.
  2. Substances produced by society must not systematically increase in the biosphere.
  3. The physical basis for the productivity and diversity of nature must not be systematically deteriorated.
  4. There must be fair and efficient use of resources with respect to meeting human needs.

type of joint is critical where the bamboo is acting in tension, and has enabled structures with very large spans to be built.

When loads are well distributed throughout the structure, bolts can be used without mortar. An example is the traditional design of a house in Japan. A gable roof consisting of rafters spaced closely together is topped by purlins running perpendicular to the rafters. In this manner, many joints are created with relatively small loads at each.

Simón Vélez, standing in one of his roof assemblies, shows the scale of the bamboo members and roof systems. At age 50, Vélez has completed more than 60 projects using the local palette of bamboo, mangrove wood, concrete, woven palm mat lathing (under plaster), and clay roof tiles. (Photo courtesy of Simón Vélez.)

A simple vault design can consist of pieces coped to fit together purely in compression. Because gravity holds them together, no bolts are needed under gravity loading. But if this vault is placed on top of vertical walls, a horizontal collar tie must be added to prevent splaying. This piece must be bolted and mortared in place. If there is an additional lateral load, like wind, then the design needs more triangulation to hold the compression-fit pieces together.

Where members of the truss come together at angles and tension forces are anticipated, a steel strap is placed to bridge the pieces and the joints are filled. This is especially necessary for center-bearing trusses, where tensile loads can be quite large. For all of the dozens of structures built by Vélez, he says, "I have never seen the bamboo fail, only the steel straps have failed under load testing."

EARTHQUAKES AND WIND There are two different strategies for withstanding lateral forces. One—represented by the recent engineered Latin American structures—relies on the shear provided by mortar on both the bamboo-lathed walls and the roof. In April 1991, 20 houses constructed under the instruction of Jules Janssen in Costa Rica survived a 7.5 earthquake near the epicenter. The relatively lightweight bamboo structures generated far smaller lateral forces than the surrounding masonry buildings, which sustained significant damage.

Steel scaffolding is used to service construction of a bamboo roof system in Colombia—the reverse of the Asian practice of using bamboo scaffolding to work on steel buildings. (Photo courtesy of Simón Vélez.)

A structure in Colombia with bamboo roof designed by Marcelo Villegas. (Photo by Darrel DeBoer.)

The other approach depends on the flexibility of the lashed, pinned, or bolted joints traditional in both Asia and the Americas. In these locations, much anecdotal evidence shows bamboo structures surviving unscathed when the adjacent rigid concrete structures routinely fail. Even the structures built with intuitive engineering and nonoptimized joinery take great advantage of bamboo's ability to be pushed out of shape and return once the load is removed. It is very difficult to cause failure in pure compression or tension, and bending can be quite dramatic and the structure still not fail.[8] The joints are most critical.

To overcome lateral forces in roof structures, Simón Vélez and Marcelo Villegas have created the redundant second roof. They believe it is important to design structures with redundant systems, capable of both tension and compression. A joint purely in compression under gravity loading may be pulled apart in tension in the event of high winds or earthquake. Thus, a second set of rafters, which triangulates with the first, can carry those temporarily opposing forces when needed. As a bonus, this sophisticated structure is quite beautiful. Of the buildings constructed with this new type of earthquake-resistant design, none sustained damage in the January 1999 earthquake—including a 60-foot tower (a pole structure linked together with a bamboo roof frame) built almost on the epicenter, about 20 miles from the city of Armenia, where two-thirds of its concrete and brick buildings collapsed. A few very old, traditional bamboo buildings failed (those using untreated poles and known to be unsafe).

BUILDING CODE APPROVAL Throughout most of the world there is no provision in building codes for bamboo construction. A model nonprescriptive code written by Jules Janssen for the International Network on Bamboo and Rattan (INBAR) is intended for inclusion in the year 2000 International Building Code (IBC) and could lead to widespread acceptance of bamboo construction. This code helps the user to analyze beams, trusses, columns, joints, and composite materials. Janssen has also written, for INBAR, the *Standards for Testing*

🔲 CHECKLIST FOR WELL-DESIGNED BAMBOO TRUSS STRUCTURE

✤ Good, solid static analysis to distribute loads more evenly among the joints and axially along the pole.
✤ Slenderness ratio of less than 50.[9]
✤ Bolted joints with solid-filled internodes.
✤ Dry poles that are still easily workable—about 6 weeks after harvest is ideal.
✤ "Good hat and pair of boots" for the building—keep the poles out of the sun and dry.
✤ Find a way to obtain lateral strength—either through creating a shear panel consisting of a mortar bed over lath or by avoiding mortar altogether and relying on redundant triangulation within trusses to distribute the forces.
✤ Refer to the engineering formulas and testing criteria developed by Jules Janssen.[10]

Bamboo, an important work because of the many varieties of bamboo and the current difficulty of comparing test results from different sources.

In the United States, Jeffree Trudeau and David Sands of Bamboo Technologies in Hawaii are currently working on a UBC standard for bamboo and have been successful in achieving code acceptance, which puts responsibility for the design and inspection firmly in the hands of the designer. The county of Kauai has required the signature and stamp of a structural engineer or architect, along with special inspection (UBC Section 306.a.14) to obtain approval. These authors have also suggested that a prescriptive code be developed by the local design community to "provide uniformity in submittals."

As the United States is expected to adopt the International Building Code, Janssen's model code will likely be applicable here, and the step-by-

This modest bamboo house, also designed by Vélez, uses bamboo for walls as well as the roof. (Photo courtesy of Simón Vélez.)

An experimental bamboo truss structure built recently in Hawaii. Interest in bamboo has taken a foothold in Hawaii, where it already exists as part of the culture. Many tropical varieties can be grown here. (Photo by Darrel DeBoer.)

step prescriptive code will also be useful to facilitate approvals in jurisdictions with little experience in structural bamboo.

Cultivation

Think of bamboo differently. The strength of a bamboo plant has less to do with the visible portion above ground and much more to do with how well the rhizome below ground has been storing energy throughout the year. Once a new shoot appears, it will never get any larger in diameter. As the plant becomes established, subsequent years' shoots are larger than the "parent" culms.

The fiber lengths vary across the culm wall: shorter in the interior and exterior than in the center of the wall. The shortest fibers are near the nodes, where cracking is likely to first appear.[11] Fiber strength is significantly greater at the exterior. In sharp contrast to the way wood grows in trees—from the cambium layer outward to form bark and inward to form annular rings—bamboo, as a grass, grows by stretching. The structural and design implications of that cellular difference have yet to be fully explored.

SPECIES SELECTION

For structural use, several species stand out and lend themselves to specific climatic selection. Although wind should be avoided and water ever-present, the most significant factor in site and species selection is temperature. Several large, strong temperate *Phyllostachys* running bamboos can withstand 0 degrees Fahrenheit and appreciate hot, humid summers: *P. bambusoides, P. heterocycla pubescens (moso),* and *P. nigra henon.* One tropical clumping variety spans the range between the temperate runners and the tropical clumping types: *Bambusa oldhamii,* hardy to 15 degrees Fahrenheit

Giant, tropical clumping bamboo in southwest China. Bamboo craft begins in the bamboo grove. The best groves are well watered and sheltered from winds and thus produce tall, straight poles. Knowledge of yearly harvesting cycles has been refined to the point that the raw material has a high natural resistance to insect and fungal problems. (Photo by Darrel DeBoer.)

(although its new shoots occur in November and December, so even though the plant may not be killed, it can be set back a year with a hard freeze).

The tropical clumping varieties, like the other *Bambusas, Guadua,* and the strong *Dendrocalamus* species (*D. brandesii* and *D. asper*), barely tolerate freezing. Extensive research with interplanting between trees, avoiding frost pockets, and using mineral salt fertilizers is being carried out in Australia.[12] Using these strategies, the researchers have observed tropical bamboos resisting temperatures 9 degrees colder than the species was previously seen to tolerate. Unfortunately, in the Northern Hemisphere, the vulnerable shooting season is the coldest time of year.

Daphne Lewis of Seattle has detailed species selection for growers in the U.S. Pacific Northwest, covering the *Phyllostachys* varieties related to size, hardiness, quality of poles and edible shoots, forage quality, and flowering dates.[13]

USE OF SMALLER POLES In this country, by far the most common variety is *Phyllostachys aurea*, or Golden Bamboo. Next is *Phyllostachys nigra*, or Black Bamboo. These two are readily available, but rarely more than 15 feet in useful length or more than 1 inch in diameter. The fibers are strong, though somewhat prone to splitting. A great challenge is to find use for these common varieties. One solution is to make structural members out of split bamboo, which offers the further advantage of being able to form curves.

HARVESTS AND GROVE MAINTENANCE Propagation of small plants is labor-intensive and slow, especially for the running bamboos. So, the price of plants in the United States will probably remain at a premium. Planting a grove of timber bamboo in loose soil like sand and mulch allows for easier propagation and digging of rhizomes. Bamboo needs a moist, well-aerated soil in order to spread rapidly and produce high-quality poles. Bamboo groves can be a high-value use for wet, nonsalty soil, but only if the rhizomes still have access to oxygen. Water-tolerant bamboos—*P. atrovaginata, P. heteroclada, P. heteroclada* solid stem, *P. nidularia,* and *Arundinaria tecta*—all have air canals in the roots and rhizomes as an adaptation for wet soils.

The major work season in a grove should be scheduled for the plants' dormant months. Ideally, poles should be harvested about six weeks before use, as they will shrink to their expected size during this waiting time, while maintaining the working characteristics of green bamboo. In early spring, walking in the grove should be avoided once the shoots are expected. Several varieties, notably *Moso*, push moisture ahead of the new shoot, so a very clear wet spot can be seen on the ground just before the arrival of the new culm.

New plantings of bamboo are extremely fragile, as the rhizomes may have just been cut—somewhat like an umbilical cord. Until the canopy grows high enough to shade the ground, which usually takes several years, the new plants are subject to quickly drying out, and can die quickly if water is forgotten. A mature grove will spread in a radius equal to its height. Thus, a 60-foot-high *Phyllostachys henon* grove will expand to a 120-foot diameter and find stasis and comfort in the microclimate it creates for itself.

Once the grove is mature, culms selected for strength and durability should be in their third to fifth season of growth and harvested when sugar levels are at their lowest. In the tropics of South America and Asia, this corresponds to the dry season. On the West Coast of the United States the best time for harvesting the running species may be late spring, once shooting has ended (see Table 11.4).

Bamboo Preservation and Curing

There is a vast amount of stored knowledge in traditional bamboo cultures—a living school kept by craftsmen following methods that have proven effective over generations. In one of the most rigorous systems, kept alive today in Kyoto, each species of bamboo requires a slightly different harvest and curing procedure, with key seasonal factors dictating once-yearly maintenance to clean up groves and produce useful by-products.

Table 11.4. Bamboo Harvest Seasons in Western United States

	Running Bamboos	Clumping Bamboos
Edible shoots	March–June	October–December
Rhizome divisions	Just before shoots begin	January–February
Plant divisions	Avoid heat of summer	January–February
Leaves as animal feed	Any time, but especially after pole harvest	Any time, but especially after pole harvest
3 to 5-year-old poles	April–June	January–February

Strategies to preserve bamboo culms for light construction are grouped in two categories. The first includes methods to cultivate strong, healthy plants; this requires a clear grasp of local growing conditions that can be manipulated to give optimum results. The second category is methods of drying. Because there are no radial fibers in bamboo, the hard outer layer does not transpirc and, if dried too quickly or unevenly, will split because of its watertightness. Three to four months of air drying is sufficient to slow-cure the poles. These months of curing should be cool, the area well ventilated and free of organic material that could harbor pests. In regions where bamboo thrives, winter temperatures seldom stay below freezing long enough to create the superdrying effect found in Europe and the eastern United States. Most important, the culms should be stored horizontally, out of direct sunlight, to prevent uneven curing.

LONGEVITY   Much of the information on the longevity of bamboo has to do with local conditions: the more rain and humidity—the more tropical the climate—the more quickly bamboo will return to compost. Some species of bamboo also have a higher starch content that makes them vulnerable to earlier attack by fungi and insects. In tropical rain forests, bamboo houses without preservative treatment are expected to last only 3 to 5 years, rarely longer than 15 years. The whole structure is rebuilt in a matter of days as part of an accepted rhythm.

On the other hand, in Bogotá, Colombia, one of the stronger, longer-lasting species, *Guadua angustifolia,* was used untreated for posts, beams, flooring, roof structure, and lathing for plaster walls in houses that are now more than 90 years old. This is possible only with the use of culms that are at least 3 years old when cut. Plaster on the ceiling and walls also protects the bamboo from beetle damage.[14] In Japan, running bamboos that grow in temperate climates have been seen to last hundreds of years when used as ceiling structures that have received the smoke from unvented cooking fires (see the following section on curing).

Cracking in a bamboo pole can be the first step toward breakdown, as insects are then given access and the strength of the tube is lessened. Exposure to the sun and dramatic changes in humidity combine to increase the frequency of cracking. Both the interior and exterior shell of a pole are watertight, and the appearance of cracks on a hot, dry summer day is accompanied by a loud popping sound as the pressure finally equalizes in the pole. Some timber species, like *P. nigra bory,* seem predisposed to cracking more than others, but the splits can be minimized by both avoiding the sun and keeping the humidity high, or at least constant. For fences and gates with minimal structural loads, the common technique is to punch out the nodes with a steel rebar or to make an incision along an unseen side to encourage splitting to happen there.

TRADITIONAL JAPANESE CURING METHOD

In selecting bamboo as an interior ceiling frame it is wise to keep the members well away from destructive moisture. The bamboo ceiling frame in traditional Japanese farmhouses lasted several hundred years and was highly resistant to insects. The smoke wafting up through the ceiling from open fireplaces coated the entire surface with a creosote-like film, which slowly dried and preserved the bamboo framing. Traditional untreated bamboo structures that lack smoke curing are said to last for only 10 years before the bamboo is infested by powder post beetles.

Thus, an abbreviated curing process was developed that takes only one to two weeks. First the bamboo poles are selected from those that have air-dried for several months and have begun to lose their green color. Then, according to ultimate use, they are matched in size and evenness of internodal spacing, determined by the culm's location within the grove. Near the center of a mature grove, the culms have fewer branches and must stretch to reach the available sunlight. This characteristic also creates very evenly spaced nodes without branches for several meters from the ground, and these poles have somewhat thinner walls, which makes them dry uniformly from bottom to top.

After careful selective grading, the bamboo is transported to a special curing yard set up with peeled *Cryptomeria japonica* drying racks. At this point the partially aged poles are pale green in color. Apprentices are set to work leaning the poles vertically in the sun—the season is usually dry but without intense sunlight or high winds. During the secondary curing the poles are turned constantly—each pole must be turned two or three times a day to keep sunlight from splitting the culm.

After having been spun approximately 20 times, the poles are ready for the final process of quickly smoking the poles to preserve them with the creosote naturally contained in the smoke. Charcoal pits are constructed at ground level, 2 to 3 feet wide by the length of the poles—12 to 15 feet. Each pole is suspended over the pit by two workers, one standing at each end of the pit, who slowly spin the culm between them until the light green color begins to vanish. This final process takes less than five minutes.

It is important that a pole has received enough time in the yard prior to this final processing. If the culm has been dried too quickly, the baking cure will result in a blotchy finish. The Japanese expression for the final curing is *roketsu-dosu*, which refers to removal of the exterior culm wax that naturally occurs on the outer surface of bamboo. The wax is quickly wiped off while the pole is still piping hot, leaving a highly polished surface. While still warm, the poles are flexible and can be straightened. As soon as the poles have been wiped and cooled in a well-ventilated rack, they are stored upright in tidy storage barns, ready for use. At this point the color is an even tan, which will very slowly develop a patina over time.

OTHER CURING METHODS A method of boiling bamboo culms in caustic ash solution has recently been used in Japan. After the drying process described previously has been followed for three months, the culms are boiled for 25 minutes. This method is not as time-consuming nor expensive, but does not leave a satisfactory polish on the culms, and the finished product brings a lower price. In Costa Rica culms are boiled in caustic soda to remove the wax so their surface will darken evenly during flame finishing.

Insects are most interested in the starchy fibers closest to the interior of the culm. But neither the smoking nor the boiling method always reaches those fibers. Holes can be drilled into the internodes to allow some penetration, but because there are no radial fibers in bamboo, full treatment is not always possible, depending on species. In tropical regions especially, it is most advisable to treat the poles with a pressure-injection system, which fills the capillaries with solution, or to use another method that fills and coats the interior walls.

PRESERVATION WITH BORATE SOLUTION The "Boucherie" method was pioneered by Walter Liese of Hamburg University. This technique replaces the starchy sap of the just-harvested bamboo with 1 kilogram boric acid (1 kilogram boron to 50 liters water)[*] solution to make the cellulose and lignin fibers indigestible to insects and microbes. By far the most inexpensive material to use in solution is the disodium octaborate tetrahydrate derived from a fertilizer, like Solubor. The culms selected must be mature, and the machinery must be near enough for them to be processed in the first day after harvest while the sap remains fluid and capable of capillary action.

---

[*] Based on field trials in Columbia, Costa Rica, Indonesia, and Vietnam, all by different investigators.

Pressure-feeding bamboo culms with preservatives immediately after harvest. Costa Rica. (Photo by Darrel DeBoer.)

The dye-marked solution is pressure-fed into one end of a pole until it is seen at the far end of the pole. Borates are used as fire preventatives in various insulations, but, to our knowledge, no testing has been done on the fire-resistance of bamboo poles treated with borates. Information on this method has been disseminated through the INBAR bamboo research network in India. A simplified method—more effective at reaching the sugar-laden inner wall fibers—has been developed by the Environmental Bamboo Foundation of Indonesia. This approach consists of drilling holes through all the nodes in a pole except the bottom one, and filling the culm with the same borate solution. This method may not be effective on all species, so localized tests are in order.

ENVIRONMENTAL CAUTIONS Most bamboo used in construction is untreated. Some of the more toxic treatment systems, relying on creosote, arsenic, and chromium, can be avoided by using more benign solutions, such as borates, or smoke, as discussed in the preceding sections. The advantage of borate is that it is already used as a fertilizer, and thus is compostable. The only major health concern is that it can be injurious if it gets in a person's eyes.

By far the most common glue used in composites, such as flooring, oriented strand board, and woven mats, is urea formaldehyde, which is highly sensitizing, is generally accepted as carcinogenic, and should be avoided.

⊠ CHECKLIST FOR OBTAINING CONSTRUCTION-QUALITY POLES

✤ Age—3- to 5-year-old culms are best, depending on species.
✤ Starch content—harvested at right time of year to minimize beetle/fungus attack.
✤ Appropriate species for the intended use.[12]
✤ Sufficiently adapted to local humidity—especially for interior use.
✤ Stored out of direct sun.
✤ In the running bamboos, use the bottom 5 feet or so for other purposes, as it is usually crooked, has nodes too close together, and its density characteristics are different from those of rest of the pole.
✤ Treated for insects and fungus if used outside.

Significantly better are the polyurethane, isocyanate, and aliphatic "carpenter" glues, which are nonreactive after a day. The realization of a sustainable future is brought closer with the recent release of soy-based glues.

Notes

1. H. E. Glenn, *Seasoning, Preservative and Water-Repellent Treatment and Physical Property Studies of Bamboo* (Clemson, S.C.: Clemson Agricultural College, 1956).

2. Jules Janssen, *Building with Bamboo* (London: Intermediate Technology Publications, 1995). Jules Janssen, *Bamboo: A Grower and Builder's Reference Manual* (Keaau, HI: HCABS, 1997). (HCABS, P.O. Box 1390, Keaau, HI 96749.)

3. Klaus Dunkelberg, *IL31: Bamboo as a Building Material*, Institute for Lightweight Structures (Stuttgart: Karl Kramer Verlag, 1985).

4. Ibid.

5. American Institute of Architects, *AIA Environmental Resource Guide* (New York: John Wiley & Sons, 1997).

6. Ibid.

7. Walter Liese, Oscar Hidalgo, Gib Cooper, et al., *Bamboo and Its Use—International Symposium on Industrial Use of Bamboo* (Beijing: International Tropical Timber Organization, 1992).

8. Janssen, *Bamboo.*

9. Oscar Arce-Villalobos, *Bouwstenen 24: Fundamentals of the Design of Bamboo Structures* (The Hague: Eindhoven University, 1993).

10. Janssen, *Bamboo.*

11. Liese et al., *Bamboo and Its Use.*

12. Victor Cusack, *Bamboo Rediscovered: Growing and Using Non-Invasive, Clumping Bamboo* (Trentham, Victoria, Australia: Earth Garden Books, 1997). (Available in the U.S. through Chelsea Green Publishing Company, P. O. Box 428, White River Junction, Vermont 05001.)

Shopping for bamboo poles. In countries where bamboo is part of the culture, such as Japan, the client buying the bamboo is usually a craftsman, and the whole bamboo industry is supported by the continued demand for highest quality. (Photo by Darrel DeBoer.)

13. Daphne Lewis, *Hardy Bamboos for Shoots and Poles* (self-published, 1998; bambuguru@earthlink.net).

14. Oscar Hidalgo, *Manual de Construcción con Bambú* (Bogotá: Estudios Técnicos Colombianos, 1981).

12
Earthen Finishes

Earth Plasters

Carole Crews

Mud, the combination of earth and water, is almost too basic and elemental a material to discuss. For as long as humans have had hands, we have been playing with mud. Judeo-Christian mythology even claims that God created man from clay. The fascinating thing about mud is that it lies somewhere between a liquid and a solid, combining the best properties of each. It is the oldest building material on earth and continues to be used by people who live ecologically viable lives throughout the world. Humans have always been clever enough to use what is available to them and to experiment by combining ingredients and trying out new techniques. Unfortunately, manufactured goods are becoming more available than the dirt under our feet, as that becomes increasingly covered over with concrete and pavement. Natural builders can feel good about reviving the use of Mother Earth's gifts and we can give thanks by using them in beautiful ways that embellish the lives and spirits of all.

Alternative Construction: Contemporary Natural Building Methods, edited by Lynne Elizabeth and Cassandra Adams ISBN 0-471-24951-3 © 2000 John Wiley & Sons, Inc.

Exterior of Carole Crews' home outside Taos, New Mexico, with relief dragons sculpted by Carolee Pelos, showing the pigmented finish. (Photo by Carole Crews.)

The integrity of a natural building is maintained most successfully by using ingredients in the plasters that are compatible with the wall materials. This lesson was learned at the famous San Francisco de Asis Church in Ranchos de Taos when it was coated with cement plaster in 1967. The plaster cracked, allowing moisture to go deep into the adobes and stay there, and it was unable to dry out again. Large sections of the buttresses had to be rebuilt, and the community has now gone back to annual renewal of the mud plaster, which not only keeps the church building in beautiful condition but strengthens neighborhood ties as well. Most earth-plastered buildings do not need such constant attention, but the parishioners are fond of a particular color of local earth called *tierra ballita*, which contains a lot of silt, making the plaster vulnerable to weathering. This, plus the fact that there are protruding buttresses and no overhanging roofs, means that ongoing care is necessary.

When most people think of plasters made from earth, their mental image is of brown, dusty walls in some poor foreign country. Yet this is not a fair picture because the properties of clay itself, one of the most abundant components of the earth's crust, can be rich and varied. The particle size of clay is extremely small, and the molecules are shaped like plates. Clay comes in a wide variety of colors, ranging from white kaolin to the lavender of the painted desert, reds containing iron, grey ball clay with a higher percentage of organic matter, desert browns and pale shades of peach, French green, dark maroon, yellow, and more. If these beautiful colors are not available locally, pigments can be added to white kaolin to achieve any tone imaginable. The surface can be burnished to a smooth dust-free luster, especially with the addition of fine mica.

In an earth plaster, clay coats the particles of sand, silt, and straw and binds them together. Clay feels slippery to the touch when mixed with a little water and will cause your fingers to stick together slightly when pinched. If your earth does not stick together when wet, it does not contain enough clay to be suitable for building or plastering, so you will have to add some from elsewhere. A shake test can be done to determine proportions of sand, silt, and clay by filling a jar one-third full of sifted soil, adding water nearly to the top, then shaking well. A teaspoon of salt can be added to speed settling, which will leave the heavier layer of sand at the bottom, silt in the middle, and clay on top. If you have at least 20 percent clay, your soil will most likely work for a plaster. It is usual to start with a coarse plaster and progress to a finer texture with each layer, so the fineness at which you start plastering depends on the smoothness of the bare wall you are faced with.

An easy way to mix mud plaster is to fill a wheelbarrow or mud boat with water one-third to one-half full. Sift the soil into the water until the surface is covered, and leave it for at least five minutes to be absorbed through capillary action. Then add straw, which will help to stiffen the mix and give it body. It is not usually necessary to add sand to a sticky base coat, but it is used for brown coat and finish plasters.

Always try out an earth plaster with which you are unfamiliar by making a patch test of several square feet. Clays differ in their shrinkage rates, and if the plaster cracks too much, add sand and more straw. Some plasters dry with cracks—like those found on the bottom of a dry lake bed—and fall off the wall. This happens more often when the plaster is troweled, because there is less surface area to release moisture than when it is left rough. If the dry plaster is very soft when you scratch it, it has too much silt or sand and needs more clay, manure, and/or flour paste, additions that will also improve a long-straw plaster for use on the outside of a building.

In India, the Middle East, and elsewhere, it was discovered long ago that old stinky mud works better than fresh mud, so the mud is mixed in a pit with animal manure and other wastes and left to "ripen." This process causes the molecules of clay to line up as closely as they can, improving wet plasticity and dry toughness. The same principle is well known in pottery making. Rotting straw is not so good, however, so when using this method, add the straw just before the mud is to be used. In Africa and India, fresh dung is added to clay to make a plaster called *litema*.

When plastering straw-bale walls, it is easiest to spread a thin layer of mud, with high clay content and no additional sand or straw, on the bales first to lock into the straw and provide a surface to which the next layer can adhere. If you use this technique on tight bales, you can avoid using stucco netting. Even when chicken wire or lath is used, the smooth mud will penetrate the metal and leave no air spaces to cause future cracks. You do not have to wait for this to dry before adding a thicker layer of plaster containing straw.

On rough cob or adobe or for the next layer on straw bales, use a plaster with a high clay content and as much straw as can be mixed in to achieve a slippery, easily spreadable consistency. This can be applied with rubber-gloved hands to a dampened wall and is very good for filling in depressions and adding decorative shapes or bas-relief sculptures. You might add extra chopped straw and a binder such as wheat paste to make a stiffer sculpting mix. The condition of the walls and how much shaping they need will determine whether to use long straw, chopped straw, or a combination. The thicker the layer must be, the more long straw it should have. Do not trowel this layer down smooth, but get it as flat as you can with your hands and let it dry out thoroughly. It will produce lots of little cracks and provide a perfect surface for the next layer to adhere to.

When plastering the outside of a building, keep in mind the following characteristics of the plaster components. The sand, when magnified, resembles little boulders stuck in the side of a cliff, which have a tendency to weather off unless sealed in by a powerful binder. The clay swells and resists moisture penetration beyond a certain point. When a thick coat of clayey mud with lots of straw and wheat paste is applied, the outer layer of clay will eventually be washed off, exposing some of the straw and creating a shaggy surface that will protect the wall by deflecting rain. If a smoother finish is preferred, the traditional way to apply a high-clay-content plaster is in thin layers, to prevent cracking.

The basic "brown coat" interior plaster (one that would go over the coarse straw-filled layer on straw-bale walls, and typically beneath a fine interior finish plaster) can be applied directly onto evenly stacked adobe walls, well-trimmed cob, rammed earth, pumice-crete, or light-clay walls. It is made by sifting the appropriate soil through quarter-inch or even win-

Exterior angel relief on free-standing wall is the work of Lori Lawyer and Carole Crews. (Photo by Carole Crews.)

dow-screen-size mesh into the water-filled mixing container. To an earth containing approximately half clay, add an equal amount of plaster sand and enough chopped straw to produce a workable consistency that does not crack when dry. This material should be sticky enough to adhere well to the dampened wall and wet enough to trowel on easily, but not so wet that it is hard to pick up. Always work with a test area to determine the proper proportions for your particular ingredients. Robert Laporte recommends adding a small amount of cooking oil to the mixture to make it slide on more smoothly. If you have gathered a purer clay that is in lumps, it is best to soak it and make a slip. Add sand to the thick clay slip until it is the right consistency—up to 70 percent, depending on the clay—and then add the straw to it. The more finely sifted the ingredients, the finer the plaster will be.

It is most effective to apply this coat of plaster with a trowel. Trowels with curved corners are good for the bulk of the wall, and pointed ones can be used for edges. Japanese-style trowels combine a pointed top with straight sides and work very well. Disks of plastic cut from the tops of coffee cans or yogurt containers are useful tools for going around curves and "bullnosing" around windows and doors. Many people like to hold the plaster in a hawk, deftly scooping some up with the trowel and sliding it onto the wall (for finish plasters this is definitely the best way), but if this technique proves to be difficult with a brown coat plaster and the hawk

Pigmented relief by Peggy Mabry along an interior stairway. (Photo by Carole Crews.)

weighs too heavily on your arm and shoulder, work from a bucket of mud instead. Cover about a square yard at a time without worrying unduly about smooth perfection, then go on to the next square yard. When the second area is done, go back to the first area, which has had time to "set up" a bit, and smooth it out with a damp tile sponge or a float. A sponge can leave a perfect, fine-grained "tooth" for a clay slip finish coat to bind with. If you want to leave the plaster without a clay slip, you can trowel it hard and smooth at this point.

A very fine finish plaster can also be made without straw, to be applied in a thin layer to a wall that is quite flat already but has a texture or dark color to be covered up. If clay is hard for you to find, you can purchase it at a pottery supply store. White kaolin, used for slips, can also be made into plasters. If you are less concerned about color, planning to put a slip over the surface anyway, ball clay of a gray color is very good in plaster because it has greater plasticity and dries somewhat harder than kaolin. Either mix the fine dry sand (70 percent) and clay (30 percent) together first and then add water, or mix the clay and water into a slip first and then add the sand. The splash of oil added for workability is optional. Proportions may vary somewhat, depending on the sand. The level of perfection you wish to achieve, as well as the surface you are starting with, will determine the number of coats of plaster necessary. One-coat plasters of the kind described in the preceding paragraph are possible over well-shaped cob, adobe, or light-clay using finely sifted ingredients and chopped straw in a half-inch-thick coating.

Clay slip (*aliz*) is most successfully made from white kaolin, ground mica, and a binder. A small amount of fine sand is often used, especially in the first coat to make it thicker and fill in any small irregularities in the plaster surface. Use cooked flour paste as a binder in the proportion of 20 to 25 percent of the liquid. To cook the flour paste, boil one quart of water for every cup of flour to be used. In a mixing bowl, whisk together somewhat more than one cup of cold water per cup of cheap white flour to make a mixture the consistency of pancake batter. When the pot of water comes to a rolling boil, pour in the flour and water mixture and stir it well with a whisk. The paste should thicken immediately and become somewhat translucent. Do not continue cooking it or it will scorch. Rye flour paste is often recommended as an addition to exterior plasters. Instant flour paste can be purchased as wallpaper paste, but tends to lump when mixed with water. It is quite useful for adding to plaster mixes, however; a box will be enough for two wheelbarrow loads. Milk products also make good binders, buttermilk being best and requiring half the amount of wheat paste in the *aliz* recipe. White glue has been used to strengthen mud too, but it makes the material stiffer and less workable.

To mix the *aliz* you will need a container at least as large as a 5-gallon bucket and a big whisk or a paint-mixing attachment on the end of a drill. Start with three parts water to one part cooked flour paste in the bucket, to approximately three-fifths full. Use a saucepan, coffee can, or other container for a scoop and start adding the ingredients proportionally. Recipes vary according to the surface to be finished and whether or not people like mica. Generally, however, use three scoops of white kaolin, two scoops of ground mica, and one scoop (more or less) of fine sand. Sand is used mainly for the first coat or if mica is not available. Second coats are best with half clay and half mica. If mica is not available, use up to 40 percent very fine sand in the first coat, but less in the second. Be careful not to breathe in the fine particles of dust and mica, or wear a dust mask while mixing. Keep adding these ingredients until the mixture is the same thickness as unwhipped heavy cream. You may have to add a bit more water to achieve this consistency. Sodium silicate, an ingredient used in slip casting to keep the particles of clay afloat in the water, is useful in this context as well. A very small amount is required. It will also thicken the mixture somewhat, as will powdered milk, which also makes the final product a little tougher. Thin washes of kaolin clay with a small amount of binder may be applied

Framed earthen relief art. Notice imbedded seashells and mica glimmering in wall behind. (Art and photo by Carole Crews.)

without sand or mica and need not be polished. If this finish cracks, you are applying it too thickly. Colored clays or pigments may be added to create different colors. The colored clays would replace kaolin in the recipe, and pigments should be soaked in water, if not actually ground, to prevent color spots from showing up. The earth pigments, some in liquid form, used to color concrete and grout are quite suitable for our purposes too. If mold may be a problem, add a little dissolved borax powder. Larger chips of mica are a beautiful addition to the mix if they are available, as are bits of chopped straw, seed husks, and other decorative material.

Ground mica is inexpensive and obtainable through industrial mineral suppliers, but shipping costs are high. The V-115, the largest grain currently available, gives a visible sparkle to a surface. The 1117, the next largest grade, has a more subtle sheen and does more for the texture of the mixture. Mica 200 is finely powdered and offers no sparkle at all, but improves the texture and thickens the mix. If it is easily available, you might try mica, as it is a lovely addition to finish plasters as well as clay slips and adds to their workability. Mica is like a molecularly flat sand, smooth instead of gritty. Mica is also used as a lubricant when oil wells are drilled and is a major ingredient in joint compound.

You will need a few tools to apply the *aliz* to your wall: a 3- to 5-inch-wide brush, a 1-inch brush for edges (natural bristles are best), two small buckets, and a fine-grained tile sponge. Sheepskins also work, but I find them slippery to hold onto, and the coat of slip they deliver is not as evenly applied. Do not moisten the wall first. Make sure it is completely dry, because damp plaster will leave water stains. Cover the floor with drop cloths, and start brushing at the top of the wall so that you do not mar your fresh work with drips. If the wood of the window edges and lintels is rough, tape it first to save yourself cleaning work. If the wood is smooth and painted, the slip can be wiped off easily while it is still wet. Use the small brush for edges. To fill cracks that often form along edges when thicker plaster dries, mix some of the slip with extra sand in a small container and apply it to the edges with a palette knife. Most walls require two coats. Make sure that the first one is completely dry before applying the second. When the second coat starts to dry and look mottled, becoming "leather hard," use a sponge and a bucket of warm water to polish the surface; use circular strokes with the squeezed-out sponge. Rinse and squeeze out the sponge often so it will cleanly polish the mica flakes and bits of straw (if you have chosen to put them in). Polishing smoothes out the brush strokes and gives

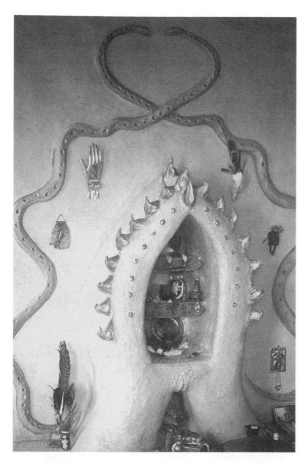

*The home altar of Carollee Pelos, close friend to Carole Crews and aficionada of earth architecture. Carollee recently died, but her book, **Spectacular Vernacular**, which she produced with her husband Jean-Louis Bourgeois, remains a great inspiration (see "Books" in Appendix A). (Finish plaster and photo by Carole Crews.)*

the surface a finer texture. You can dry polish it again later with a rag to remove any last grains of sand and bring the mica to a greater sheen. Wear rubber gloves for the wet work. When you are finished with your job, save the leftovers by drying them into "cookies." These may be stored, then soaked and used to repair any small damages.

When a tougher, more water-resistant finish is desired, casein washes are a good solution. Additional color may also be applied this way. Natural paint specialist Reto Messmer has shared this recipe for basic casein paint medium. To one liter of warm milk, add 2 teaspoons of sour cream (or yogurt). When the mixture thickens (usually overnight in a warm place), heat it gently until the curds separate from the whey. Pour it through a cheesecloth-lined colander. You may wish to save the whey for baking bread. The curds, called *quark*, are mixed with 1 tablespoon of borax dissolved in ¾ cup warm water. If possible, use a blender to make a smooth mixture. The casein will keep for about two weeks refrigerated. Thin it out with three to seven parts water, add a bit of pigment, and brush onto the dry *aliz*. It will give a glossier finish and deepen the color somewhat. However, it is best left off if you want pure white. Casein paint may also be made from the undiluted base by adding pigment and an extender such as whiting, kaolin, or chalk.

In most discussions of plasters, the question of sealers often arises. Commercial products such as water sealants and clear acrylic sealers peel off in a short time. A petroleum-based "adobe protector" will work for a while if hail does not hit the wall and nothing ever brushes against it, but it has an odor and is expensive. Asphalteum is a time-tested additive that does keep out water, but it has a tendency to stain darkly and it is difficult to get anything else to stick to it. Linseed oil darkens the mud and eventually breaks down in the sun, but can be replenished annually and is certainly good for floors. If you live where prickly pear cactus grows and are willing to go to the trouble to prepare the juice, you may try this treatment; it is said to work well. Lime washes and plasters are traditional means of protecting against weather, and in Afghanistan and Australia sheep fat is mixed with these sealants to cover domes. Coats of flour paste, especially with mica added, offer some protection, as do casein washes. The best solution is to learn to enjoy the process of occasionally renewing your home's surfaces and to be rewarded by their elemental beauty.

Earth Floors

Bill and Athena Steen

Little known in the modern world, earthen floors are a perfect complement to a home built of natural materials. They are comfortable, soft underfoot, and lack the coldness commonly associated with concrete floors. It is not an exaggeration to say that they are warm in winter and cool in summer. When finished and sealed, they often have the luster of worn leather.

Like so many aspects of earthen construction, mud floors were once reserved for the poor, who could not afford wooden floors. In generations past, the routine was to tamp down the earth inside a finished dwelling and treat the surface with animal blood, typically of sheep or oxen, to help it resist abrasion and moisture.

Today earthen floors are becoming fashionable in many a custom home in the southwestern United States. The once traditional blood-finished floor has been replaced by mixtures of aggregate, clay soil, straw, and oil sealants. Moreover, earthen floors are by no means limited to the dry Southwest. They were once common throughout Europe, where the climate is wet and cold, and they can yield a hard, durable finish that is sufficiently strong for most domestic purposes.

Earthen floors are not limited to residential use. Here poured adobe spans the large spaces of a church interior. (Photo courtesy of Bill and Athena Steen.)

If construction of an earth floor is undertaken without hiring additional help, it can be very inexpensive; typically, the only cost is for the oil and solvents used to seal the floor. Simpler floors can also be constructed with no cost at all. Traditional floors in many homes were often no more than compacted soil that was regularly sprinkled with water and then swept. Yet floors of this kind are by no means limited to the homes of poor peasants, as it has become fashionable in Japan to install this simple flooring in some rooms of very expensive homes that have conventional furnishings.

One of the most critical factors in installing a mud floor is to make sure that the substrate is free of moisture and movement. In many parts of the Southwest, earthen floors are constructed right on top of the existing ground. However, if moisture and ground movement are problematic, the substrate should be improved to create a dry and solid base. In many cases, removing 3 to 12 inches of the existing soil and replacing it with gravel, pumice, or a sandy, gravelly mixture will be sufficient. In extreme cases, it may be necessary to go deeper. If an actual barrier against moisture is needed, several inches of bentonite clay beneath the gravel will often suffice. Pumice has the advantage of providing insulation as well as drainage, and perlite and vermiculite can be mixed with a clay slip and used as an insulating base.

If gravel, pumice, or other coarse material has been used for drainage, a covering layer will be needed to help seal the porous spaces. A thin layer of straw mixed with a clay slip, burlap coated with clay, or a similar material will work.

Next, lay down several inches of a soil that has a good distribution of clay and aggregates. The soil should be dampened and well compacted either with a hand tamper or by mechanical means; this provides a good

The feel and hand-textured appearance of earthen floors provide a warm contrast to manufactured flooring. (Photo courtesy of Bill and Athena Steen.)

solid base for the finished floor. If you plan a radiant floor heating system, it is placed just below the top of this layer.

This layer should have just enough clay to bind it and yet not enough to make it crack. Too much sand will prevent the soil from compacting. These sublayers can be placed anytime during the construction of the house. The weight of foot traffic will help to compact them even further. However, the finished floor should be one of the last things constructed in the house.

The final coats are two or three layers of a mixture of dirt, clay, and straw, each ½ to ¾ inch thick, hand-troweled to a smooth finish. The trick to achieving a durable, crack-free floor is getting the right mix.

Many attempted earthen floors develop extensive cracking because they are installed in a manner that imitates a concrete pour. Using this approach, a wet mix of 4 inches of clay soil and sand are poured into a floor area and then screeded. The thickness, wetness, and too much clay result in the cracking. Typically, the cracks are opened and filled to create a look that resembles flagstone. The only problem with this approach is that the finished floor is often weak and susceptible to further damage.

If the base is compact and has been properly laid, the finished floor layer can be very thin, 1 inch or slightly more. The thicker the material is, the harder it is to control.

Probably the most important part of the entire process is conducting tests on samples before placing the floor. Initially, these can be 6-inch squares, ½ to ¾ inch thick, of varying combinations of clay soil, sand, and straw. Some will crack, some will be too weak and crumble. The clay typically gives strength, and the sand and straw regulate shrinking and cracking. After identifying what appears to be a workable test sample, a larger test should be conducted—a square meter at the thickness of the final layer. The

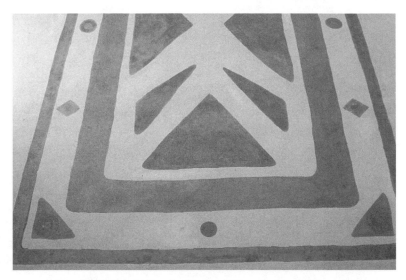

Pigmented linseed oil was applied in a pattern to this earthen floor in Austin, Texas. (Photos by Frank Meyer.)

test mixtures are usually composed of soil and sand that have been passed through ¼-inch screen.

Stabilizers, such as starch paste, casein, or synthetic glues, can be added to make a harder floor if good clay is not available or if the floor is being installed over anything other than an earthen base and additional adhesion is required. Portland cement (5 to 10 percent) can be helpful in stabilizing sandier soils.

Wood floats are good for applying ½-inch layers of the floor, and longer (3 to 4 foot) 1 × 4 or 2 × 4-inch boards can be used to achieve a more level surface. The process is very much like plastering a horizontal wall. The first coat should be left rough to ensure good adhesion of the second coat. The second coat can be applied when the first coat is still damp, but solid enough to walk on. When leather hard, it can be metal troweled to create a less porous and smoother surface.

The final step is to harden and waterproof the surface with several coats of any oxidizing oil, like linseed or hemp oil. Although it is possible to buy very expensive high-grade varieties, linseed oil (boiled or other linseed oil that contains lead-free driers) is typically the least expensive of the oxidizing oils. An advantage to the more expensive oils is that the higher quality can result in better penetrating ability and may be preferable for anyone with sensitivity to the cheaper types.

Adobe floors can be oiled and polished to a lustrous finish. (Photo by Frank Meyer.)

The oil is best applied warm, under the warmest conditions possible. The oil can be heated, but very cautiously and to low temperatures (not smoking), as combustion is always possible. Caution must be exercised if any rags are used, as they can also spontaneously combust.

The first coat should be of oil only; the oil can be diluted gradually with a solvent over several additional coats. The final dilution can contain as much as 75 percent solvent to 25 percent linseed oil. This combination also makes a good maintenance coat. Almost any solvent, ranging from inexpensive odorless mineral spirits to turpentine and other more expensive citrus-based thinners, can be used. Budget is usually the determining factor. These coatings are applied most easily by brush.

The oils and solvents should penetrate enough to waterproof and harden the floor. Apply each coat to the point where it soaks in and the material cannot absorb any more. Continuing beyond that point will leave an oily residue or film on top of the floor.

This approach will usually produce a simple floor that is resistant to wear. It is also possible to create floors that are even more finely finished, but with additional effort and detailing. Changes in color can be accomplished by applying colored clay slip over the final layer when leather hard, and then burnishing it with a metal trowel. Mottled tones can be accomplished by applying a variety of colors. If a glossier finish is desired, an oil-based varnish can be used as the final coat. This provides a good surface for applying wax; the final layer must be very smooth and free of small depressions, which tend to collect cloudy deposits of wax.

A reminder: It is an excellent idea to conduct several practice sessions before attempting a finished floor. Mixtures of clay soil, sand, and straw can be applied to a level piece of ground as practice. Once dry, they can be torn up, remixed with water, and used again. This is a valuable way to gain experience, to perfect techniques, and to confine to the practice field most of the mistakes that are common to any beginner. A few trials are usually suffi-cient to provide the confidence required to create a beautiful floor.

part III

APPLICATIONS

The contemporary case studies in Part III represent but a sampling of the variety of reports one can now find with regularity, not only in the natural building journals such as *The Last Straw* or *Joiners Quarterly*, but in mainstream architecture and construction periodicals, and with surprising frequency in consumer magazines. Included here are international examples as well as domestic, hybrid applications and single-systems use. The projects range in scale from humble hermitage to university campus, in technique from experimental to tried-and-true, and in methodology from primitive to sophisticated.

In every case there have been hurdles to overcome, particularly in introducing a construction technology that differs from the local norm, and the authors share the lessons learned. Their own lessons, too, have much to do with on-the-spot innovation to adapt to geography, resources, or culture. Witness here the natural builder or architect as inventor, diplomat, teacher, student, and strategist.

13

Integrated Systems with Rammed Earth

Charles Sturt University, Thurgoona Campus, New South Wales, Australia

Marci Webster-Mannison

Charles Sturt University has three main campuses in regional New South Wales, at Bathurst, Wagga Wagga, and Albury. The 87-hectare site of the new Thurgoona campus is about 10 kilometers outside Albury and bounded predominantly by roads and by a creek to the north. In the mid-1800s the area was cleared for farming. Today the once-rich flood plain is badly degraded and infested with weeds, and the creek is severely eroded. The opportunity to implement an environmentally sensitive design, from the broad scale of site planning to the detailed selection of nontoxic finishes, is unique in Australia for a project of this size. Here ecodesign has been put into practice, through the "deep green" ethics of the designers, offering a model for the future that

Alternative Construction: Contemporary Natural Building Methods, edited by Lynne Elizabeth and Cassandra Adams ISBN 0-471-24951-3 © 2000 John Wiley & Sons, Inc.

has interesting implications for government bodies and large corporations that wield great power over the built environment.

From the outset, design principles were established for the Thurgoona campus that addressed the concerns of building users, owners, designers, and community, as well as environmental consequences. The resulting physical form comprises buildings sitting comfortably on the site's natural contours connected by a pedestrian spine that rings the hill.

The principles considered a wide range of issues related to building development and to the particular requirements of this project. Recognizing that building development can contribute to the increase of greenhouse gases, breakdown of the ozone layer, deforestation, production of waste and associated waste disposal problems, soil, water, and air pollution, damage to ecosystems, and loss of species and habitats, the challenge of limiting environmental damage had to interface with the project requirements.

This design approach recognized that certain environmental principles always apply, but that they need interpretation through specific design criteria for particular applications and circumstances.

The establishment of environmental principles and interpretation techniques also involved the development of a value system for making appropriate design decisions. Through the process three unique solutions evolved for: (1) buildings, (2) water management, and (3) landscape design.

Buildings

The first building constructed and completed in May 1996 was the Student Pavilion. It became a prototype for the design of future buildings, laying the foundation for acceptance of an environmental design approach.

The thermal mass of the rammed earth walls stores the sun's heat in winter and keeps the building cool in summer. Solar panels produce hot water that circulates through the slab for space heating. Cooling is by natural ventilation, assisted by fans and spray misting of the external areas. High levels of daylighting, integrated with flexible switching arrangements, reduces dependence on artificial lighting. The windows are double hung, and frames are recycled timber. Other environmentally friendly aspects of the design include the use of recycled and plantation timber, natural materials such as wool insulation and nontoxic paint, collection and reuse of rainwater, and the ponding and reuse of storm water.

An office building for approximately 100 staff and postgraduate students, specialist teaching space for the School of Environmental and Information Sciences, and a herbarium were completed in 1998. Construc-

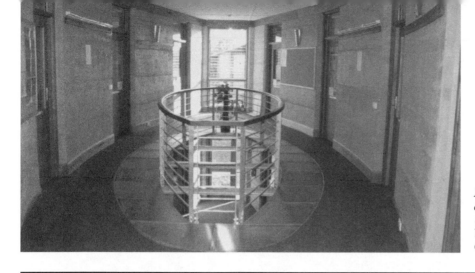

Environmental Design office building interior. (Courtesy of Marci Webster-Mannison, Charles Sturt University.)

⊞ ENVIRONMENTAL PRINCIPLES FOR CAMPUS DESIGN*

Low Energy

❖ Maximize the passive use of the building form and fabric to collect, store, and distribute energy.

Resource Management

❖ Minimize the depletion of natural resources, especially nonrenewable resources.
❖ Ensure social equity in the distribution of the costs and benefits associated with the use of resources.
❖ Maximize the health, safety, and comfort of building users.

Environmental Impact

❖ Minimize interference with natural systems.
❖ Minimize pollution of soil, air, and water.
❖ Maintain and, where it has been disturbed, restore biodiversity.
❖ Enhance opportunities for conservation of ecosystem, habitat, and species.
❖ Increase awareness of environmental issues.

*Courtesy of Marci Webster-Mannison, Charles Sturt University.

tion of a network center, a lecture theater complex, and residential accommodation for 32 students followed in 1999.

The requirements of space planning and the optimization of daylight, solar access, and natural ventilation determined the building forms. The offices include two stories, north facing; long, thin building wings connected by a staff room on the first floor, and an open breezeway on the ground floor. The breezeway clears the in-ground services aligned along the secondary path system that connects the building cluster's main pedestrian spine, roads, and parking. Thereby, it provides covered access between

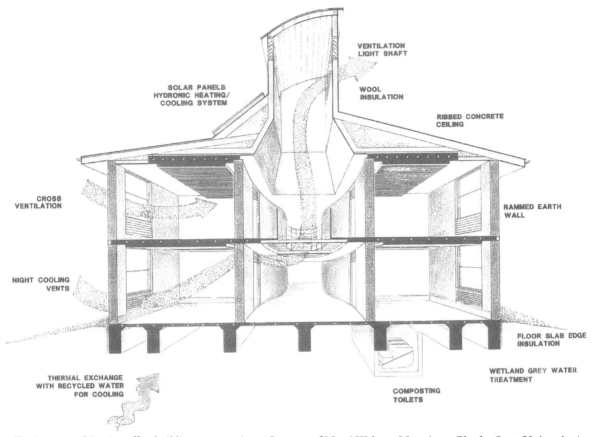

SOLAR PANELS
HYDRONIC HEATING/
COOLING SYSTEM

VENTILATION
LIGHT SHAFT

WOOL
INSULATION

RIBBED CONCRETE
CEILING

CROSS
VENTILATION

RAMMED EARTH
WALL

NIGHT COOLING
VENTS

FLOOR SLAB EDGE
INSULATION

THERMAL EXCHANGE
WITH RECYCLED WATER
FOR COOLING

COMPOSTING
TOILETS

WETLAND GREY WATER
TREATMENT

Environmental Design office building cross section. (Courtesy of Marci Webster-Mannison, Charles Sturt University.)

Environmental Design office. (Courtesy of Marci Webster-Mannison, Charles Sturt University.)

⊞ECOLOGICAL BUILDING FEATURES

Thermal mass of the rammed earth walls and concrete floors and first floor ceiling slab acts as a heat sink to stabilize room air temperature. Ribbing of the underside of the first floor and ceiling slabs in the office building increase the surface area of the thermal mass and thus improve heat transfer between the slab and occupied space.

Insulation of the ceiling with wool reduces heat loss in the winter and heat penetration in the summer.

Shading of all direct sun, designed specifically for each orientation, reduces heat gain in summer and adjusts daylighting for comfortable use.

Air circulation by ceiling fans assists homogeneous distribution of the air temperature in a room and has a cooling effect.

Cross ventilation is critical to admit fresh air and to exhaust heat. Air movement also provides a cooling effect. The windows are double hung, and frames are recycled timber.

Night cooling through automatic operation of low- and high-level louver vents reduces the temperature of the thermal mass and spaces in summer. Low-level vents are below the windows, and high-level ventilation is by large shafts central to each office wing and clerestories in the teaching building and herbarium.

Heating and cooling are by the circulation of water through the slab. In winter, the water is heated by solar collectors and stored in large tanks. The tanks are above the central staff room of the office building, and in the tower over the entries of the teaching building and herbarium. There is gas backup for heating the water. In summer, the system operates in "reverse," and cooling is achieved by thermal exchange, with the water pumped up from the reservoirs to the top supply dams. The reverse operation involves circulation of water through the solar collectors at night, when the temperature is generally cool.

Light and ventilation shafts in the offices and clerestories in the teaching building and herbarium provide daylight penetration to internal zones and assist cross ventilation via the stack effect.

Destratification tubes in the office building collect hot air from the ventilation shafts and redistribute the heat. This process can also operate in reverse for summer conditions.

Lighting design, based on energy conservation and flexibility of individual control, suits variable user requirements by the integration of daylight and supplementary, direct, and indirect task lighting. Large, fully shaded operable windows make the use of daylight possible for a wide range of tasks and provide good views to the outside. Individual switching of general and task lights allows the user to control the light level and direction to suit a particular task.

Rammed earth is the choice wall construction, because of its thermal characteristics and because it is the most environmentally sound, thermally massive material locally available. Rammed earth construction involves the compaction of soil in form work to produce walls that are very durable, require little maintenance, and are resistant to fire, termites, and weather. The color of the finished wall depends on the mix of soil; our walls are a soft earthy pink.

The soil was obtained from a nearby quarry, as the soil on-site was not suitable. The raw material was stabilized with off-white cement to achieve the required erosion standard, compressive strength, and dry density. The unreinforced rammed earth walls are load bearing and support the concrete slabs at the first floor and ceiling levels, and take the roof loads. The specialist rammed earth contractors used prefabricated plywood form work and replaced it on-site as necessary.

Adaptability and flexibility of servicing was a design issue. Communications and electrical cabling conduits were cast into the slab and rammed earth wall to feed skirting ducts throughout the buildings. The skirting ducts were recessed into the rammed earth with the use of blockouts, as were strip-lighting wall recesses, picture rails, and wall niches.

Recycled materials include timber and some interesting material from the dismantling of the Dixson stacks at the State Library of New South Wales in Sydney. Recycled timber is the preference throughout; however, where this is not practical, plantation timber is substituted.

The stacks, originally installed in the New South Wales library in 1930, are an integral structural system of glass flooring and library shelving. Installation of the 24-millimeter-thick glass flooring around the voids and in the corridors in the office building enhances daylight penetration. The original steel beams support and frame the panels. The shelving components are also reassembled for reuse in the offices.

Nontoxic materials, such as natural fiber wool and linoleum floor coverings, wool insulation, nontoxic paints and timber treatments, mesh protection from termites, as well as an attractive, comfortable atmosphere, contribute to the making of a healthful building. Minimization of the use of polyvinyl chloride (PVC) is achieved largely by the substitution of high density polyethylene (HDPE) conduiting and pipework and clay pipes where possible.

the two office wings and forms a vista though the site to a significant stand of remnant trees and the reservoirs.

Early decisions favoring passive techniques were critical in developing a building envelope responsive to temperature variations. The high thermal mass of the concrete floors and rammed earth walls, a high level of insulation, the optimization of windows for heat loss/gain and daylight, and the utilization of natural ventilation were the key features planned to stabilize internal conditions.

The herbarium building required air-conditioning in the storage area and in some work areas, mainly to ensure positive pressure for insect control without reliance on chemical sprays. The work flow and its direct relationship to insect control determined the layout of the herbarium, as a primary user requirement was to minimize the operational need for highly toxic chemical control of insects. The herbarium can operate in a mixed mode, with or without air-conditioning, to ensure future flexibility for changes in operational behavior or use.

Rammed earth exterior of Thurgoona campus lecture theater, south view. (Courtesy of Marci Webster-Mannison, Charles Sturt University.)

Water Management

A total water harvesting strategy is being implemented in all design works with the storm water recycling system completed in early 1996. Reservoirs, diversion banks, swales, and wetlands recycle water from the catchments initially affected by development, and planning includes progressive extension of the remaining subcatchments. The treatment of gray water in artificial wetlands and the use of dry composting toilets obviate the need for connection to the town sewage system.

STORM WATER MANAGEMENT

Waterways collect the storm water, and through a system of connected wetlands, deposit the cleansed storm water in retention basins. The wetlands provide aeration by dropping the water over rocks, sedimentation due to the shape of the ponds, and filtration through reed beds. A windmill and solar power pump water from the retention basins to two supply reservoirs located on the top of the hill. These reservoirs release water back into the top of the waterways and supply irrigation to the campus. The constant recirculation of water keeps the system alive and prevents the creation of mosquito breeding grounds.

An overland flow drainage system, combined with a minimum of pipe work, harvests all the runoff and transfers it to the artificial wetlands.

Tanks store rainwater collected from the roofs of all the buildings. The predominant use of the rainwater is temperature control in the buildings.

GRAY WATER

Gray water cleansing uses artificial wetlands for treatment and aeration, with final disposal by subsurface irrigation. A piped drainage system collects

Construction of rammed earth walls for student pavilion at Charles Sturt University, Thurgoona Campus, New South Wales, Australia. (Courtesy of Marci Webster-Mannison, Charles Sturt University.)

gray water, which is water generated from washbasins, showers, baths, kitchen sinks, and laundries in each of the buildings. The gray water flows through a grease trap and is then treated in intermittently loaded, gravel-based, artificial wetlands planted with locally native emergent aquatic plants. The wetlands empty into evapotranspiration mounds for final disposal. Any overflow is to ephemeral wetlands planted with trees that will tolerate both dry and wet conditions.

The wetlands have a modular basis, allowing additional modules to be added as the volume of the effluent to be treated increases with growth in the size of the university. Once it can be demonstrated that the treated water from the wetlands is of consistently high quality, the water may be released into the storm water system for recycling.

COMPOSTING TOILETS Conventional toilet systems contaminate high-quality water, and the subsequent piping and sewage treatment they require is expensive, resource-intensive, and environmentally unsound.

The installation of dry composting toilets for the offices, teaching building, and herbarium overcomes these problems without the need for the relatively large infrastructure required for an on-site sewage treatment system.

POTABLE WATER Town water supplies every building for drinking, cooking, and washing. Demand for the town supply is low because of the use of composting toilets, the use of drainage runoff for irrigation purposes, and the use of rainwater tanks to collect building roof runoff.

Landscape Design

The landscape design uses plants indigenous to the Murray Valley region and is therefore not reliant on permanent irrigation. Installation of the reticulation system for irrigation of plants, drawing water from the supply reservoir, is not necessary until the arboretum is planted. The ironbark eucalyptus and subalpine regions of the "living herbarium" along the waterways, as well as the stringy bark eucalyptus hilltop plantings, are complete.

The landscaping of the site emphasizes low maintenance and low water use. The landscape character of the campus reflects both the rural and social environments of the region and is laden with teaching and demonstration opportunities for the Environmental and Park Management courses offered by the university.

Restoration and erosion control are planned for the creek, which is part of the Regional Parklands network. The hilltop is revegetated with stringy barks, which, although indigenous to the locality, are substantially threatened.

The plants of the waterways and wetlands help to filter the water and remove nutrients. The large reservoirs, reinstatement of planting indigenous to the region, and the restoration of the creek have provided wildlife habitats, and an increase in bird and frog populations has already been observed.

PHYTOGEOGRAPHIC ARBORETUM

The landscaping of the campus includes an arboretum, presenting a phytogeographic tour of the world. Trees and shrubs from the vegetated regions of Eurasia and Gondwanaland, that are ecologically similar to the Murray Valley, will be planted along the main pedestrian spine.

MURRAY VALLEY ARBORETUM

The vegetation of the major climatic regions in the Murray Valley is being established on the waterways that collect storm water. This represents a live version of the collection in the herbarium, which is also being made available electronically on the Internet.

Eco-Solutions

A pattern is emerging from the examination of environmental philosophy and the views of some contemporary proponents of environmental design. One difficulty is a lack of integration and recognition of ethical propositions as the performance base of design strategies.

Therefore, a strategy that can provide clarity of purpose and fluidity in the process is a basis for implementation of realistic courses of action. Designers must balance environmental ideals and the need for radical

Student pavilion interior with natural linoleum flooring. (Courtesy of Marci Webster-Mannison, Charles Sturt University.)

change in society with practical limitations. Inherent in this balancing act is the need to address the web of interrelationships between the issues. Ecodesign necessitates a change in life-styles that the designer can only facilitate. For ecologically sound buildings and systems to operate successfully, they must be maintained and fine-tuned, and people must think more about their interaction with their surrounds.

The design of the Thurgoona campus demonstrates outcomes of a new approach to building design. The environmental benefits are local, national, and global and help stem global warming, ozone depletion, loss of genetic diversity, and air and water pollution.

The project further illustrates that the architectural expression of a new environmental rationale inspires a new aesthetic. The resulting architectural interpretation, based on a response to climate and responsible resource management, generates a pleasing synthesis between architecture and land use, and architecture and living.

Acknowledgments

Thanks to Cliff Blake, vice chancellor, Charles Sturt University, for having the courage to pursue a new direction in design; to Chris McInerney, architect, Charles Sturt University, for creative energy; and to David Mitchell, water management consultant, Johnstone Centre, for commitment to ecological principles. Thanks also to Che Wall, Advanced Environmental Systems, for the modeling of thermal systems; to Branco Engineering for making these systems work; and to Andre van Egmond and Mark Orton, construction supervisors, Charles Sturt University, for putting the ideas into practice.

14

Straw, Clay, and Carrizo

Obregon Project, Northern Mexico

Bill and Athena Steen

After completing our book, *The Straw Bale House*, in the fall of 1994, we came to spend a significant portion of our time, in the southern part of the Mexican state of Sonora, in a city named Ciudad Obregon. Traditionally, the area was once the domain of the Yaqui Indians, who concentrated themselves alongside what is one of northern Mexico's major rivers, the Rio Yaqui. This region has always been a land of great abundance. Frequent flooding of the river continually renewed fertile soils, making farming highly productive. The nearby ocean and mountains provided a wealth of additional resources, and the deceptively barren desert annually produced a variety of cactus fruits, sweet beans from legume trees, and seasonal wild greens.

After the turn of the century, European and American entrepreneurs, well aware of the region's agricultural potential, gained control over the Rio

Alternative Construction: Contemporary Natural Building Methods, edited by Lynne Elizabeth and Cassandra Adams ISBN 0-471-24951-3 © 2000 John Wiley & Sons, Inc.

Yaqui and began extensive hydraulic development, constructing dams and a large irrigation network. Within a short time, the region became the birthplace of the Green Revolution of modern agriculture, which increased the production of wheat from 1,545 pounds to more than 5.4 tons per hectare on the same amount of land. However, this overabundance of wheat also led to an overabundance of straw, which, when burned annually, leaves a gray haze over the city and detrimentally affects the health of the local population.

This region of the Sonoran Desert is dry most of the year. At times it is unbearably hot, and with the influence of the summer monsoonal rains, *las aguas*, the humidity can rise to make what some call the "not-so-bad" dry heat even more unbearable. Relief comes in the form of afternoon thundershowers, which typically last no more than one to two hours. The rains are considered a priceless blessing, except when they are influenced by tropical storms that bring torrential driving rain and high winds. These major storms usually destroy large numbers of the marginally adequate shacks, leaving many without shelter.

The northern branch of Mexico's Save the Children organization, based in Ciudad Obregon, had been attempting to find solutions to improve living and health conditions, including housing. Its members had heard about straw-bale construction and so held a workshop with the help of the Farmer to Farmer program of the University of Arizona. In spring of 1995, a small demonstration building was put up as a meeting room for a local women's group in a neighborhood called *Aves de Castillo*. It was a modest one-room straw-bale structure that introduced locals to this new technology, but it relied heavily on techniques and materials being used in

The 5000-square-foot Save the Children office was built to demonstrate inexpensive methods that use local natural materials. The exterior walls are straw-bale, with a structural system of concrete. Interior walls were constructed of light-straw-clay blocks. (Photo courtesy of Bill and Athena Steen.)

the United States, such as milled lumber, which in this and other timber-scarce environments, is usually of poor quality and very expensive.

When Save the Children was able to secure low-interest loans for anyone living in the *colonia* of Aves del Castillo to build a straw-bale house, there was only one taker. Teodoro Lopez was in need of a home and had enough vision to recognize the potential. He persuaded his reluctant brother, Emiliano, a builder of 30 years, and together they built an 800-square-foot, two-bedroom house. Emiliano agreed to help only if he would not be expected to build a house for himself. Ironically, he has since emerged as one of the main supporters and builders of straw-bale construction in the area.

Combining their talents, Emiliano and Teodoro built a house that was better matched to the region's resources, climate, and style than the first women's building. One of their major innovations to fill cardboard boxes with loose straw and use them in conjunction with concrete in a waffle-like pattern to form the roof. They significantly reduced the amount of concrete in the roof and added much needed insulation. The house gained greater acceptance when a hurricane destroyed many homes in the area and more than 50 people sought refuge in Teodoro's home.

Shortly thereafter, the director of Save the Children, Jorge Valenzuela, contacted us to inquire whether we would be willing to volunteer our time to help the organization develop a straw-bale building program. We readily accepted. Upon entering Aves de Castillo for the first time, we were confronted with intense poverty that was, at times, uncomfortably close to basic survival. Shacks pieced together with black asphalt panels, cardboard, and scraps of metal and wood densely lined the rutted streets. An open sewage canal ran through the middle of the community. We held our breath as we crossed the canal on a rickety bridge, with nightmarish visions of having to retrieve one or both of our two young boys from its foul depths. Barefoot children, dusty and big-eyed, came flocking, in what seemed like hundreds, to encircle our Suburban, their noses pressed against the glass. They were curious to know who we were and what we might have. We often wondered how anyone could exist in such bleak and oppressive conditions, yet we soon discovered in the midst of it all a people, gracious and generous, rich with life, who would eventually make us a part of their world.

Initially we were introduced as "the American experts," which gained the immediate respect of some, generally the higher classes. Yet in the people with whom we would be working, it provoked a cautionary distance.

We were politely greeted, listened to, and watched, frequently with no response. But we kept returning, often to what seemed like looks of surprise. Many volunteers who enter communities such as theirs, wanting to "help," typically come once or twice and never again return. Most likely, it was our continuity that began to change politeness to friendship. Eventually, after long days of working together, discussing problems and solutions over simple meals of tortillas and beans and cups of sweet black coffee as our children played together, a mutual trust and respect began to grow.

Save the Children wanted help to improve a building program it had started, which involved the construction of 12 to 15 houses. The labor was to be provided by each family, with Teodoro as supervisor. The idea was good, and everyone was sincere, but with one person in charge of directing the construction of so many houses going up all at once, by people who had no straw-bale building experience, the program had little chance of success. Marital difficulties, drinking, and a slow rhythm of work further complicated the scenario. The final blow was a devaluation of the Mexican peso that cut its value in half, making it very difficult for anyone to complete his or her house. In many cases, the bale walls were left unprotected and susceptible to rain, and running down to the local store to buy tarps or plastic to cover them was not a possibility. Today most of these houses are crudely finished, at best, many of them having sustained permanent damage to their walls.

In Mexico, cement runs a close second to religion. A house built from concrete is, in Spanish, said to be a house of "material" (meaning the same in English and Spanish). *Concrete* and the word *material* are synonymous. For most of the straw-bale houses going up at the time, a large amount of cement was needed. It was being used in the foundations, supporting columns, roofs, floors, and the thick plasters. It is the dominant building material in Mexico, as it is in most of the third world, because of the scarcity of lumber. The wood that *can* be obtained is typically expensive and of poor quality. After the inflation that followed the devaluation, one sack of cement became, for many, equivalent to one day's wages, raising its cost to exorbitant levels. As unfortunate as this situation was for most, it created an opportunity for us to introduce and explore new materials and techniques that otherwise may never have been considered or accepted.

There were many resources in the immediate area that could greatly reduce the amount of cement and other costly materials needed for construction. Carrizo and clay were two of the most important. Carrizo

Carrizo (a local reed) was used to reinforce straw-bale walls. Placed opposite one another, these poles are tied together, sandwiching the bales between. Notice the original shack behind the structure going up. (Photo courtesy of Bill and Athena Steen.)

(*Arundo donax*), a bamboolike reed, grew in abundance along the irrigation canals, and clay soils, which contractors had excavated from building sites, were dumped in great piles at easily accessible locations on the outskirts of town. These two materials became as essential to all our future buildings as they had been for the local Yaqui and Mayo Indians. We were also inspired by the building traditions of other countries that used the same or similar materials.

Much of this early experimentation culminated in the building of several simple one-room homes. We realized that we could build a structure that would cost little more than the temporary shacks that most people built, yet provide a substantial increase of comfort and, given a little upkeep and maintenance, could be expected to last much longer. The first house we built was called a *casa provisional* so as not to threaten the family's dreams for one day owning a house of "material." We used sandbags for the foundation and bales of Johnson grass for the walls. We eliminated the common rebar pins and used instead carrizo poles that we placed on the outside of the bales, opposite one another and tied together—a method now called "external pinning." The roof was made from small-diameter poles that were too small for milling. Mats, or *petates*, made from split carrizo were used for the ceiling, which supported an insulative mix of straw lightly coated with clay. A highly reflective and lightweight concrete coated with white cement, ground marble, and an acrylic sealant made up the roof. All the plasters were clay, sand, and straw mixes, as was the floor. The finished cost of this house was $350. The windows and doors, made by a local carpenter, comprised at least half of that cost. These houses have not only provided comfortable shelter for families in need, but have served as great learning tools, offering chances to try our new ideas without a huge investment of time, energy, or money.

After experiencing some of the problems involved in using bales in a country where detail and waterproofing are not always high priorities, we started to experiment with straw-clay mixes that proved to be more water-resistant than straw by itself. Several houses were built in an attempt to use the European light-clay/straw method of packing long lengths of straw coated with a clay slip into forms attached to a wood framework. In place of the wood post-and-beam structure, as used in Europe, concrete columns were substituted. Emiliano, Teodoro's brother, with his wife Juanita and their five children, built a house using this technique. Despite their ingenuity in using recycled washing machine panels for the forms instead of expensive plywood, the complexity of the technique was difficult and soon grew tiresome.

Athena Steen works with volunteers and the women of Aves de Castillo, mixing straw with clay slip to make light-clay blocks. This simple building material was quickly understood by local families familiar with adobe. More than 10,000 blocks were made for the Save the Children office building. Anyone from the community, including children, could join in to make blocks and be paid one peso per block. (Photo courtesy of Bill and Athena Steen.)

What did prove successful, however, was taking the same mixture of straw and clay, packing it into oversized block forms, and sun drying the material to make large lightweight, insulative blocks. This system drew on a skill that had always been a major facet of Mexican life—making adobes. Being more accustomed to that type of work, the people could therefore accept it more easily and learn it quickly. Moreover, the only tools required were a wheelbarrow and a few simple forms. Because they were lightweight, these blocks were easy to handle, and because they were smaller than straw-bales, they were more convenient to use for interior walls. Across the street from her newly built light-

Straw-clay blocks are coursed and mortared with the same method used in making an adobe wall. (Photo courtesy of Bill and Athena Steen.)

clay and straw house, Juanita began supervision of a small straw-clay block factory that was to produce approximately 10,000 blocks over several months. It was open to anyone who wanted to come in and make blocks. Women, men, and children alike were paid one peso for every block they made.

Some of these blocks were to go into a three-house project begun by Save the Children. The organization having learned from past mistakes, the families were given small loans and were required to work along with two experienced workers who were in charge of the construction of their houses. The houses were of about 800 square feet with two bedrooms, an indoor bathroom, a kitchen, and a living room. Two had straw-bale exterior walls, and another was built entirely of the new straw-clay blocks. They all shared the same common cardboard box and straw roof system that Teodoro and Emiliano had conceptualized in the first house. The houses were plastered with clay and straw and finished with lime. Their cost ranged from $3,500 to $4,000 (US dollars.) The least expensive prebuilt concrete block housing, referred to by the locals as "pigeon coops" because the homes are built in long continuous rows, all looking the same, typically costs $20,000 for 400 to 600 square feet. In comparison, the straw-bale, straw-clay houses were of much higher quality, larger, more comfortable, and one fifth the price.

These three houses were basically well built. Yet the idea of what constitutes a good house is a disputable point. For many of these poor people, the notion of living in a house of mud or straw is what they have been running from all their lives. Their dream house is of concrete, like the houses they see belonging to the rich. Even if they know they will sacrifice comfort and coolness during the heat of the day, or that they will never be able to afford that dream house, they do not care. They will wait for the day when they can have "a real house." We were constantly confronted with this problem, especially by the men. We had to show them something that was "of the rich," built with the same materials and methods that were being proposed to them. We needed a project we could take to its full expression without being aborted by the owner's fear of clay and straw. We needed a building that was official, natural, low-cost, do-able with few or no tools and that took full advantage of the natural creativeness of the Mexican people. It had to be something people could come to see and about which they would say, "This is beautiful—we want one!"

Our wishes came true. Wanting to create a demonstration center for alternative and sustainable technologies, Save the Children acquired 10 acres with a small lake on the outskirts of Obregon. We were asked to design and oversee the construction of a new office building using the methods and techniques we had been developing over the last several years. We found the opportunity irresistible, even though it meant committing large amounts of volunteer time. We began work in the fall of 1997. Emiliano was in charge of the main construction and a work force of men and women, several under the age of 18.

Our only construction drawings were a hand-drawn floor plan faxed back and forth from us to Emiliano. Emiliano performed all structural calculations on-site with little more than a visual assessment, when needed. The rough floor plan was approximately 5,000 square feet, including an interior courtyard that would function as the heart of a passive cooling system. The entrances were sized and placed to capture the summer breezes that blew across the lake. Smaller windows were situated on the east and west to minimize the impact of the intense summer sun. High ceilings with vents were placed in every room to allow for the exit of hot air. The thatched portal roof, circling the courtyard, was also vented to allow for the exit of heat. Densely planted trees and bamboo were used to shade the exterior walls.

The portal roof around the inner courtyard was thatched with palm leaves from a nearby mountain village. The pavers in the garden were poured in place and stained. (Photo courtesy of Bill and Athena Steen.)

Approximately 20 species of bamboo were planted around the office building to help determine which species respond favorably to the climatic conditions in and around Obregon. The initial plantings included species from Asia, India, South America, and Mexico. The Sonoran desert is dramatically undersupplied with timber-quality trees, yet there is sufficient land bordering the rivers, as well as irrigated agricultural land, where timber bamboo can easily be grown. Once plants are established it takes only three years for a bamboo pole or culm to reach maturity, which can then be used as part of the structure or roof of a building. Within the first year, two species from India, which grow to 6 inches in diameter and up to 60 feet in height, were outperforming the others. The successful establishment of several species in the area can help relieve the heavy dependence on concrete and provide a versatile resource that can also be used for food and furniture.

The size of the office, along with the weight of the roofs, necessitated the use of concrete columns for structural support, with straw-bales and straw-clay blocks as infill materials. The straw-bales were used for all the exterior walls, and both the walls around the courtyard and the interior walls were built with straw-clay blocks. The first coat of plaster was a mixture of clay with large quantities of straw. The finish plasters and paints for the interior walls were varying combinations of clay and wheat paste; the exterior of the building was coated with lime, slaked and prepared on-site. Some rooms have earthen floors that are sealed with linseed oil and waxed, and the majority of heavily trafficked areas are concrete, scored and acid stained to resemble tile and stone.

Flexibility, change, and innovation were ever present during the building process. Wooden poles were placed in the straw-bale walls at the height of the window and door openings, allowing their locations and sizes to be easily changed or determined later. The same kind of poles also functioned as part of a compression system for the bale walls. As the walls were completed, room by room, the roofs were decided. We stood in each room, evaluating its shape and span, and then decided on a form

Floors are of concrete, scored and colored with acid stains. Ceilings are carrizo reed held by cement viguetas or beams stained with ferric nitrate. (Photo courtesy of Bill and Athena Steen.)

Colored clay from nearby mountains was used for the plaster and paint for this room. The floor was created by pouring concrete pavers in place and grouting the joints between them with exposed aggregate and ceramic tile. Notice the roof of multiple vaults. (Photo courtesy of Bill and Athena Steen.)

that would demonstrate a varying combination of the materials we had available. The result was a collection of highly original roofs that, in many ways, play a large part in defining the beauty of the building. Carrizo vaults, flat ceilings of herringbone carrizo between small concrete beams, smaller vaults, and a dome plastered in clay emerged from this process. All used a mixture of straw-clay for insulation, except for a section of the building that demonstrates the use of concrete with cardboard boxes and loose straw. In the hands of women and children, the same materials, straw, clay, and carrizo, were artistically molded into bookshelves and seating.

Not following any preestablished style, the building's form emerged from the integration of the place, its materials, and the imagination of those involved. The result was a rustic and organic elegance in which the human touch is more visible than the precise machinery of economic efficiency.

The office building serves as an inspiration to both poor and rich. Its presence has influenced several projects, one of which has been a joint effort between Save the Children and the city of Obregon, to build 50 low-income houses in poor communities using many of the same techniques and materials.

These built-in bookcases are made of carrizo, clay, and straw. (Photo courtesy of Bill and Athena Steen.)

However, the most exciting project for us was one that grew out of the enthusiasm and determination of a group of 15 women who decided that they wanted to improve their lives. They were from a rural area, Xochitl ("flower"), on the outskirts of Obregon, where they lived in marginaly adequate shacks of scrap without water or electricity. Their distance from the city and lack of nearby access to public transport made their lives even more difficult. Because they were unable to qualify for the city's new housing program owing to lack of title to their land, the group contacted Save the Children. They wanted to build their own houses and asked Save the Children whether they could be supplied bales.

After consulting with Jorge Valenzuela, the organization's director, we met with the women and notified them that they could have all the bales they wanted. They looked pleased, but asked, "How are we going to build? We do not know how. Maybe if you helped us build one, we could learn." Teodoro and Emiliano were ready for their Easter vacation, but they knew, as we knew, that there was no choice. The women quickly decided among themselves which of them was to receive the first house—the one most in need. Chosen was a young woman with four small children whose husband was in prison and who had no other family in the area.

We began the next day, a Monday afternoon, collecting pieces of broken concrete from the sides of the road to recycle for use in the foundation and

A house under construction in Xochitl. The entire family participates in the construction. Primitive by our standards, but affordable for the local families, it was a great improvement over their original dwellings. (Photo courtesy of Bill and Athena Steen.)

Left: *Mud plaster goes directly onto the bales without any wire mesh. The first coat is always applied by hand to ensure good adhesion. (Photo courtesy of Bill and Athena Steen.)*

Right: *Unlike most conventional plasters that use sand, this mixture relies on large amounts of straw mixed into clay slip. It makes a strong, water-resistant base coat that can be applied thickly. (Photo courtesy of Bill and Athena Steen.)*

floor. On Wednesday the foundation was built and soil was screened, to be used later for plasters and the roof. On Thursday the bale walls went up and plastering was begun. On Friday, the carrizo straw-clay roof was started, as plastering continued. By Saturday, most of the roof was finished, and broken pieces of concrete were laid like flagstone to make a new floor. Sunday was Easter, a day of rest.

The following week, with vacation over, we had to return to our construction projects in Obregon and could be in Xochitl only at the end of the day to show and instruct the women in what to do next. Emiliano and Teodoro's two 12-year-old boys were hired for five dollars each to supervise and help finish the roof. In the evenings of the following days, we helped build a bed, closet, and seating, using straw-clay blocks, carrizo, mud, and soil fill.

As of this writing, the summer of 1998, the women are on their fourth house. They are planning to build as many as 15. All are one-room dwellings, small and simple, but a major improvement over what they had had. Because the clay used in the rough plasters and roofs needs time to dry, the women are completing several houses at a time before returning to apply the final lime finishes. The lime plaster that covers the straw-clay roof is being polished and sealed with alum and soap, as were the roofs of many Mexican colonial buildings. The women are working hard, learning as they go, and except for occasional assistance from one of the Save the Children builders or us, they are doing it all themselves.

A way of building is gradually evolving, in small steps, over a long winding path of friendship, innovation, and, at times, frustration. It is not happening all at once, all in one building, but with time, or perhaps, lack of time. As we build relationships with the place and the people who live there, each of us becomes more than we once were. In the process, we learn patience and the capacity to adapt and endure. We learn to make tortillas, to fish with our hands, and to polish lime-coated roofs with river stones. As we build more simply and more beautifully, building becomes more an activity than a product.

The process of mutual learning and working together has brought promise and hope to people whose lives often seem little more than an endless struggle against overwhelming odds. We have watched change come, not with a grand transformation, but rather carefully and quietly—often without notice. To be part of that, we continue.

Light Clay House Additions

School/Residence in Wisconsin

Lou Host-Jablonski, AIA

This case study describes two house additions within a few blocks of each other in urban Madison, Wisconsin. Both projects are additions to 90+-year-old homes and part of the larger context of working within my existing neighborhood.

My efforts, as part of a nonprofit architectural and planning office that is one of some 40 Community Design Centers in the United States, are focused in the city, reusing and adapting existing buildings, supporting use of the existing infrastructure, and sustaining healthy community. Ever since my days in architecture school I have been seeking integrated systems of natural building. I am somewhat conservative with my clients when it comes to proposing new construction practices. I want to be sure that they will perform satisfactorily, and most owners are reluctant to risk building something they have not seen before.

Alternative Construction: Contemporary Natural Building Methods, edited by Lynne Elizabeth and Cassandra Adams ISBN 0-471-24951-3 © 2000 John Wiley & Sons, Inc.

The School Addition

The light-clay addition project came about because my partner Vic needed a low-toxin new "school space." She has been caring for children professionally more than 20 years and runs a family child care center in our home. Her state-licensed program includes children of various ages, disabilities, and stages of development, coming from families of diverse ethnic, racial, social, and economic characteristics. Vic had become aware that the children were noticeably affected by their physical environment and wanted a warm, sunlit space with fresh air and no chemicals.

For my part, I wanted to explore how far I could go toward creating, in a cold Midwestern urban setting,

School addition—view from the east. (Photo by Lou Host-Jablonski, AIA, architect.)

an all-natural structure that met my criteria. It had to function without a host of problems, be energy-efficient, durable, beautiful, and lend itself to good, buildable detailing.

The addition we built is passive-solar heated, low-toxin, and wheelchair accessible, with a backyard designed to integrate the interior and exterior teaching and living spaces.

Large portions of the addition were built by students as part of several workshops to teach natural home building. The workshops were sponsored by Design Coalition and led by Robert Laporte and Steven Vessey of the Natural House Building Center. We were also blessed with enthusiastic support from the families served by the child care center, or "school," as the children call it. The cohesive community of parents and volunteers at the school is one of its unique strengths, and several of its members spent many hours helping with construction or providing meals in an exchange arrangement.

Bill's Addition

At the same time we were working on the school, a friend and neighbor in Madison, Bill, decided to build an addition and took our workshop. Despite my cautions about the pioneering aspects of this method of construction, he enthusiastically dived in and has been a wonderful, hard-working support on our project as well.

As you will see, Bill chose somewhat different materials and techniques, which make an instructive comparison. His addition may be called "sun tempered," in that it is properly oriented and designed with the sun but does not have added mass for thermal storage.

Green Construction as I Practice It

The range of "green" or "sustainable" construction practices is a continuum. As we design projects, we are always asking the questions, "How green *is* it?" and "What does that mean, for this region, this project?" It is helpful to think of the following broad categories:

1. Energy-efficiency. Most people and building codes recognize at least some elements of energy-efficient construction.

Examples: Careful thermal envelope design, including good windows and exterior doors; airtight detailing and construction; heat-exchange ventilation; superinsulation, selection of energy-efficient heating, cooling, and lighting equipment. At the outer edge, this category also includes passive solar design with thermal mass; solar panels for space heating, electrical generation, and/or water heating.

2. Resource-efficiency. The construction industry and building materials manufacturers are slowly becoming more resource-efficient.

Examples: Use of salvaged and recycled materials; use of products made in part or wholly from recycled materials; timber from sustainably harvested sources; minimization of construction waste; design for low water use.

3. Low-toxicity or "healthy house." In its pure form, this is specialized construction. It is appropriate for a small but growing percentage of the population with chemical sensitivities and for persons with specific disabilities. Of course, *everyone* benefits from low-toxin construction.

Examples: Use of specially selected paints, adhesives, and sealants; extensive use of hard-surface materials, such as tile or wood floors instead of carpet; elimination or isolation of plastic-containing products and VOCs; rigidly controlled construction site practices; and low-EMF wiring and appliances.

4. Natural or low-impact construction. Beyond the aforementioned objectives, the goal of "natural" construction is near-zero pollution and low levels of embodied energy. The goal is elusive, of course, but the principle guides the search.

Examples: Climate-appropriate building technologies such as light-clay, adobe, rammed earth, straw-bale; use of only locally made and renewable "least-processed" materials; plant-based (not chemical-based) coatings; no petrochemical-based materials; no PVC; integration of natural systems within buildings, such as "living machines" to biologically clean and recycle wastewater, natural building cooling and ventilation, and substantial indoor plantings.

OBSERVING THE CONTINUUM

Generally speaking, and as currently practiced, each step forward on the continuum includes those preceding it. Each step yields more sustainable results and takes us further from today's building norms in the United States. That is, steps taken in category 1 are closest to conventional practice and those in categories 3 and 4 are farthest from it.

These broad categories are not rigid, of course; some of the techniques overlap. For example, imagine a wall section using 12-inch-deep Larsen trusses (instead of studs) made of recycled lumber, sheathed in low-toxin fiberboard, with locally manufactured recycled cellulose insulation to produce superinsulated construction using less wood. It should be noted that although categories 1 through 4 are all viable as construction approaches, the techniques are not necessarily applicable in all situations. Some techniques can have negative implications for affordability, at least until they become more common. Others may require a longer learning period, or encounter greater resistance or difficulty with building codes.

SUSTAINABLE BUILDING TAKES MORE

Clearly, a more *sustainable* project is more demanding of everyone involved. More research must be done, more time allotted to sourcing materials, more thought given to the design phase to examine new details and materials, more attention to communication among the disciplines involved to integrate new techniques, more care taken during construction.

When you are planning to build, you need to find out where *you* see your project on the continuum. The whole project team—the owner, the designer, the builder and tradespeople, the consultants, the materials suppliers—must be communicating to work out how far the project is to travel on the path of innovation and sustainability.

Primary Construction Materials and Techniques

FOUNDATIONS

The foundation of the school was made from a composite recycled wood and cement masonry product called Faswall. This kind of material has been used in Europe and Canada for decades and is available through a few franchised factories in the United States. In Canada a nearly identical product is called Durisol. The primary United States producer in New Jersey has recently developed a system that can be used with common block-making machinery. A greater number of producers means that shipping distances, and thus shipping costs, will be lower, enabling more widespread availability.

Faswall forms are composed of about 85 percent (by volume) shredded wood pallets, and 15 percent portland cement. Substitution of this product for concrete block can help reduce the use of cement (a material requiring energy-intensive manufacture), while offering features such as thermal mass, sound control, fire resistivity (four-hour rating), insulation value (R-22 for a 12-inch wall, with insulation inserts), and durability. This product was approved by the manufacturer for use as a foundation.

FOUNDATION OF BILL'S ADDITION

Bill's house used dry-stacked concrete block (without mortar), coated with a glass fiber-reinforced parging that waterproofs and provides lateral strength roughly equivalent to that of standard mortar. Bill found this assembly to be more labor-intensive than he had expected; conventional concrete blocks are not precision manufactured to be accurately square, and much shimming was required. On the plus side, a block wall uses less concrete than a poured concrete wall, and he avoided the damage to his site and trees that a concrete truck would have caused.

FRAMING THE SCHOOL

The load-bearing structure of the school is of Wisconsin white pine timber, harvested a few hundred miles away. This heavy timber frame was hand-crafted with traditional pegged mortise-and-tenon joints. The horizontal braces are in the Japanese style.

Frames built this way have endured for centuries. Typically, large timbers are air seasoned, not kiln dried, so less energy is used in processing. In addition to the heavy timber, there is also "blind framing" of standard lumber that supports the straw-clay and provides nailers for electrical fittings and wood trim.

The roof trusses were built, on-site, of glue-nailed plywood gussets, water-resistant glue, and reused short lengths of 2× lumber. About 40 percent of these components was reused material. We engineered the trusses following the glue-nailed model truss plans published years ago by the Illi-

Teamwork—lifting the frame. (Photo by Lou Host-Jablonski, AIA, architect.)

Glue-nailed roof trusses of reused lumber. (Photo by Lou Host-Jablonski, AIA, architect.)

nois Small Homes Council. Our structural engineer pronounced them stiffer laterally and straighter than the factory-made metal plate versions.

STRUCTURE OF BILL'S ADDITION

Bill used 2 × 6 (40 × 140 millimeter) balloon framing, which accommodates a straw-clay wall that is thicker than the framing. The top-plate configuration was slightly modified from that used in standard practice to allow a better straw-clay installation; one plate (a 2 × 6) is turned vertically, and the other (a 2 × 4) is laid horizontally. This configuration allows a continuous blanket of insulation and eases access to the stud cavity to place and compress the straw-clay.

WALLS

Both additions used straw-clay (aka "light clay"), which in this case was mostly wheat straw mixed with clay. The straw insulates, and the clay acts as binder, fire-retardant, and heat-storage medium (see Chapter 9, "Light Clay").

The walls of the school are 9½ inches (24 centimeters) thick, with an insulation value approximately the same as that of the rest of the house, about R-13 unadjusted for the mass effect. In Bill's addition, the walls are about 8 inches (20.5 centimeters) thick.

Mixing clay slip is the first step in making a straw-clay wall. For the workshop we first taught hand-mixing. Because clays and earth composition

Mixing the straw-clay (aka light clay) with pitchforks. (Photo by Lou Host-Jablonski, AIA, architect.)

in different locales vary considerably, successful straw-clay construction depends on developing a feel for the consistency and stickiness of the mixes. We later switched to a mechanical mortar mixer for volume production.

We coated the straw with clay by pouring buckets of slip onto opened bales of straw on a plywood mixing floor and tossing it with pitchforks—exactly like making a giant salad. The mix is ready when a tightly squeezed handful of straw-clay holds its shape, and the goal is to use as little water and clay as possible to achieve this condition. We delivered the straw-clay to the form work in large hand-rolled, cigarlike armloads and compacted it with our feet and tamping sticks. Our form work required special attention to detail, with its splayed window jambs and blocking for baseboards and wood trim, and the form builders had to work fast to stay ahead of the straw-clay workers. The forms were stripped immediately after compacting the straw-clay (and reused elsewhere) to avoid surface mold and allow the wall to begin to dry.

This method of light-clay construction, as often taught in workshops, is manual and labor-intensive. However, the process lends itself to mechanization with conventional or slightly modified farming equipment such as balers and mixing auger/conveyors. Much of the research on straw-clay and its variations (straw-clay blocks, wood-chip and clay, etc.) has been carried out in Germany in the last two decades, and good resource books are available in German (see Appendix A, "Books").

Straw-clay walls directly earth-plastered are the key to this inte-

Form work and straw-clay edge. After packing with straw-clay, the forms are immediately stripped. (Photo by Lou Host-Jablonski, AIA, architect.)

grated building system. An exterior wall of this type requires no vapor barrier, no drywall, no wall sheathing product, no metal lath. Although the wall does not allow air to pass through it (this would cause a real air-infiltration problem), it does possess the ability to absorb, rerelease, and transpire moisture because of the hygroscopic property of the clay. A straw-clay wall can help to moderate indoor moisture variations while eliminating, unlike most hollow-wall systems, moisture problems within the wall itself.

ROOF INSULATION We used the same straw-clay mix for the roofs as for the walls, except that we installed it "fluffy" (loosely packed) and up to 22 inches (50 centimeters) thick for high insulation value. A portion of the school ceiling has recycled cotton insulation.

EARTH PLASTER Clay, sand, and chopped straw fibers, with a small quantity of linseed oil, were added to the final coat for durability. We used a finish coat of white silica sand mixed with white kaolin clay, then colored with a little potter's iron oxide. There was no cement, lime, or gypsum included.

Initially, I was concerned about the durability of this finish material, especially because children tend to be hard on finishes. The earth plaster is, however, performing well. In both the school and Bill's addition earth plaster was applied as a veneer over gypsum drywall in a few areas that make a transition to the existing houses.

EARTH FLOOR IN THE SCHOOL The spaces in our home are radiantly heated by hot water coils embedded in the floor (via the house's existing efficient gas boiler). The floor is adobe, of essentially the same materials as the wall plaster, but in slightly different proportions (see Chapter 12, "Earthen Finishes"). With both flooring and wall plaster, we find it more successful to mix the clay and sand dry before adding water. Mixing is faster and more precise, and the sand helps prevent the clay from forming unmanageable clumps.

We wanted warmth at the floor where the children are, and I wanted to experiment with the durability, appearance, and thermal conductivity of earth as a flooring material. We are pleased with all aspects of the flooring. It has a texture and feel underfoot that cannot be matched by a concrete slab, and I now consider earth to be a premium flooring material that is worth the extra effort and time it takes to create.

We colored the final layer of earth with red iron oxide powder and sealed it with two applications of boiled linseed oil, poured on hot. This application is basically the same as an oiled finish on wood furniture. We made small samples of the finish layer to test colors. A surprisingly small quantity of iron oxide powder is needed to obtain a rich color, as the linseed oil deepens the hue and enhances slight variations in texture. When the linseed oil cured after a few weeks (and the odor dispersed), the surface of the earth floor was remarkably hardened. The odor could have been much reduced if we had used a clarified (and more expensive) form of linseed oil such as Livos. The earth floor is waxed and maintained much as a linoleum floor is.

BILL'S FLOOR Bill chose conventional maple strip flooring, on floor joists and subflooring, which he will finish with a low-VOC, water-based acrylic urethane.

REUSED MATERIALS IN THE SCHOOL We collected lumber from curbsides, lumberyard castoffs, and materials discarded from other projects. The roof trusses were hand-built from short lengths of two-by-fours and scraps of plywood. The fir ceilings are resurfaced old-growth Douglas fir, scavenged from several local porch demolitions. The old glass doors are inexpensive finds. The sun space was designed around a large sash that had been misordered by the contractor who had built a house for Vic's mother.

PATIO Both additions have patios of limestone pavers on limestone fines. There are no decks, no railings, no toxic-treated wood, no energy-intensive concrete or brick pavers.

The "fines," fine, sand-sized particles that are left over from the operation of crushing limestone into gravel, are locally called "choker." The material is compacted and used as a kind of natural paving for trail sys-

Bill's addition. (Photo by Lou Host-Jablonski, AIA, architect.)

tems. It is also excellent for bedding unit paving (that is, bricks, stones, etc.) because it spreads and levels easily, but, unlike sand, it then compacts into a semihard layer.

In fact, much of our patio remains choker because Vic finds it to be a softer surface for young knees to fall onto than stone pavers.

SOLAR DOMESTIC WATER HEATER The school's solar heat collectors were acquired for reuse from the Packerland array in Green Bay, Wisconsin (this company underestimated the efficacy of the solar contribution, and had to sell off half of its collectors). An antifreeze solution of nontoxic glycol circulates through the collectors to prevent freezing in winter. The heat is transferred to the water in a basement storage tank via simple double-wall heat exchange tubes.

What We Learned

Now, in subsequent projects, we do the following:

Build thicker walls, using 12 inches (30 centimeters) as a functional maximum for this type of wall, because straw-clay tends to mold in this climate before drying if it is much thicker. At this thickness the insulation value is about R-20, and the thermal storage capacity even greater.

Use some pea gravel aggregate in the base courses of an earth floor. This ensures quicker drying, less cracking, and somewhat better passive solar thermal storage and better hydronic heating conductivity.

Insulate more highly at the edge of the slab, and consider 1-inch (2.5 centimeter) rigid insulation under the gravel subbase. In the cold Midwest, the energy embodied in construction—and the first cost—is dwarfed by the heating energy consumed during the life of the building. Extruded polystyrene is one of the few plastic materials in our addition,

Cross sections of both timber-framed and light-framed straw-clay walls. (Illustration by Lou Host-Jablonski, AIA, architect.)

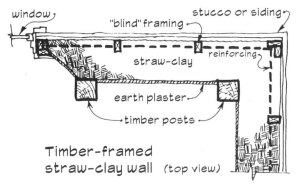

Timber-framed straw-clay wall (top view)

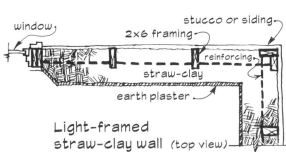

Light-framed straw-clay wall (top view)

a precious material that is used sparingly and only in the locations where nothing else will do the job adequately.

Consider using tile or fiber-cement roofing. The cedar shingles we selected are a renewable resource, longer-lived than conventional asphalt shingles, and a beautiful, all natural material. However, they are made from old-growth trees, which means that some habitat destruction results from their manufacture.

Consider using cellulose ceiling insulation with a fire retardant. We had a nasty scare when I was sweating the copper pipes for the solar panels. Despite what I thought were proper precautions, the propane torch ignited the loose straw-clay. The smoldering fire quickly burrowed out of sight and was extinguished only by thrusting the garden hose deep into the insulation.

I am, however, very comfortable with the dense straw-clay walls, which have very good fire-resistive characteristics.

The Context of Community

In our region there is great concern about the rise of "factory farming" and the accelerating disappearance of family farms to development. There are many forces driving this trend, an important one of which is the increased value of land created by willing purchasers of homesites. Land sold for development brings in large amounts of immediate cash, and the pull is strong for farmers to sell and create a nest egg for their children or for retirement. Even lands far beyond the fringe of development experience this kind of pressure.

This situation has led me and many others to question the ecological wisdom of increasing the number of new buildings in nonurban areas. The most fundamental treasures of our earthly community are its soil and its water. These are threatened by sprawling developments, which increase road building and habitat destruction and create demand for expensive and ever-more far-flung public services like schools and fire and police protection. Sprawl divides us and robs us of the joined human energy we need to heal the places—and the hearts—that we have damaged. Development tends to create more development, just as road building creates traffic. Even "green" sprawl is still sprawl.

Valuable hands-on experiments in home building are possible in ex-urban settings, because here construction requirements, zoning, and building code enforcement are usually less stringent than in other areas. It is time, however, that the fruits of this research have greater influence within cities. A return to the city would save both city and countryside.

Final Remarks

Construction, people, and environment are not separate entities, especially in ecological building. The excellence of Vic's child-care program and the parents' commitment to it, the quality of design supporting the childrens' nurturance, the choices of materials and techniques appropriate to this project, the concern for both the childrens' well-being and the environment by all those involved in the building process, and our personal and collective growth in all this—these do not exist separately, but constitute a worthy whole.

Many hands . . . .
(Photo by Lou Host-
Jablonski, AIA, architect.)

16
Variations on Earthbag

The Honey House—Moab, Utah

Kaki Hunter and Doni Kiffmeyer

After a one-day workshop with Nader Khalili at Cal Earth in Hesperia, California, we returned home eager to build our first earthbag project. We started with simple linear, buttressed exterior walls, graduated to serpentine garden walls, progressed to a small dome, and have now finished a larger dome with a vaulted entry way and big sunny arched windows. This last project turned into a casual workshop, as we invited people to learn earthbag construction with us.

And learn we did! We made ease of construction our priority. As long as the work was fun and simple, it went quickly and the results were solid. We learned that when the work became in any way awkward, frustrating, or slow, it was time to stop, change tactics, or simply go to lunch. Returning refreshed, we often found a clear answer to the problem.

Alternative Construction: Contemporary Natural Building Methods, edited by Lynne Elizabeth and Cassandra Adams ISBN 0-471-24951-3 © 2000 John Wiley & Sons, Inc.

Earthbag construction is a free-form version of building with rammed earth. Because the bags act as a flexible form, it allows the architectural design of curvaceous, sensual, monolithic structures. We have the ability to mold, bend, writhe, and swoop sculptural forms inspired by nature's artistic freedom of expression, while providing structural integrity. Thus, a whole house, from foundation to walls to roof, can be built using the earthbag technique.

The procedure is simple. The bags are filled with moistened soil right on the wall being built, and laid in a mason-style running bond. We use No. 10 coffee cans for scooping and filling; this eliminates any heavy lifting. After a course of bags is laid, the course is rammed with hand-tampers. Between every course are laid two strands of 4-point barbed wire, which act as a "Velcro" mortar, holding the bags in place. This allows the courses to be stepped-in to create self supporting corbeled domes and other unusual shapes. Arched windows and doorways are built with the use of removable plywood forms until the keystone bags are tamped in place.

The bags we used for our construction are woven polypropylene "misprints." The companies that manufacture these bags sometimes have imperfections in the printing that render them unsuitable to their clients. Rather than discard them, the manufacturers may sell them at a reduced

The second-story loft made excellent scaffolding for continuing construction of the corbeled dome. An adjustable sliding compass delineated placement for each row of bags. (Photo courtesy of OK OK OK Productions.)

Corbeled dome of Honey House, Moab, Utah, shown here after removal of arch forms and before addition of cob and plaster finish. The structure used a combination of short and long bags. (Photo courtesy of OK OK OK Productions.)

price. A comprehensive list of bag manufacturers can be found in the *Thomas Register* at your local library.

The material we filled our bags with was "reject sand," obtainable from any gravel yard. Reject sand is the by-product of separating sand and clay fines from gravel. This reject material often has the best ratio of clay to sand for rammed earth construction, and, it is dirt cheap. We paid $1 per ton; some places will give it away.

To keep construction easy, we developed a few specialized tools that enhanced the precision and quality of the construction. They include (these are our names) bag stands, sliders, tube chutes, full pounders, quarter pounders, sliding compasses, fans, halos, and chicken wire cradles.

During the construction of the Honey House (a 16-foot corbeled dome), an average of four people working five to six hours a day moved 40 tons of earth with coffee cans to complete the bag work of the structure in 19 days. In another 7 days, we moved 7 tons more in the form of cob or sculpted adobe onto the roof to provide a 6-inch base for Bermuda grass. Living in a climate with an annual average of 8 inches of sporadic rainfall, we are experimenting with what we call a "living thatch roof." Our idea is to keep the earth alive rather than stabilized. Because the dome is very steep and Bermuda grass is dense and droopy, it is likely to shed water nicely.

Tossing cobs for the first layer finish on the flexible-form rammed-earth Honey House dome. The cob was built up to a 6-inch depth to accommodate a living Bermuda grass roof. Because annual rainfall averages only 8 inches, a drip irrigation system was built into the roof as well. (Photo courtesy of OK OK OK Productions.)

Although building codes in our area have yet to accept monolithic earthbag construction, we have been able to raise awareness of this alternative kind of architecture by building structures that do not require a permit. Allowing people to see, feel, and understand this alternative is the first step toward acceptance.

Our material costs to date for the Honey House (to the nearest $5 US) are as follows:

Professional backhoe excavation (2 ft deep × 16 ft diameter)	$150
40 tons reject sand (delivered)	150
1,000 bags (delivered)	250
Homemade tools (compass, stands, pounders, etc.)	75
4-point barbed wire (2 rolls)	90
Plywood arch forms (reusable)	150
Straw for plaster/cob (20 bales)	35
Chicken wire	20
Total	$920

Residences in Arizona and Wisconsin

Dominic Howes

My grandmother, Shirley Tassencourt, a life-long potter and sculptor, was inspired by the earthen building technique she had learned at Cal Earth in Hesperia, California, with Nader Khalili. The day after her weekend workshop she invited me to her land in Dragoon, a high desert region in southeastern Arizona, to build her a home using a method similar to the way a potter sculpts a pot. In the spring air, under a warm Arizona sun, we constructed the first of four buildings with compacted earth in sandbags.

The first structure was a 15-foot-diameter dome. The next was a 25-foot-diameter by 30-foot-tall two-story dome, her main living quarters. Finally, we built two smaller rectilinear buildings to be used for storage and as a study. The rectilinear buildings accommodated standard windows, doors, and furnishings, which made them considerably less time-consuming

Domed earthbag residence of ceramics artist Shirley Tassencourt, Dragoon, Arizona. (Photo courtesy of Dominic Howes.)

and expensive than the custom work required for the domes. The inventiveness of the domes, however, is unsurpassed.

The following year I built a 625-square-foot, square-shaped and hip-roofed home for my grandmother's neighbor, Allegra Ahlquist. We attempted to streamline the construction process, using long tubes of compacted earth—a newer method developed by Nader Khalili, which he calls "superadobe." The material was aesthetically beautiful, affordable, and energy-conserving.

This newer tube-bag, linear wall system demanded some innovations. We used horizontally placed rebar to strengthen the walls and a steel-reinforced concrete bond beam to cap them. That created a level surface on which to erect the roof structure. We installed rigid insulation board under the exterior plaster and wire to help reflect the hot summer heat and retain the warmth in winter. We also sandwiched burial grade Romex electrical wire between the courses and laid the plumbing against the interior walls under the plaster.

I have used many of the same techniques to complete a variety of structures in the last two years. The first was a small domelike prayer and meditation chapel for a Dragoon-area nun. The most recent structure was a 1,500-square-foot sandbag, timber frame hybrid house in rural Wisconsin. This home is the first of its kind in such a wet and harsh winter climate.

Although I have, to some degree, moved toward a mainstream approach with this process, using standard doors and windows or applying insulation to the exterior of the bags, I have also developed an intimate knowledge and

Residence of Allegra Ahlquist uses long earthbags in a rectilinear structure. Earthbag buttresses add stability. Rigid insulation board under the exterior plaster gives protection from the desert sun. Dragoon, Arizona. (Photo courtesy of Dominic Howes.)

Timberframe and Superadobe earthbag systems were combined in this recently constructed passive solar house in rural Wisconsin. Rigid foam was added to insulate the stout earthen walls against winter cold. (Photo courtesy of Dominic Howes.)

feel for the earth I build with. In the beginning I often struggled with tape measures and other devices to keep the walls in balance. I gradually realized that my artistic and intuitive senses were my greatest tool in forming these structures that are more like sculpture than generic walls. My hope is to continue to be fortunate enough to have opportunities to fuse the artistic and the practical to create homes that inspirit life. I suspect my grandmother would have it no other way.

Sandbag House, Rum Cay Island, Bahamas

Steve Kemble and Carol Escott

When we took on the challenge of building on a remote island in the Bahamas, we observed that the current construction trends relied on the importation of almost all building materials. In keeping with our work in the United States, we wanted to explore more sustainable building options and use our construction project as a demonstration of appropriate earth-building techniques. It was our hope to effect a shift toward more ecological development on the islands. After five years of winter trips to this island and

Before plastering begins, Carol Escott and Steve Kemble pose under the structural Egyptian arches of their coral sandbag residence in Rum Cay Bahamas. (Photo courtesy of OK OK OK Productions.)

researching the availability of materials and the costs of importing, we decided to build a home using woven polystyrene bags filled with sand and crushed coral. Although it is too soon to tell whether the systems we are using will be accepted and adopted, we have attracted a lot of attention thus far and the islanders have started viewing our project as an unusual, yet workable, solution to their housing situation.

Design Constraints

Rum Cay is a 6-by-12-mile island with a population of 55 native Bahamians and about seven American families with winter residences. Building in such relative isolation tested our faith in the universe to provide and forced us to be inventive. The islands of the Bahamas are subject to devastating hurricanes each summer and fall. Any lasting structure must be able to withstand these destructive winds, from any direction. There are termites aplenty on the islands, and all wood products are subject to attack. It also rains here regularly, and moisture levels stay high for weeks. Anything built of untreated wood is probably unusable within a year. Pressure-treated (PT) wood is used extensively and lasts longer, but the intense summer sun can deteriorate exposed PT wood in as short a time as five years.

The Bahamas have developed very little industry other than tourism. Most building materials (with the notable exception of cement) must be imported from the United States, the closest industrialized nation. Material costs in Nassau are typically two to three times U.S. prices because of import duty, stamp taxes, and shipping fees. Prices increase even more when these materials must be further shipped by mail boat to the outer "family" islands. A concrete block weighing 31 pounds may also originate in Trinidad or Puerto Rico (the locations of the only two block plants in the Caribbean), be shipped to Nassau, and then shipped to the family islands like Rum Cay, 200 miles away, bringing the individual block cost to $1.70— about the same it would cost to import (in bulk) a $0.65 block from Miami via Nassau.

The climate is subtropical: hot and humid. The ocean breezes are needed for cooling. Our particular building site is surrounded by bush up to about 8 feet tall. Most people clear-cut the bush on their property to allow the breeze to blow through and to eliminate hiding places for mosquitoes. We decided to leave a lot of bush around our site for overstory protection of new plantings, shade, and moisture retention, to maintain biodiversity and habitat for beneficial creatures, and to provide personal privacy.

We needed a home that would be durable, reliable, and easy to maintain, inasmuch as it might be unoccupied during times of the most harsh weather.

Local Resources Available

Because of thin soils and relatively little overall rainfall, most of the vegetation is low bush and trees for lumber are virtually nonexistent. Casuarina pines can be found at the edge of the beaches and bogs. Stripped casuarina logs are resistant to termite attack and will last many years if kept out of contact with the ground. However, this is an extremely heavy hardwood species that grows in remote locations, making the logs difficult to transport to the building site in quantity. In addition, they are seldom very straight, tend to split, and are so hard that working with them dulls and breaks tools quickly.

Coconut palms were too few on the island to consider using the trunks as a building material. Limestone outcroppings were located either on the beach, where gathering in quantity would deteriorate the beach bulkhead, or in the dense bush, making gathering difficult. The most abundant and easily collected natural building resource on the island was sand. Where dredging for a marina had occurred, we found piles of sand mixed with

Doni Kiffmeyer collects the lime-rich local coral sand that had been dredged for a marina. Pure coral sand had never been used before for an earthbag structure. When dampened and rammed in long-bag wall forms, it "set" to become cementitious. (Photo courtesy of OK OK OK Productions.)

crushed coral, all available for the taking. The lime in the coral acts as a natural binder with the sand, creating a cementitious composite.

An equally valuable resource were a number of strong young men on the island, eager for an opportunity to earn money, use their building skills, and learn something new.

Historic Island Building Methods

Exploring the ruins of previous settlements built on the island over the past three centuries, we discovered that the primary building method was what is referred to as "rubble stone," or rough, uncut limestone set in a soft lime mortar and plastered with lime plaster. According to some of the island's elders, the plaster was made by burning and crushing conch shells. The dwellings, with walls ranging from about 14 to 24 inches thick, were usually one-story structures with wood-framed roofs, floors, lintels, windows, and doors (the wood was presumably imported by boat).

The Design

After looking at the innovative work done in earth-filled bag construction by Iranian architect Nader Khalili in southern California, the inspiring homes built with this method by Dominic Howes in southeastern Arizona, and the beautiful earthbag domes and walls built by Kaki Hunter and Doni Kiffmeyer in Moab, Utah, we decided to employ this building method for our Bahamas project. Kaki and Doni have developed site-built hand tools and a process for building that simplifies, makes neat, and eases this normally labor-intensive construction method. We sensed that the project was blessed when Kaki and Doni agreed to come to the Bahamas to help us get started. Here are their comments on the design:

> We are convinced this form [earthbag] of architecture is well suited for the Bahamas. Corbeled dome designs would excel in grace, beauty, and resistance to hurricanes and termites. We imagine roofs that resemble giant seashells with spiraling gutters to catch precious rainwater. We'd like to try draping a polypropylene fishing net over the exposed bag work of a dome as a substitute for chicken wire, then plaster the surface with a coral pink stucco made from white lime and crushed conch shells. Stucco works well in the Bahamas, as there are no freeze/thaw cycles to promote cracking, and cement and lime are two of the few locally manufactured building materials.

A tube chute holds open a long polypropylene bag. On Rum Cay Island a chute was improvised from metal roofing, pounded into a cylinder and duct-taped along the seam. In the United States, a sono tube and bungie cord do the job. (Photo courtesy of OK OK OK Productions.)

We designed the home to be a two-story dwelling to collect the breezes, with an attached deck on the second level. The shape is six-sided, a cross between a square and an octagon, 24 feet in diameter, with a hip roof to shed heavy hurricane winds and rain. The first level is made of polypropylene bags filled with sand and crushed coral along the perimeter, shaped around six peaked, eight-point Egyptian arch openings (one arch per building side). The arch apertures will be left open for breezes, and the covered lower level used for utility, storage, and a shop. The center core of the lower level is a 3,000-gallon concrete block rainwater cistern, which doubles as load-bearing structure for the second level.

The sandbag walls use salvaged (misprinted) 100-pound rice sacks for the bottom 4 feet, resulting in a 20-inch-wide wall up to that level. The top 4 feet of the sandbag walls use continuous poly tubes, resulting in a 14-inch-wide wall at that point. We ran two lengths of 4-point barbed wire between each two courses of bags for reinforcement and to prevent the bags from slipping over each other. The arches were temporarily formed with scrap plywood, two-by-fours, and concrete blocks (because they were available to borrow). We hand-mixed and poured a 6-by-10-inch continuous reinforced concrete bond beam across the tops of the sandbag walls, and another across the tops of the cistern walls, all level to each other. Next, the bag walls were covered with a layer of 1-inch chicken wire, secured to the walls by galvanized tie wires laid between bag courses as the walls were constructed. We strapped the bond beam down to the bag wall using both the

Completed sandbag structure with concrete bond beam. Note the rainwater cistern in the center of the building. (Photo courtesy of Sustainable Systems Support.)

chicken wire and poly strapping, which was ratchet-tightened before stuccoing.

Kaki Hunter describes a bit of the adventure of working on a remote island:

> We arrived with snorkel gear instead of bag stands, pounders, and cans. Steve and Carol had brought the bags, but the barbed wire (with the rest of their building materials) was stuck with bureaucratic red tape in a cargo container in Nassau waiting for customs approval. So like the Bahamians, we improvised and went shopping at the dump. We found empty bleach bottles, bed springs, and rabbit fence, cut casuarina pine limbs to make tampers, and metal roofing to turn into sliders. We scavenged plywood from abandoned hurricane-ravaged houses to build our arch forms and borrowed cement blocks from Bahamian friends to use as the lower rectilinear portion of the arched doorway forms.

> The most ingenious discovery was turning a food serving tray into a collapsible bag stand that just happened to fit the 100-pound bags we were using. Surprisingly, No. 10 cans turned out to be the most elusive commodity we needed. We asked everyone to save the big cans for our construction project and would order pizza at the marina restaurant in hope that they would open a can for tomato sauce!

The completed bag walls were then coated with a scratch coat of cement/lime/sand stucco, to protect them from ultraviolet (UV) rays, and were later finished with a colored coat of the same for a durable, maintenance-free surface. We bolted down plates to the top of the bond beam, then framed (with two-by-sixes) a floor across the sandbag walls and cistern.

Left: *Plastic bleach bottles, bed springs, and rabbit fence scavenged from the local dump are prepared by Kaki Hunter for pouring cement hand tampers. Limbs of the island's indigenous pine trees make sturdy handles. (Photo courtesy of OK OK OK Productions.)*

Right: *Discarded folding trays were inverted to make improvised bag holders. (Photo courtesy of OK OK OK Productions.)*

A local island crew helps to hand mix and apply a stucco scratch coat. Chicken wire attached to bags helps the stucco adhere. (Photo courtesy of Sustainable Systems Support.)

View of arch opening after plastering. This completed first floor of the Rum Cay Island residence has already withstood one major hurricane. (Photo courtesy of Sustainable Systems Support.)

The 500-square-foot enclosed living space will be built on top of the floor structure, utilizing light wood framing. The roof will incorporate a built-in solar batch water heater, photovoltaic panels, ventilation, and gutters to keep the cistern filled. A sun deck will be added at the second floor level to extend the living space outdoors. The deck will be supported on the perimeter by casuarina pine poles harvested on the island.

Other features of the entire home system will include many varieties of fruiting and medicinal trees and plants (watered by a gray water system, cistern overflow, and drip lines), coconut palm, timber bamboo, and casuarina pine stands; a perennial herb garden and an annual vegetable garden; a small toilet-only septic system with low-flush toilet; and propane cooking and backup hot water.

Conclusion

After the completion of our ground floor, we are very pleased with the results and have started calling the dwelling our "sand castle." The islanders have accepted our house, stating that it looks like the old rubble stone structures around the island, whose walls are mostly still standing. It feels sturdy enough to take on the worst hurricane, relentless sun, and regular windblown rains. It cost a fraction of what "conventional" building options would have cost, and it used as the primary material something readily available on this remote island. Although it took a substantial amount of manual labor, this is a do-able method, requiring only minimal hand tools, and we had fun coordinating our teamwork into a smooth process. The three young men we trained are considering building houses for themselves with this method and are even discussing becoming contractors for other people.

The Value of Indigenous Ways

Habitat for Humanity International

C. Wayne Nelson

To the privileged few who have grown up in the United States or another of the more developed areas of the world, indigenous building systems may seem strange, not worthy of use as our primary means of shelter. Yet people in many countries are familiar with such systems and appreciate how they function, interacting with nature, sun, heat, humidity, and rain. Much of world's population is keenly connected with God's creation within their daily lives, in touch with the earth when riding animals or walking, waiting for the rain and sun to grow the plants that provide food and shelter. Indeed, the closer to and more dependent one is on nature the easier it is to understand God's provision of materials for housing.

To better understand the constraints of low-income communities, imagine yourself having volunteered to be sent, as I once was, by a nongovernmental organization (NGO) to a community halfway around the

Alternative Construction: Contemporary Natural Building Methods, edited by Lynne Elizabeth and Cassandra Adams ISBN 0-471-24951-3 © 2000 John Wiley & Sons, Inc.

Building materials, bamboo and thatch, at market in the Philippines. (HFHI photo by Ellen Patton.)

world. You are given two assignments. The first is to live in a way that the indigenous people could aspire to at their economic level (average annual income is $126). The second is to help families in the community to build houses, using volunteer labor and a no-interest loan for materials.

After traveling for days, you arrive at the room in which you will stay for the next few years and head for market to get something to eat. The market consists of a few tables and woven grass mats spread on the ground, displaying dried fish, roots, rice, and a few types of vegetables. You find that the paper bills of money you have brought from the bank in the capital city (each bill worth US $1.50) are too big for any market person to change. So they *give* you food, which makes you feel bad because you are taking food from really poor people.

As you head back to your room, you try to think of how you can build a house with these people. One corrugated metal roofing sheet could cost more than six months' wages. To avoid burdening them financially, you will need to use no more than 20 percent of their income. That could be less than $2.00 a month, amounting in more than 10 years' payments to less than $250 for the total cost of the house. The only answer is the same answer for much of the world's population—build with whatever natural materials are available in the area.

HFHI work preparing for the foundation of a house in the Philippines, where coconut tree lumber has been used to build wood-framed houses. Coconut lumber has been used for walls, floor, roof structure, windows, and doors. Coconut palms must mature at least 30 to 40 years to be durable, although they reach full height in just a few years. Nipa thatch has been used for the roof. Traditionally, thatch is tied on with cords made of natural vine, but now it is often tied with cords made from used tires. (HFHI photo by Branson.)

Ways of the Grandparents

The "ways of the grandparents," which have often been passed on for generations in many parts of the world, are being abandoned for higher-technological solutions to housing. However, there is much knowledge that has not yet been lost, because a large portion of the world's population still lives in rural areas without dependence on the industrialized, energy-intensive construction systems so prevalent today. Even in modern cities, older buildings may provide examples of more natural building techniques.

Here you may find structures that used unsawn round poles or rafters from split poles. One old method to preserve the wood is to char the poles in a fire, especially important when they are to be placed in the ground. Such poles can last for hundreds of years. In the Pacific islands, for example, it is common to find poles passed on for several generations. Much of the resistance to the use of unsawn timbers is caused by poor construction methods and lack of timely maintenance, rather than the fault of the natural materials.

Usually, when wood is sawn, the grain is cut across. At each cut the exposed grain will take in moisture, rot more readily, and result in cracks under load. In contrast, round and split timbers, whose grain is cut across only at the ends, utilize the full strength of the original material. Natural preservatives, such as a mixture of melted beeswax and oil, should be used on cut ends to help prevent decay when they are exposed to weather.

Round poles, long overhangs that protect the walls, and steep pitches on thatch roofs that help the thatch last longer are traditions passed on from the grandparents. (Photo courtesy of HFHI.)

In traditional cultures of West Africa round poles are often cut from the branches of trees that are hundreds of years old, a practice that leaves the old root structure in place and produces wood much more quickly than cutting the whole tree and starting again from a seedling. Cropping branches may be one of the best practices of sustainable construction with wood.

Usage and Availability of Material

Many natural materials have been used improperly, creating doubt about their durability. For example, woods and bamboo that are in short supply are sometimes harvested before they are mature and strong. It is also common today to see thatch roofs with very little pitch in areas where metal roofing sheets have been used. A slight pitch on a metal roof is fine, but the steep traditional pitch of thatch is a major factor in its longevity. Thatch on a low-pitched roof retains water longer and deteriorates more quickly, thus creating the false perception that it performs poorly. At the same time, because the required reed or palm is now not as readily available, thatch will be laid in fewer layers than in previous generations, significantly lowering its performance and longevity.

In considering the use of natural building materials, it is important to weigh the cost of maintenance as well as the initial cost. In the case of a thatch roof, the next roof may be locally grown. If the tradition is kept up, local artisans can install it, which has the benefit of keeping money within the local economy. This is a key factor in the well-being of a community. A sustainable community finds ways to provide for most of its needs through the services and goods of those in its society.

Building with local materials provides a community with a sustainable future. Thus, there is a great need to encourage the planting of building materials to keep up with future demand. Although governments have made many unwise choices, such as encouraging monocrop farming (with the use of herbicides that kill the plants used in traditional crafts) or bringing in cement factories (instead of employing indigenous solutions such as bamboo farming) for construction materials, some governments are now providing tree seedlings for their citizens. In many countries where Habitat for Humanity International (HFHI) is working, trees are being planted. In Brazil, HFHI home owners have tree nurseries, and in Ghana teak and *odum*, a native hardwood, are being replanted.

Traditional Design

Housing has become so dependent on industrialized systems that nature's solution displayed in a traditional building is often overlooked. A natural arch, for instance, is a great construction method; it disperses the stress that often forms cracks at the top corners of windows built with horizontal lintels. A simple brick arch can easily span an opening for a window or door, yet because we have become accustomed to the concrete and steel lintels of modern construction, many people are without good houses.

The design of older buildings derives significant shading from long porch roofs. In addition to keeping the house cooler, these overhangs also protect the walls, keeping them in sound condition longer. These designs may also have placed the greatest amount of window to the south (or to the north in the Southern Hemisphere) to allow for warming from the winter sun. Despite the simple logic and economy of these ideas, they are often abandoned with the use of mechanical heating and cooling systems.

Cultural Acceptance

Probably the major reason that traditional and simpler construction methods are abandoned is that residents of less-developed regions are trying to make their houses look more modern. At one time, while living in Ghana, West Africa, I suggested building my bed frame of bamboo. People began to laugh at me. They knew I could easily afford a bed manufactured by the most sophisticated methods and found it absolutely silly that I should make a rustic one.

The round house has been a traditional style in much of the world. It is easier to build with nonindustrialized material, and it is stronger than a house with corners because it disperses stress throughout the structure. This second factor is particularly important in areas experiencing high winds, earthquakes, or shifting ground resulting from moisture or freezing.

Despite these advantages, it is difficult to find people willing to purchase a round house, which is perceived as "the old way" or as "the poor person's house," inasmuch as it appears that all wealthy people live in rectangular houses. The low-income person most likely will have just one chance in life to buy or build a home and does not have the freedom to experiment.

It will always broaden acceptance for those of affluence to use natural, environmentally friendly methods of construction. In Travandrum, India, Laurie Baker has been putting this principle to work, getting the rich to build with fired brick, using no plaster inside or out. He does this specifically to show those with low incomes, who cannot afford plaster, that it is fine to have exposed brick.

Some Habitat for Humanity International houses in the United States are being built of earth blocks, a technique previously believed affordable only by the wealthy. Habitat for Humanity is employing this method to build affordable houses with low-income families by using donated labor and keeping the house modest in size. It is also building homes with ecological, natural systems that require less heating and cooling, thereby reducing utility bills and making the homes sustainably affordable.

Good looking, durable Habitat for Humanity housing has been built in other countries using natural materials such as locally fired brick and compressed earth blocks. In Malawi, Uganda, and other countries, HFHI has combined local resources with a bit of industrialized technology. Many houses are built with locally fired clay bricks. Sometimes bricks are fired with agricultural waste such as rice husks, the ash of which can also be used as an additive to cement. Foundations can be made with local stone. Mud mortar is used, and on the outside it is sometimes pointed with a cement mortar to make it more durable and weather-resistant. Combinations of methods and materials, with only a little of the industrialized materials such as cement, can reduce the need for maintenance.

An earth block press, made regionally, produces uniform, strong blocks that are easier to make and use. These blocks can be pressed even when it is too rainy to make sun-dried bricks. (Photo by Wayne Nelson.)

Modern Engineering

The use of modern engineering knowledge in combination with natural materials can enhance safety. After an earthquake in northern India, Laurie Baker wrote a rudimentary book showing why houses fell apart and how to build with traditional materials and timber reinforcement. The use of this simple picturebook conveyed the necessary knowledge for families to build their own houses.

Traditional construction in Kyrgyzstan uses a slip form adobe wall, which now incorporates a modern chicken wire mesh for attaching plaster. The concern for earthquake resistance can be addressed by using chicken wire on both the inside and outside face of a wall, so that it contains the earth in a basket. This method has been tested on shake tables at universities in California and Peru. Results have shown that although the walls may crack, the chicken wire keeps the house from total collapse. Present building codes require steel-reinforced concrete with an infill of fired clay brick, a method not affordable to many.

In Kenya, the standard wattle and daub is modified to place the poles on the outside of the wall rather than buried in the wall. This leaves them exposed, so that if they deteriorate it can be easily detected. Poles can be replaced without disturbing the entire wall. Using the poles on the outside and inside of the wall adds a supporting frame around the wall itself, again improving strength and earthquake resistance.

Challenges of Technology Transfer

There is a need to further the application of modern technological understanding in the use of available local materials and traditional building systems. A major reason that this approach is not being used extensively is that there is no profit incentive to work with low-income people. Another reason is that the fields of engineering and architecture are so specialized and industrially dominated that the designers of buildings have rarely had an opportunity for hands-on experience with natural materials. The only way to fully understand how the materials can be used is through field experience. Ideally, designers can learn by working alongside master crafts people. Because women have been the backbone of much of the building in indigenous communities, they are rich sources of information on local techniques and can be helpful in introducing new methods of construction.

Besides the difficulty of learning different building processes, there are many other problems with technology transfer. Many development projects that use alternative materials or construction methods have outside funding for start-up, yet often, when that funding is spent, the new methods are abandoned. Occasionally, economic situations change so that the product is no longer affordable; at other times, the money needed to continue the work is stolen; in still other instances, the one person who was really making the system work moves on, quits, or dies, and so does the new system. For technology transfer to be successful, it must be well absorbed into the community by many of the people.

Several organizations have helped to advance the worldwide transfer of technology. Many of these groups' efforts have been published by the European organizations of SKAT, ITDG, GATE, and CRATerre (see Appendix B). The United Nations Center on Human Settlements has gathered and distributed information and has also implemented residential construction. These groups have training centers where methods and manuals are developed, and skills and theory are taught, but they also send teams to the construction sites to analyze and conduct hands-on training. Often, these groups coordinate their efforts with local universities so that they can better understand the local culture and needs and facilitate training. Mission and para-church organizations such as Habitat for Humanity International and World Vision International, as well as many religious denominations, have also played a role as they work with communities to improve housing. Yet some of the best improvements are generated by the people within the community, as they hear about or see methods from other places in the world and work with local artisans to adapt these new technologies to their own purposes.

In Galilee, Ghana, I was surprised to find a family making concrete roof tiles, using a wooden mold the local carpenter had made for them. The total cost to start this business was a few dollars. Concrete roof tiles have been promoted by international aid groups for years, but the setup of manufacturing often costs more than $10,000. In addition, equipment repair needs to rely on parts from other countries, which can take months to receive.

Concrete tile roofs are becoming prevalent in some areas because of their comparative durability, low cost, and the fact that this roofing relies on small-scale local manufacturing.

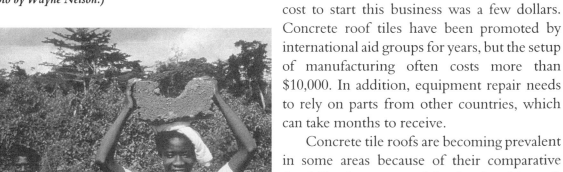

This girl carries a pressed-earth block that will form the chimney for a fuel-conserving stove. The load is cushioned with a cloth. (Photo by Wayne Nelson.)

A concrete roof tile form made by a local carpenter in Galilee, Ghana, allows a durable roof to be made locally with minimal input of industrialized building

The durable concrete roof gives good protection to the earth walls of the traditional formed-in-place swish *style. (Photo by Wayne Nelson.)*

Although the cement itself most often is imported and uses a highly industrialized, energy-intensive technology, there are some places with small-scale local cement manufacturing, particularly in Asia.

Standardization

Much can be gained from the use of jigs and standardized sizes. The ease and speed of initial construction, along with the ability to replace standardized parts, is valuable. Compressed earth blocks, which have been used for many decades, can be made with the use of either hand-operated presses or hydraulic machines. One of the more recent technologies (developing since 1979) is the use of compressed earth blocks, which can be stacked like children's Lego blocks, eliminating the need to mortar between courses. Instead, mortar is poured down through holes in the blocks, which seals the wall by joining the blocks together. A significant advantage of this method is that it is easy for unskilled labor. Because learning to bed blocks well in mortar, on the other hand, is often a time-consuming process, modularized materials such as compressed earth blocks increase the speed and accuracy of construction and reduce waste. Similar processes have been developed in Auroville, India, where an entire town has been built with the use of pressed soil block produced by a hand-powered press called the "Auram."

With a modular block system only two kinds of block are needed: a full and a half. The half-block is used at openings. The simple system also allows space within the block to run electrical and plumbing lines.

Another advantage of modular construction is elimination of custom fitting windows and doors to irregular apertures, because window and door units are made to specific sizes equal to the length and height of a given number of blocks. Modular wall bricks also save the extra time and mortar needed on traditional walls, in which bricks from different sized molds are used in the same assembly.

Local Materials Policy One of the first decisions of Habitat for Humanity International in 1976 was to use local materials as much as possible. This will be the basis for sustainable housing for the world. An example is provided by Habitat's work in Zaire, where, amid civil strife and loss of transportation services, the most critical housing components were still available to the people, including locally made fired clay roofing tiles.

In Zambia, a house in the rural areas costs $500, whereas the same size house in the city costs $1,500, because of transportation costs, a factor often ignored. Transportation alone can constitute 25 percent of the cost of a simple house, even with the use of materials available in country. Yet another transportation cost to be considered, besides that for construction materials, is the expense to the family for simply getting to basic services and daily tasks, such as work, school, and market, which points out the importance of siting housing within fully functioning pedestrian communities.

Industrially manufactured construction materials will continue to be unaffordable for many. Communities need housing systems that are not dependent on outside help in terms of labor and technology. For these reasons, it is most important to continue researching natural building methods and to find ways to make them more efficient to fulfill Habitat's goal— that all people have a simple, decent place to live.

Where there are no machines to do the work, people work in community to move heavy loads and perform construction tasks. Boats are often used to move building supplies to construction sites. (Photo by Wayne Nelson.)

18

Building Technology Transfer

Raising Straw-Bale Housing for Farmworkers in California

Kelly Lerner, Dan Smith, and Bob Theis

A modest residential straw-bale project was developed as a prototype for a series of units to house farmworker families at Frog Hollow Farm in Northern California. A grower of organic produce for the East Bay region of San Francisco, Frog Hollow Farm is situated in Brentwood, a part of eastern Contra Costa County. The cottage itself was located along a busy highway and did successfully demonstrate, if only informally, the ability of straw-bale walls to block sound and provide comfortable living quarters despite the less-than-ideal setting.

The project was designed and built in 1996 by our architectural firm, Dan Smith and Associates, which has developed a specialty in straw-bale, and its construction was managed by the foremost straw-bale building contractor in this region, Skillful Means Builders. The structural engineer was

Alternative Construction: Contemporary Natural Building Methods, edited by Lynne Elizabeth and Cassandra Adams ISBN 0-471-24951-3 © 2000 John Wiley & Sons, Inc.

This compact cottage for a farmworker's family was designed by the authors as a prototype for a housing cluster at Frog Hollow Farm in California's Central Valley. (Photo courtesy of Dan Smith and Associates, www.dsaarch.com.)

Richard Hartwell. We selected it to serve as a training ground for the many people wanting to learn this new technology. A number of people, such as Matts Myhrman and Judy Knox of Out On Bale, offer courses on the process of managing workshops, but in this case study we focus primarily on construction techniques. We should also note that the methods described here represent only one set within a diverse and ever evolving construction genre.

Raising the Walls

The walls of the first cottage at Frog Hollow Farm were raised during a weekend workshop in June 1996. In preparation, Skillful Means Builders formed and poured a concrete slab-on-grade the day before. While the participants assembled, the wall base of mudsills, with gravel between them, was placed.

The workshop group included the typical selection of mostly middle-class people: a few with building skill, most without; a few with strong muscles, most without. It is usually necessary to do some cutting and nailing during a wall raising, so we had a framing carpenter, along with appropriate tools and materials, on hand.

We showed the assembled group the basics of stacking the bales and how to tie half-bales. The group then divided into wall teams, with one person on each team in charge of quality control, making sure that the bales were sound, straight, and laid up plumb, especially at the corners.

The wall teams were guided by bale elevations we had drawn up, showing how to arrange the bales at the corners and where to locate the half-bales and custom bales for door and window openings.

As these openings rose two or three courses high, the rough buck for each opening was set in place and loosely pinned. At the third course, we

Men pounding rebar with a sledgehammer. As the walls are laid, the bales are pinned to the courses beneath by driving in wood, bamboo, or rebar stakes . (Photo courtesy of Dan Smith and Associates.)

began driving two 48-inch pins of ½-inch rebar into each bale at roughly the first and third quarter points. We have also used 48-inch lengths of ½-inch bamboo and even sharpened 1½ -inch-diameter saplings for pinning.

Over in the shade we set up a custom bale workshop, where the shorter bale lengths were created. No matter how much we try to keep our openings to a 24-inch module, so that in theory only half bales are needed, in practice we always seem to require three-quarter-bales, one-third-bales, and so on. "Jamming in" the final bale of a course can often push the bales at the nearest corner or opening out of plumb. Far better to use a shorter bale and stuff gaps (up to about 6 inches wide) with a "flake" of straw from a cut bale. For increased fire safety, we now tend to stuff with light clay, which is loose straw tossed in a clay slip (see "Light Clay," Chapter 9).

When the bales have to be raised higher, the work slows. Having a sufficient number of extra bales for stair step scaffolding makes getting the bales up there much easier and reduces falling hazards. Even with the bales pinned together, a straight wall of bales more than 6 feet high is very flexible, so the bracing effect of bale scaffolding is useful.

Box Beam

A major reason we like to use box beams on top of bale walls that are to receive a conventional joisted roof is that they are so great at taking the floppiness out of a wall. In addition, they are strong enough to span most domestic-scale openings, so we do not have to bother with lintels.

When all the courses had been laid and pinned, the sections of the box beam were passed up, set on the walls, and fastened together. The walls were given their final adjustments for plumb and then pinned to the box beam. Although rebar was used on this project, we now specify 2-foot-long stakes sawn from two-by-twos set through square holes cut in the base of the box beam.

At this stage we checked on the levelness of the box beam. Here is where the flexibility of bale construction came to our aid. We measured around the house and recorded the various sill plate–to–box beam heights. We gathered several heavy people on one spot on the box beam to see how far down we could easily squeeze it. We then arrived at a judicious average height. We cut grooves in the walls at our column locations (a small chain saw works well, but a traditional hay saw is a lot more fun; that is, safer, less noisy, and, surprisingly, equivalent in speed).

We cut all the columns to the average height, inserted them in the grooves, and nailed them to both the embedded hold-down straps at the foundation and the box beam. In a few places, the bale wall was not tightly up against the box beam. Although a tight fit is not structurally necessary when columns are taking the load, it means a firmer surface for the finishing work. We used 12-inch-long wood wedges cut from two-by-fours ("vampire stakes") in a middle course to wedge the wall tightly to the box beam. If we were erecting a straw-bale without columns, we would suggest this as a method to level the wall plate.

Above the box beam, the roof construction was fairly conventional stick framing. The rafters were deepened with two-by-fours hung on plywood gussets so that bales could be laid within the roof structure as insulation. We believe that a bale-insulated roof is worth the additional weight and structure; it is the appropriate extension of the snug enclosure begun by the walls.

Prep for Plaster

The following weekend, we prepared the bale walls for plastering. First, all loose straw was clipped off the walls with razor knives, hedge clippers, and weed whackers. The window bucks were leveled and plumbed, then pinned into the adjacent bales. The bucks were flashed with Tyvek on the sides and head, and the sills waterproofed with Bituthene. The columns and any other wood elements under the stucco were wrapped with Tyvek. The first course of bales was protected from ground splash with a layer of Tyvek, tucked at the top between the first and second courses and lapped at the base over a weep screed nailed to the mudsill 6 inches above grade (the code-required termite barrier). The windows—which were ordered with stucco stops—were set into the wrapped buck and flashed at their heads.

The shear straps, 12 gauge by 3-inch-wide coil straps, were fastened to the foundation by bolting the straps to hold-downs, which were bolted onto anchor bolts cast at the strap angle into the foundation.

As the straps ran diagonally down from each corner of the building, we could bolt both straps at each upper corner to each other and a corner plate that prevented the straps from crushing into the wood. (All these connec-

This construction photo of the Santa Sabina Hermitage (see Chapter 10, "Straw-Bale") shows preparation of walls similar to that done at Frog Hollow. Notice the expanded metal lath at the window buck, the judi- cious use of Tyvek wrap, and the diagonal metal shear straps . (Photo courtesy of Dan Smith and Associates.)

Masking of window frame prior to plastering. Notice the construction felt on the lower sill that provides extra protection against moisture. (Photo courtesy of Dan Smith and Associates.)

tions have since been refined and modified in subsequent projects, but we have included them here to illustrate the approach.)

When all the bracing, flashing, and waterproofing was in place, we covered the exterior with 17 gauge galvanized stucco netting, nailed off at the box beam and the sill plate with eightpenny nails at 6 inches on center, so that, should any strap fail, the stucco can effectively transfer shear loads to the foundation. In addition, the netting was fastened back to the straw with "Robert pins"—large bobby pins about 12 inches long made from 12 gauge wire. Two such pins per bale are sufficient.

On the interior walls, the electrical boxes were installed in niches cut for that purpose, and the wires were pushed 4 inches deep between the courses. Niches are great—don't miss the opportunity to carve into a wall. What other wall material allows this?

The splay at each window was created by using expanded metal lath nailed to the window buck, stretched about 5 inches beyond the interior corner of the bale wall and pinned with Robert pins, and stuffed with loose straw. We try to create a splay that is about 60 degrees from the plane of the wall. This keeps the window from looking like a slot in the wall and mitigates glare.

As we have become familiar with metal lath, we use it to shape elements for the interior. On this job we created sloping seat backs at a window seat in the living room. Consider the possibilities in this medium: shelves, light sconces, benches?

A word of caution: At all phases of construction, but especially as other tradespeople work in and around the house, be vigilant about the potential for fire. The bale walls may not burn well, but more than one serious fire has occurred at a job site when sparks or flames generated by construction activity, such as grinding, soldering, or welding, ignited the ample amounts of loose straw that had piled up.

Flexible Details

As we gain experience with the informality of bale walls, we have come to favor trim elements that adapt to the walls. Wood trim elements, unless they are extremely flexible, do not follow the informal contours of plaster on bale walls. As a result, we detail doors and windows with plaster returns and use tile, molded in place stucco, or simply paint for baseboards. We are still looking for flexible outlet and switch plates.

A child at the front door of the cottage at Frog Hollow Farm. Even a simple house becomes a home with ornamentation like this sculpted plaster on the door surround and inlaid tile in a wall niche. (Photo courtesy of Dan Smith and Associates.)

The substantial interior sills created beneath windows require design attention. We try to avoid plastering them, contending that plaster soft enough to be a comfortable wall is not tough enough to be a good sill.

At Frog Hollow, Skillful Means Builders cast in place 2-inch concrete slabs during the finish preparations. Once the forms creating their lips were removed, plaster could simply die into the sill all around.

In general, the more relaxed character of less-than-straight edges and less-than-planar walls is one of the sought-after qualities of straw-bale construction. What is pleasantly informal for us may be unpleasantly wacky or coarse for you, however. Given our lack of vocabulary to describe this range of finish levels, the best idea may be to seek out built examples (pictures are not usually textural enough); these can be used as reference points for the sought-after standard, as well as for what is *not* desired.

Introduction of Straw-Bale Construction to Mongolia and China

Kelly Lerner

Mongolia
Given straw-bale construction's low-tech, sustainable approach, it may seem odd that most straw-bale buildings are built in the richest, most resource-consumptive country in the world. But straw-bale construction is slowly moving into other parts of the world, where it may revolutionize the construction of small insulated buildings. What makes an area appropriate and receptive to straw-bale construction? What are the most effective methods for moving this technology from the United States to a developing country? A look at the growth of straw-bale construction in Mongolia and China begins to answer these questions.

About twice the size of California, Mongolia sits on a high plateau, on average 5,200 feet above sea level, completely landlocked between China and Russia. With vast grassy steppes, high mountains in the north, the Gobi desert in the south, and no large bodies of water, the climate is quite extreme. Temperatures in the Gobi average –20 degrees Celsius during the six-to-eight-month winter and can reach more than +40 degrees Celsius in summer. Annual precipitation varies from 75 millimeters in the southern desert to 250 millimeters in the wettest areas. This is an unforgiving land, but, luckily, the steppes produce a wealth of grass for livestock and wheat is grown in the north central province of Selenge.

Although the Mongol empire once covered most of Asia and eastern Europe, Mongolia's current economy is struggling to make the leap from communism to free-market capitalism. Loss of Soviet subsidies and collapses in national manufacturing after a bloodless democratic revolution in 1990 provoked an economic depression three times worse than the Great Depression of the United States. By the late 1990s, with foreign aid and foreign investment flowing into the country, Mongolia's nascent market economy is growing and living conditions are improving, but the country is still marked by scarcity of building materials and low labor costs.

As Mongolians struggle to improve their standard of living, they are challenged by the centralized systems of production and an aging infrastructure left by the Soviets. All heating and power in the city are provided by coal-burning hot water/electric plants. Mostly built since the 1950s, all large buildings are of uninsulated masonry or tilt-up concrete. About 60 percent of the urban population lives in four-to-nine-story walk-up apartment blocks. In the *ger* (felt-covered round tents, called "yurts" by Russians) communities that surround the city centers, each *ger* is heated by its own small coal-burning stove. Air quality in the cities in the winter is so bad that visibility is reduced to 200 meters.

Traditionally, Mongolians have been subsistence nomads, living in *gers*, moving three to five times a year in search of good grazing for their herds. With the influx of Soviet technology and industry, small cities sprang up, to which many moved for industrial jobs. Currently, more than 50 percent of the 2.4 million population lives in urban areas. The World Bank estimates that 50,000 housing units will be needed over the next five years in the capital, Ulaan Baatar, alone. Construction of more uninsulated houses and apartments will mean even more pollution from the archaic coal-burning heating

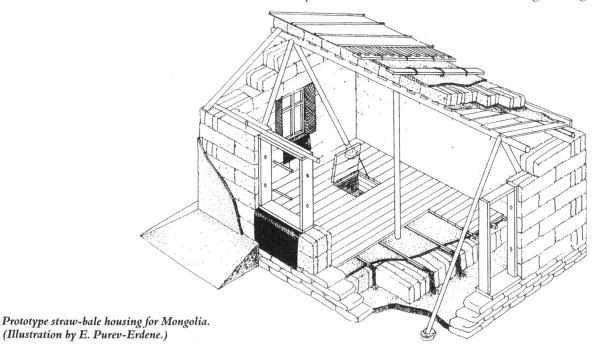

Prototype straw-bale housing for Mongolia.
(Illustration by E. Purev-Erdene.)

plants. Mongolia has experienced some of the largest earthquakes on record in the world, and those living in concrete tilt-up and unreinforced masonry apartment buildings are at great risk.

The cold, dry climate, environmental challenges, seismic threats, and housing shortage in Mongolia make this region an excellent candidate for straw-bale construction. When Scott Christiansen, of the Adventist Development and Relief Agency (ADRA) came to Mongolia in 1994, he saw the physical and economic effects of extreme cold—families living in *gers* and rural social service agencies that spent more than 50 percent of their yearly income on heating. After convincing the government officials that straw-bales *are* legitimate building materials, two small load-bearing prototypes were built in Ulaan Baatar with design and teaching assistance from Bob Theis, Dan Smith, Matts Myhrman, and Steve MacDonald.

These two prototypes featured a concrete-reinforced straw-bale floor and bale-insulated ceiling. Their construction was not without its challenges: The loose two-string bales had to be recompacted; wet, rough-sawn wood made for heavy trusses; a Masonite board ceiling provided little fire protection; and the designs lacked an air-lock entryway. Wire mesh was not available, and the Mongolians substituted their usual wooden lath under the weak concrete plaster. Subsequent inspections have shown significant cracking around windows and moisture condensation on the masonry ceiling board at the intersection of wall and ceiling where there is less insulation in the ceiling. Overall the prototypes were well received, but as private residences they did not have enough continued exposure to support replication.

In 1996, ADRA teamed up with another nongovernmental organization (NGO), World Vision, to construct a more public building—a kindergarten in a *ger* neighborhood. Straw-bale builder Jon Ruez traveled to Ulaan Baatar to construct the post-and-beam building. Although the straw-bale aspects of the kindergarten were successful, the occupants have been plagued by uneven heating from a masonry stove and leaky single-pane

The foundation of this straw-bale kindergarten and community center was built on short piers to address frost heave. (Photo by Kelly Lerner.)

windows. Over the summer of 1998, with more imported building materials becoming available, ADRA installed new windows with weather stripping and storm windows, and wall fans to move hot air into rooms isolated from the masonry stove.

In 1997 straw-bale construction in Mongolia "hit the bigtime." Based on Scott Christiansen's vision and the success of the first three buildings, the United Nations Development Programme (UNDP) in Mongolia launched a $2 million project to build 90 social service buildings in rural communities. Although UNDP is not in the habit of funding buildings, its research showed that, as mentioned earlier, rural social service providers (clinics, schools, and cultural centers) often spent more than 50 percent of their annual budget on heating fuel. With straw insulation reducing heating bills by 50 to 70 percent, schools and clinics can provide better services to more people. The Adventist Development and Relief Agency, the only NGO with straw-bale building experience, served as technical consultant and subcontractor for the project.

The UN Development Programme's project focused not only on constructing straw-bale buildings, but also on training (capacity building) and introducing other energy-efficient technologies: passive solar design, photovoltaic arrays, improved windows and heating systems, greenhouses, and retrofitting existing buildings with straw insulation. To meet the project's goals of training and capacity building, Steve and Nena McDonald spent the summer and fall of 1997 teaching Mongolian builders and site engineers in the techniques of straw-bale building. To develop local professional expertise, three Mongolian architects traveled to Berkeley, California, to study straw-bale building design with Dan Smith, Bob Theis, and me, and

Framing goes up for the second floor of the women's center in Ulaan Baatar. Straw-bale walls, energy-efficient windows and passive solar design significantly reduce the energy needed for winter heating. Notice the ger, traditional housing of Mongolians still being used, in the foreground with a stovepipe projecting from the roof. (Photo by Kelly Lerner.)

Front view of the women's center. Wooden lathing was used under the exterior plaster, as metal mesh is not made in this country. Traditional ger is in the left foreground. (Photo by Kelly Lerner.)

then on to Tucson, Arizona, to get hands-on experience with Jon Ruez. While in Berkeley, they helped us design Mongolia's first two-story straw-bale, passive solar building, a training center and shelter for the Mongolian Women's Federation.

I first traveled to Mongolia in the fall of 1997 to supervise construction of the two-story building and design energy-efficient windows, heating systems, and a prototype greenhouse. While there I also had the opportunity to design the UNDP social service buildings, write training manuals with Steve, and build the first permanent straw-bale vault in the world. Construction on the two-story building began the day after I arrived.

The salt-box design included a new balloon framing system with 50-by-100-millimeter studs built at 1,000 millimeters on center, with a 32-by-150-millimeter ribbon joist and diagonal wire seismic bracing. Paired 50-by-200-millimeter floor joists bear on the notched-in ribbon joist, and a

Rear view of the women's center. The standing-seam metal roof being assembled here was cut and bent on-site. (Photo by Kelly Lerner.)

three-layered truss sandwiches the columns. After the grade beam and insulated floor were poured, we put up the first-floor columns and built the second floor with the columns running long—500 millimeters above the floor level. The second-story columns were assembled as part of the three-layer trusses, with 32-by-150-millimeter chords and 50-by-100-millimeter webbing, and tilted into place, sistered to the first-floor columns.

The framing system worked well, but with heavy, wet larch, basic hand tools and a chain saw, even rough framing was a challenge. All lumber had to be milled from rough slab sides, and it was impossible to find tongue and groove decking (or even dry decking) for the second-floor diaphragm. Interior partition walls were framed with 50-by-100-millimeter studs at 1,000 millimeters on center and stuffed with straw under a cladding of lath. All wire mesh in Mongolia has to be imported from China, so we used wood lath instead, nailing it horizontally to the vertical frame. The standing seam metal roof was hand-fabricated on-site from 1,000-by-2,000-millimeter sheets of 20-gauge steel.

Again, the most difficult aspects of the building were unrelated to the straw bales. Freezing and cracking plaster (applied in late September), warped windows manufactured from wet wood, and a malfunctioning thermo-siphon hydronic heating system were the greatest trials. But the two-story

To insulate a Mongolian ger for the bitter winter, we improvised a temporary wrap of straw bales. Notice the masonry building being constructed in the background. (Photo by Kelly Lerner.)

framing system was enough of a success to be replicated during the following building season in a nearly commercial building.

In one of our lowest-tech experiments, we wrapped an existing *ger* with hay for the winter. We stacked and pinned four courses of bales, curving each bale slightly as it was placed. The fifth course was turned 90 degrees and bore half on the bale wall and half on the *ger* roof. Over the winter, fuel consumption was reduced by about 50 percent and the inside temperature was much more stable. Hay is available and cheap at the time of the fall harvest, but expensive in spring. A family could easily buy hay bales in the fall, insulate their *ger*, save fuel all winter, and the sell the bales for a profit in the spring.

A STRAW-BALE VAULT I had been experimenting with load-bearing vaults for three years and promoting them as an "easy-to-build," more material-efficient approach to construction with straw. A straw-bale vault can provide a fully insulated shell without the inherently difficult straw-wood transition at the top of the wall, and without the skill and wood needed to build a roof. With friends and colleagues, I had built five temporary vaults previously, and I felt sufficiently confident in the technology to try a permanent one in Mongolia.

With an eye toward future replication, I designed a simple 5-by-8-meter vault with a south-facing end-wall sun space and attached water tank room. The vault was part of a larger Permaculture design for an east-facing slope with transverse swales (to increase groundwater), gardens, a greenhouse, a water catchment system, and chicken husbandry. On September 30, 1997, I naively gathered a crew of inexperienced workers and began work. We used soil, produced by digging the swales and leveling the site, for filling earthbags for a foundation.

The falsework was made of rough slab sides—remilled, after the falsework was removed, on our table saw and used in the roof and floor. The falsework was covered with a skeleton of vertical 12-millimeter rebars at 500 millimeters on center. The structure of the vault was formed by bales in a compressive arch, stabilized by an exoskeleton of paired ribs inside and outside, tied through the bale wall with 2-millimeter wire. For this vault, I also trussed the ribs with diagonals to reinforce it further, because the narrow site prevented any buttressing.

The vault went up easily, with incredible interest from the surrounding community. For the shorter ribs on the back half vault, I tried laminated lath instead of the more costly rebar. They performed well, and I had visions of low-cost vaults made only from straw bales and laminated ribs.

Left: *Temporary form work for the straw-bale vault residence was made of local rough-cut lumber. Procuring usable construction lumber was always a challenge, as most of it is uncured, and the purchase itself required persistence and a lot of haggling. (Photo by Kelly Lerner.)*

Right:*The experimental straw-bale vault house performed better than expected in testing of insulative ability and fuel consumption during its first winter. (Photo by Kelly Lerner.)*

The weather was getting colder every day. The thin coat of fireproofing mud plaster froze as we applied it. When the big day came to remove the falsework, the vault settled less than 20 millimeters, consistent with my previous experiences. But as the front sun space frame was being prepared, I noticed changes in the vault's shape. Soft bales and softer wedges were taking their toll. Despite the added trussing between the exoskeleton ribs, the vault was definitely deforming, so I installed an interior ridge beam to support some of the weight and arrest the slumping.

The roof consisted of 50-by-50-millimeter purlins running along the length of the vault, covered by a skip sheathing of lath and tar paper roofing. The joints were secured with lath strips, and the ends were held in place by site-fabricated metal rainwater gutters. The floor was insulated with bales laid in rows, with 100 millimeters between them. These spaces were filled with compacted gravel that held a 50-by-50-millimeter sleeper, which supported the wood floor. The sun space sported a concrete slab floor, insulated with a layer of fly ash. Of course, with temperatures of –4 degrees Celsius, some of the finishing work would just have to wait until spring. A family of three moved in the day I left Mongolia, exactly two and a half months after we had begun.

In spite of leaky windows and cracked plaster, the vault performed far better than I had expected over the winter. The following ADRA report summarizes the results:

Two separate fuel-use tests were conducted at the vault site; one in mid January before the arrival of monitoring instrumentation, and one in mid March after the arrival of monitoring instrumentation. The results of the tests basically demonstrate the influence of passive solar gain and warmer temperatures in March compared to the very low daylight exposure and severe weather in January. Each test lasted two days.

	Vault	*Ger*
Test 1—January 1998		
Average outside temperature	–25 C	–25 C
Average inside temperature	21 C	15 C
Kilos of coal/day	10.2	34
Kilos of coal/day/square meter	.463	2.125
Inside humidity	30%	15%
Test 2—March 1998		
Average outside temperature	–4.5 C	–4.5 C
Average inside temperature	21 C	17.8 C
Kilos of coal/day	.85	15.3
Kilos of coal/day/square meter	.039	.956
Average inside humidity	26%	15%

In the severe weather conditions of January, the vault showed a savings of about 70% over the ger. When corrected for per-square meter fuel use, the vault showed a savings of about 80%. Comfort and health levels in the vault were much higher than in the ger. In the March test, with less severe weather and more solar gain, the vault showed a savings over the ger of almost 95%. When corrected for a per-meter measurement, the savings of the vault over the ger is about 96%. Taking the fuel-use data at hand and trying to extrapolate savings over the season (with significantly more solar gain at the beginning and end of the heating season), ADRA estimates that a net savings of at least 85% can be achieved by the vault over the ger on a per-meter basis.

So, in the end, the vault was a successful experiment. It provided excellent fuel-use data and a warm home for a family. Yet over the course of construction, I ultimately realized that a vault is just too complex a structure for a typical owner-builder to intuitively understand and build. Because of the unique shape and composite construction, there are too many places where simple mistakes can threaten structural integrity. Perhaps the greatest success of the vault is that it taught me clear lessons that could be applied to the next generation of straw-bale houses in Mongolia.

1998 BUILDING SEASON Funding for international straw-bale projects can come from surprising sources. One angle that holds promise is carbon dioxide (CO_2) mitigation credits. In short, many countries throughout the world have agreed to lower their CO_2 emissions over the next 20 years in order to protect the ozone layer. Polluting companies can choose to lower their own CO_2 emissions or they can lower emissions somewhere else.

The coal-burning heating systems in Mongolia emit massive amounts of pollution with deadly consequences. More than 50 percent of child mortality in Mongolia is due to respiratory tract infections. The vault had shown that replacing a *ger* with a straw-bale house can save 7.25 tons of CO_2 emissions per year. And Ulaan Baatar would need more than 50,000 housing units over the next seven years. If ADRA or the UNDP could find money to pay for materials and training, owner-builders could be taught how to build their own houses. But to do this we needed more information: Could a design be simplified enough for owner-builders? What would a small, passive solar house like this cost?

The goal of the project was to design and build four simple houses (25 to 40 square meters) in order to analyze construction costs and ease of replication. Two of the units were load-bearing with gravel bag foundations, and two were lightweight post and beam (32-by-100-millimeter posts at 1-meter centers, notched into the outside) with concrete footings. All four featured a central ridge beam to short roof spans and a new bale sandwich roof system to avoid the heavy trusses used in previous projects. Borrowing a design archetype from the *ger*, two central posts supported the ridge beam

First prototype straw-bale house built in Mongolia in 1995. (Photo by Matts Myhrman.)

and helped to divide the main room into different areas. With straw-bale insulated wood floors, new windows with hinged interior storm windows, weather stripping, and a decorative (and time-saving) banding of rough plaster called a water table, these units set new quality standards for straw-bale buildings in Mongolia at an affordable price.

Having learned about loose bales from the vault project, I started by recompressing all the bales in a simple hand bale press. Leaving the strings in place, one person would place the bale in the press and bear down on the lever, another person would catch the string with a small stick, twist until the string was taut, and then pound the stick into the bale until tight. The bale walls were also reinforced with an exoskeleton of 50-by-50-millimeter interior and exterior paired ribs, tied through the bale wall. These stiffened up the floppy walls and provided attachment for the 10-by-20-millimeter wood lath.

The load-bearing units were designed with a ladder truss roof bearing assembly, but the post-and-beam units were to have rafters bear directly on the posts with a small 32-by-100-millimeter ribbon rim joist to secure the top of the posts. As it turned out, the site had no electricity for the first five weeks of construction, and we were not able to rip the 32-millimeter slab sides into posts when the wood frame was to be erected. Instead, we put the bale walls up first, with plans to notch the post in later (this was standard practice on many of my jobs in the United States). With nothing to secure them at the top, the eight-course-high walls danced and waved as the wind

Mongolians plastering over wooden lath of the second model of straw-bale housing. These simple houses cost even less than a new ger and half to a third as much as new brick housing. (Photo by Kelly Lerner.)

Plastering nearly complete. Notice how the walls have been gently curved into the windows. (Photo by Kelly Lerner.)

blew, and we cut the notches with a hand saw. Finally, pulling the walls back into plumb, we installed a collar tie and then used site-fabricated galvanized metal ties to connect rafter and posts. We learned a valuable lesson: If the frame is to be notched-in after the bale wall is built, a collar tie is required to hold the top of the wall in place. The collar tie can be omitted only when the frame goes up first and is braced in place (as in the two-story building).

The roof was a simple sandwich of recompressed bales placed between rafters. Ceiling purlins were hung off 32-by-250-millimeter rafters, which bore on the outer wall and the 50-by-250-millimeter ridge beam. Bales were placed between the 1 meter rafters on center, and the cracks tightly stuffed with straw, before four evenly spaced purlins were placed on top and tied through to the rafters below. Site-fabricated standing seam galvanized metal roofing bore directly on the purlins and bales. The single-walled chimney was protected by a galvanized metal box filled with fly ash, which kept it away from combustible materials. A gypsum board ceiling isolated the bales in the ceiling, protecting them from interior fires.

Although Ulaan Baatar has not experienced any large earthquakes since its construction, Mongolia is seismically at risk. By virtue of the inherent shock-absorbing qualities of straw, many believe that straw-bale construction can be seismically resistant. David Mar, a structural engineer in Berkeley, California, and I worked together to develop a system that will hold the walls and roof together in case of a big quake. Luckily, the units are so small that even with the added weight of bales in the roof, the seismic loads are relatively low.

In Mongolia, wire is imported from China and costs four times more than wooden lath, so David and I adapted wire mesh details into vertical and horizontal wire ties. In the load-bearing units, wire loops for tie-downs ran beneath a 50-by-50-millimeter purlin, sandwiched between the second and third course of gravel bags. After the wall is up, wire tie-downs go over the roof-bearing assembly and are tightened after the weight of the roof precompresses the bale wall. The walls are tied together horizontally, with ties at the second, fourth, and sixth courses, which wrap around exterior ribs.

Window and door openings are secured with horseshoes of wire mesh, as it is needed for plastering anyway. Wooden lath prevents the cement plaster, wire ties, and bale walls from knitting together as well as the solely wire mesh version, but the wire ties, notched-in exoskeleton, and wood lath all work together to contain the bale wall as it absorbs lateral forces.

Final material costs for Mongolian owner-builder replication averaged $47 per square meter ($4.37 per square foot). Compare this with material costs for the other types of new housing: A new *ger* at the market in Ulaan Baatar runs about $51 per square meter, a plastered log cabin costs about $60 per square meter, and a new brick house can cost anywhere from $80 to $150 per square meter. Straw-bale houses are clearly a better solution at a better price. These cost data, along with the fuel-use data from the vault, lays the groundwork for ADRA's CO_2 mitigation proposals, which may fund several thousand owner-built straw-bale houses over the coming years.

The immediate success of straw-bale building in Mongolia has not been achieved without encountering problems. The interest in and demand for straw-bale buildings currently outpace the experience and skills necessary to design and build them. Although many workers have been trained in basic straw-bale building techniques, architects and engineers with experience in this method of construction must still be imported from outside Mongolia. Because the first buildings were small, low-tech houses, some professionals and government officials developed an impression of straw-bale construction as low-quality, poverty housing and have no interest in pursuing the technology. The lack of well-trained workers and a few low-quality straw-bale buildings have supported this stereotype. While we seek and apply low-tech solutions, we must be careful to maintain the highest quality in straw-bale building and educate professionals (as well as workers) at every turn.

China

In January 1998, Zhang Bei County in the northeastern province of Hebei, China was struck by a 6.2 earthquake. Fewer than 100 people lost their lives, but more than 42,000 homes, businesses, and schools were destroyed, leaving families scrambling for shelter in the −30 degree Celsius weather. Scott Christensen, country director for the new ADRA office in China, used this earthquake as the opportunity to introduce straw-bale technology in China. I traveled to China for a week in July 1998 to study the feasibility, research available materials and construction techniques, find a suitable site, and, if possible, design a school building.

When introducing a new technology (like straw-bale), I find it much better to start with known, local materials and add innovations bit by bit as the area and the people become better known. The materials that were available would determine to a large degree the composition of the building.

I was relieved and amazed to find steel vinyl-clad windows and hydronic thermo-siphon heating systems manufactured in Zhang Bei, only 10 kilometers from our chosen building site in Xiao Dongpo village, where we would be adding a new three-room middle school to an existing primary school. Wood is expensive and scarce in China. The large beams and purlins used in reconstruction typically came from hundreds of kilometers away from a building site. Bamboo grows in the wetter temperate and subtropical areas of the country. So instead of a wood or bamboo truss, I chose a locally available double-angle steel truss to span the 6-meter classrooms. The trusses bear on double-angle columns, which notch neatly into the interior face of the wall. Because this is a wetter region than Mongolia (400 millimeters of rain per year versus 200 millimeters in Mongolia), a 200-millimeter layer of locally mined pumice insulated the brick floor. My research turned up only one missing building material—bales.

Land in this area is farmed by families that lease plots from the government. Each family grows a variety of crops and vegetables—beans, potatoes, corn, greens, wheat, and barley. Grain crop plots are small and most straw is used locally, so baling is not typical. In fact, no one in this region had ever seen a baler or a straw bale. We inspected wool balers to see whether they might be used, but they were the wrong size—too large and square. Finally, modifying the design of a hand baler used to recompress bales in Mongolia (originally designed by Steve McDonald), I had the local master carpenter make a hand baler. Before the end of the week, using straw from the previous year's harvest, we manufactured the first two straw-bales in China.

When I visited six months after the earthquake, professional building teams and farmers were working side by side to relocate 152 villages (averaging 50 houses per village). The collapsed earthen and stone-walled homes were being replaced with houses of unreinforced brick masonry on foundations of dry stacked stone, faced with cement plaster. Three large beams (250 millimeters in diameter) bore on walls 3 meters apart, supporting 50-millimeter-diameter purlins, and a bed of mud and straw under a terracotta tile roof. Although brick bearing walls may be better than earth and stone, heavy tile roofs and large south-facing openings put these new buildings at great risk in the event of a similar seismic occurrence. Clearly, the earthquake-resisting abilities of straw-bale construction are even more impressive than its insulating potential in these circumstances.

Applying his experience with seismic design for concrete and steel frame buildings, David Mar, S.E., has proposed a new conceptual model of

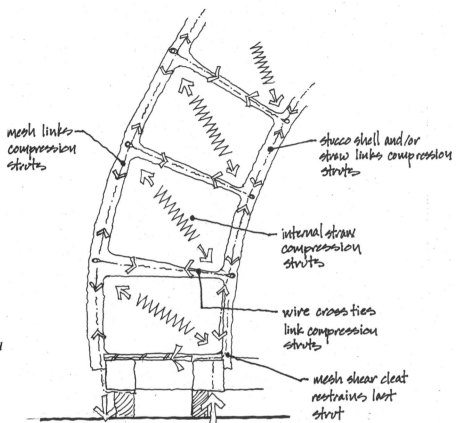

mesh links
compression
struts

stucco shell and/or
straw links compression
struts

internal straw
compression
struts

wire cross ties
link compression
struts

mesh shear cleat
restrains last
strut

*Shear mechanism developed
by engineer David Mar for
compression struts in a
vaulted straw-bale wall
assembly. (Illustration by
David Mar.)*

how straw-bale walls can perform. A straw-bale wall acts like a low-strength version of concrete, transferring lateral seismic loads to the foundation through internal diagonal compression struts.

A simple version of David's structural system had been informally tested for a vault-shaped structure with great success early in 1998. Given the high seismic risks in China, I was eager to apply the system to this building. David and I worked together to design and adapt mesh details to a hanging bale-insulated ceiling and largely open south wall. Without the data provided by formal testing, we cannot predict exactly how the structure will perform in an earthquake, but I sleep soundly at night, knowing that this building will be standing long after its brick and tile-roofed primary school sister next door.

⊠ KELLY'S PRINCIPLES FOR INTRODUCING ALTERNATIVE BUILDING TECHNOLOGIES IN DEVELOPING COUNTRIES

Bringing ecological building technologies to another country, bridging cultural gulfs and foreign languages, is full of obvious as well as hidden pitfalls. First, there are the challenges of constructing in a new place—finding materials, modifying designs, communicating with the building crew, and dealing with local authorities. With patience, humor, an adequate budget, and an experienced builder, most problems of this type can be solved over time. Changing attitudes and educating building professionals, on the other hand, is the greater challenge.

Most developing countries are anxious to modernize. In many places, a community has given up sustainable indigenous building methods in favor of "Western" concrete, wood frame, and steel, at great expense to their natural environment, culture, and standard of living. A return to local materials and low-tech techniques can be associated with a return to poverty. In order to take root, sustainable building techniques must be proven to be clearly superior—less expensive, more beautiful, warmer, cooler, or easier to build.

But changing attitudes takes more than hard evidence; it requires give and take, a localization of designs and methods, a dialogue between respected colleagues. Introducing a new technology is never a one-way process.

Toward the goals of two-way learning and planting sustainable technologies, here is a short list of principles that guide my work in foreign countries (and they work well at home, too!)

1. Learn from the locals

Ask questions, observe living patterns, investigate construction sites, build relations with people, listen to their dreams and their fears. Look at traditional architecture and how it has been translated (or ignored) in modern building. Travel. Immerse yourself in the people's lives. Check and recheck design ideas with all kinds of people. You may be a technical expert in one area of construction, but they are the experts as to their needs, desires, and resources.

2. Find and work with good local partners (individuals and organizations)

Look for partners who are curious and innovative, but well rooted in their own culture, skilled at their craft, and respected by the community. Good partners can act as cultural interpreters, open doors to hidden resources, and help implement projects in the most appropriate manner.

3. Use local materials and local techniques wherever possible

Introducing new technology can be tough—ease the process by combining the new with familiar materials and locally proven techniques. Go slowly. Introduce and fine-tune one new material or technique at a time. Moving slowly allows for feedback, learning, and local innovation.

4. Ground every project with local investment and involvement

This may take the form of sweat equity, land donation, material supply, or administration. Local involvement builds in a cultural feedback loop, eases implementation headaches, and ensures local responsibility after the project is completed.

As a result of her work teaching alternative construction in countries where she did not know the language, Kelly Lerner helped to create a simple primer, using instructive visuals like these with no text. She hopes it can be published for owner- builders in Mongolia and other countries who would benefit from this technology. (Illustrations by E. Purev-Erdene.)

Constructing earthbag footings for a straw-bale greenhouse and cold frame.

Wall and roof assembly of greenhouse and cold frame.

Wrapping a traditional ger with straw bales for winter insulation.

5. Train, train, train

Your main objective is to develop local expertise and work yourself out of job. Teach at every level: architects, engineers, government officials, building inspectors, builders, owner-builders, and the general public. Hands-on, experiential training programs are most effective. Make every building a learning and teaching opportunity. Use training manuals with lots of pictures and three-dimensional drawings to transcend language barriers.

6. Be professional; work to the highest possible standards

When you introduce a new technology, your project may be the only example of it in the entire country. Many people will be watching, and the whole technology may be judged a success or a failure based on your one building. Do your homework ahead of time, work closely with your local partners, and maintain high building and safety standards on the job site.

7. Be flexible to local custom, but do not compromise on structure or other life safety issues

Often, local materials or local methods will compromise your design ideas or your intents. Work closely with your partners and your construction crew to find innovative solutions and material substitutions, but do not compromise your standards on life safety issues, especially in regard to structure or fire resistance. Many people are watching you and assume that your way is best. Make sure that your approach is worthy of replication.

8. Find intuitive building approaches

Try to devise solutions that make sense in the local context and with the available materials. If a design approach is too costly, too difficult, or too complicated, it will be poorly replicated or left out entirely next time. The best solutions are intuitively understood and replicated without mistakes.

9. Take a long-range view

Do not try to do everything at once—you will fail miserably. Plan for a diversity of projects over time, each highlighting a different approach or application. Each project should build on the last and should have increasing input of local expertise.

Epilogue

Finding the Soul of Natural Building

Tom Bender

Do places have souls?

There is life in *all* creation. There are wombs in space that give birth to galaxies and stars. The hearts of stars sing like bells. The rocks under our feet thrum with messages from within and around the world. Trees make love with a thousand others at the same time. Microfauna in our cells create communities and transportation systems. Communities have personalities. A forest is a single organism. Planets have consciousness. And they all sing together in harmonious celebration of life.

Places, even, *do* have souls—small or great, gentle or fierce, nurturing or debilitating. Like all life, they have distinct and often strong personalities. They have auras and energy bodies. They are touched and altered by our regard or disregard of them, and they are able to move our hearts and alter our lives. They can enrich and nurture us, empower and connect us.

In their most powerful form, places connect us into, and allow us to coexist in, the nonmaterial planes of existence as well as in the material one. With them, we can individually or as a group expand our conscious presence into some of those other realms. Most simply put, making places with souls is the most central thing we must attend to in making buildings to shelter our lives and nurture our hearts.

Alternative Construction: Contemporary Natural Building Methods, edited by Lynne Elizabeth and Cassandra Adams ISBN 0-471-24951-3 © 2000 John Wiley & Sons, Inc.

We have been able to learn in recent years how to create places that can achieve such goals. When all the pieces are right, everyone who enters such places breathes a sigh of relief and happiness. Their legs get rubbery, and they want to sit down and just soak in the energy. Such places are filled with a powerful silence. They connect us to the rest of creation. They nurture us with the breath of life. They are the soul of natural building.

<hr>

Intention

There are many aspects to places with soul. One of the most important is *intention*. All of our surroundings are like mirrors, reflecting back to us the intention that has gone into their making and use—the values of their makers. If made from greed, if made to deceive, they convey that. If they come from a meanness of soul or smallness of spirit, they infuse us with that essence. If made with love, with generosity, with honoring of all life, they support and evoke the same intentions in our own lives.

Clarity, strength, and rightness of intention also bring life-force energy, or *chi*, into a place, with its ability to nurture our lives. The nature of our intention—whether in making or using a place—reflects that same energy back into our own lives, enhancing or weakening our own energy.

Even more, our intention toward a place can change the lives of others. Out of an intention of making a Head Start Center good for the kids using it, we asked ourselves what would make us feel best if we were kids coming in the door. "The smell of good food!" was the unanimous response. This led us to put the kitchen right in the middle of the building, open to all the classrooms and entry. It works wonderfully, giving immediate pleasure and a sense of rightness to those coming in the door. It also gives parents a place to stop for a cup of coffee and a chat, to peek around the corner to see how their kids are doing. It allows the cook to be an extra friend, a source of snacks and hugs for the kids, and a backup pair of eyes for the teachers.

What we did not realize until later is how much our intention totally changed the idea of working as a cook in this place. Cooking is usually a "back-room" job, tucked away out of sight in a service area near the loading dock. In contrast, putting the cook in the *middle* of everything, and in contact with everyone, made the cook a whole-person part of what went on.

An architect later asked what we would do if the center were larger and needed a bigger kitchen and loading dock. I said, "You've just defined *too big!*" A change in intention—from wholeness and people-centeredness to optimizing mechanical function—underlies our gut feeling of wrongness when something becomes too big.

Just as approaching building with positive intentions can change the lives of people using our places, it can be empowering to the people making them. Materials that can be obtained locally, and can be put together with sweat equity of the owners instead of bank loans, encourage a sense of accomplishment for the owner-builders, provide opportunity for enrichment rather than standardization, and avoid the ecological costs of transportation and industrialized processing. The intention to empower brings forth a need for the natural building materials being rediscovered and refined today.

Even in working with architects and professional builders, there are ways of empowering people doing the building if we make that part of our intention. I began several years ago to add a "1 percent for Heart" section to my specifications, to pay for suggestions from the workers on improving the spirit of the building. This has encouraged the workers to think creatively about *everything* they do in the building, often resulting in many wonderful touches without any additional cost.

Without a clear and positive intention, ugly and uncomfortable buildings can be made with natural building materials just as easily as beautiful and satisfying ones. With the goal of designing a place with a soul, we give direction and destination to our powerful engine of creativity in building. That intention guides not just choice of materials but the shapes and spaces created from them, the means used to warm and cool the occupants, the connections made with the rest of nature, and the values expressed in the building. Unconsciously or consciously, I think most people working with natural building materials are seeking that soul of place. And it is that intention as much as, and in combination with, the materials themselves to which we respond with an emotional sense of rightness.

Chi

Chi, or life-force energy, is another part of the soul of place that we are rediscovering today. A central part of the philosophy, healing arts, and operation of society in most cultures worldwide, *chi* is today becoming acknowledged in our own culture. It underlies acupuncture, faith healing, *feng shui*, martial arts, yoga, and a variety of other practices. Combined with intention, it forms the subtle energy template upon which our material world takes shape in its many wonderful variations. It is vital to supporting our physical, as well as emotional and spiritual, health. It is blocked by artificial building materials, intensive use of electromagnetic devices, and cultural practices based on taking from others.

We are learning today, in addition to identifying good natural power spots of *chi* on which to locate our buildings, that *chi* energy can be called in, enhanced, and worked with by individual intention and group ritual, and that it forms the glue that keeps a community healthy.

Diseases of the Spirit

The truly rampant diseases in our culture are not of the body, but are *diseases of the spirit*. They arise from lack of self-esteem and mutual respect, being of value to our community, or finding meaning in our lives. They find expression in rape, substance abuse, addictions, violence, crime, obesity, isolation, depression, and despair—conditions possible in any culture, epidemic in ours.

Healing diseases of the spirit requires that we nurture, not neglect, the emotional and spiritual well-being of all. This requires in our surroundings the honoring of the materials, the elements and forces of nature, the rhythms and cycles of life, and limiting our wants so as not to prevent the fulfillment of other forms of life. These are all possibilities inherent in natural building materials, used with reverence.

Giving

In a culture rooted in taking from others and keeping things to ourselves, the act of giving is a powerfully transformative deed. Expressed in the shaping and use of our surroundings, it becomes the embodiment of the spirit needed for sustainability as individuals and as a culture. Giving enriches places through what we discover can benefit other people or other life in the process of building. Shading, or giving the scent and beauty of flowers to adjacent public areas; allowing pedestrian ways to cut through large projects; providing low walls that can be used for seating, or facilities that can be used by the community when not needed by the primary users, are all gifts.

Providing habitat for birds, spiders, bats, and butterflies; restoring creeks and watersheds; providing wildlife migration routes—all are forms of giving, as is restraining our building to allow room for the rest of nature to live unthreatened. A place may well achieve that generosity of spirit in surprising ways—like a Japanese room, which is generous in space because of its *emptiness*, not because of its size. Generosity is created out of the love and energy put into making. It gives the *unexpected*.

Another and wonderful form of giving is *honoring* or *celebrating* elements of nature and life in what we make. Honoring is a giving of respect. Celebrating is a giving of thanks. A building with a soul honors its surroundings and the materials that were given to make its existence possible. We can honor the materials by allowing their history, beauty, and power to come through their use in ways that move our hearts. A building with a soul honors the skill, competence, and the sacredness of the work gone into its making.

A place with a soul can honor our inner resources as well as our material ones, and it can honor its users, as in the Japanese custom of placing guests before a *tokonoma*, giving them a sense that they and their activities are of value. In respecting building tradition, a place honors the insights and wisdom gained from the past. By planting trees, or by other means, it honors a hope for a future. It honors all life, and the power that begets it.

A building with a soul fills primal psychic needs—for protection, for warmth, for companionship, for meaning. It enfolds and gives refuge and sanctuary to all who enter it. It affirms *sacredness* and meaning in our lives and surroundings and creates places for our hearts and minds as well as our bodies. A building with a soul draws on and connects its users to power extending beyond the material world.

Joining the Community of Life

A building with a soul is enriched and given meaning through its connection with other things. It brings us into closer touch with each other, the rest of the world, and the rhythms of nature. By opening our places to sunrise and moonset, it connects us to the daily and seasonal cycles of the sun, the moon, and the stars; to the beauty of rain, fog, and snow; and to the visible and invisible universe. It adapts readily to changes in use and additions to its structure.

Every place has distinct communities, singularly tied to the qualities of that place, which have evolved through the ongoing testing of centuries. Until we learn to nestle our lives into those distinct ecologies, to celebrate the specialness characteristic of each place, we remain as awkward outsiders. We stumble around foolishly—unintentionally disrupting, wasting, and destroying through our every act while failing to receive the ease and plenitude that lie in being an integral part of a place.

Taking part ourselves in building, bringing materials from local sources, learning the real costs of what we create, brings us in touch with that local community in ways impossible with power tools, transported materials, and professional design and construction. It teaches us directly, not intellectually, the importance of nurturing what lies on both sides of our skins.

A building with a soul fits its site and makes the best use of it, making almost magic connections between location, relationships, and views. The arrangement and organization within it, outside it, and in connection with the life around it, are apt. It fits its use and users and the dreams that drive their society. It fits the capabilities, beauty, and aptness of local materials, local ways of building, local traditions of design, and local patterns of living. It chooses local wisdom for dealing with its unique climatic conditions and ways of heating, cooling, and ventilating. It touches the spirit of where it is. It helps us make where we *are* paradise.

Right Duration

Because it is loved, a building with a soul often endures beyond the needs of its makers to become a gift to future generations. A cathedral lasting 20 generations, or a bridge lasting 20 centuries can give back far more than the effort put into their making. Such endurance immeasurably alters the per-generation cost of resources and construction. Durability thus grants a generosity to the places we make that can be obtained in few other ways. A building with a soul needs to be as comfortable a thousand years in the future as it was in the past and is today. It needs to be comfortable with the changes of time, neglect, and love—mellowing and becoming enriched rather than tarnished and tattered. There is a hoary strength and a nourishing peacefulness in the timeless qualities of a building that truly fits our hearts and spirits.

Yet not everything benefits from lasting longer than its nature. A generation from now we may not wish that some of the things that we have recently created had lasted beyond their time and intention. It may be good that our homes or vehicles are durable. It may not be desirable that some of our building materials are preserved with poisons that linger and harm.

The Inuit who throws away a scrimshaw carving once the empowering act of creation is finished, or the Balinese village or Indian pueblo that returns imperceptibly to the earth when its use is done holds a rightness of duration and of material choice.

The Web of Relationships

It is important, as we learn the wisdom of material choice and the greater goals for which we assemble them into new creations, to follow far the web of significance of each of our choices and to sink deeply into the earth the roots of our own understanding. Today it may indeed be wise to use less wood and to find more locally appropriate materials in lands where trees are not plentiful.

Yet wood is a wonderful, natural material. Our demands today are limitless and beyond the capacity of any resource to satisfy. Reasonable demands may be satisfied by materials that are appropriately obtained and appropriate to such use. Building with wood in a place such as the Pacific Northwest (this author's home), which *is* wood as far as the eye can see, honors the rightful and abundant life of the place.

To ensure its well-being, let us use wood with honor. Let us use it in ways that move our hearts by its wonderful nature, love it so much that we demand and ensure the survival and health of our forests.

In Japan, for example, you can find doors made of single slabs of wood showing 500 years of growth rings. You can find buildings with whole trees, not sawn timbers, used for posts and beams. You can find a verandah made with only two planks—sawn through the center of the tree so that you walk the life and growth of the tree as you walk the length of the verandah. You find wood honored and celebrated, and used with love, care, and restraint.

Simplicity

A building with a soul takes a simple and modest route, rather than a complicated one, to fulfilling our needs. It lets nature do the work rather than machines. It finds simple answers to needs rather than complicated high-tech solutions. It knows that excess is as harmful as meagerness and discriminates between things that harm and those that nurture.

Invisibility

A building with a soul fulfills our needs without calling attention to itself, making its use respond to important inner qualities rather than superficial outer ones. It demands little in the way of resources.

It is filled with the *emptiness* of Lao Tsu's teacup and reverberates with the peace of *silence*. It is free of unnecessary possessions and mechanical noises, and open to the joyful sounds of birdsong and laughter and the sound of the wind. It draws back into shadow, letting the light and attention rest on its inhabitants and their partners in creation.

Coherence

A building with a soul is consistent and arises out of a single, whole, and clear vision of the needs it can fill and the possibilities it can unfold. It reflects a lucid and unencumbered *intention* of its owner, designer, and builder. It has sought and found the heart of the institution it is sheltering, and found ways to unfold that heart in its making. The issues it has addressed are fundamental and not frivolous, and the solutions it has created are sound.

Love

Most simply put, a building with a soul is one that is built from love.

The Soul of Natural Building

In honoring the character of natural building materials, we can create joyful places with souls—places of power, gardens for our spirits, and cities of passion. We can rediscover ways to collaborate with nature. We can relearn that simple and wise building can accomplish far more than complicated and isolated technologies. And, through natural building, we can begin to make safe places within which our own hearts can open, join with others, and flourish.

Appendix A

Recommended References

Vernacular Architecture and Modern Interpretations

1. Brown, R. J. *The English Country Cottage*. London: Hamlyn Paperbacks, 1979.
2. Brunksill, R. W. *Illustrated Handbook of Vernacular Architecture*. New York: Universe Books, 1971.
3. _____. *Traditional Buildings of Britain: An Introduction to Vernacular Architecture*. London: Victor Gollancz, 1981.
4. _____. *Houses*. London: Collins, 1982.
5. Denyer, Susan. *African Traditional Architecture: A Historical and Geographical Perspective*. New York: Africana Publishing, 1978.
6. Duly, Colin. *The Hoses of Mankind*. London: Thames and Hudson, 1979.
7. Evans, Ianot, Michael G. Smith, and Linda Smiley. *The Hand-Sculpted House*. White River Junction, VT: Chelsea Green, 2002.
8. Fathy, Hassan. *Architecture for the Poor: An Experiment in Rural Egypt*. Chicago: University of Chicago Press, 1973.
9. _____. *Natural Energy and Vernacular Architecture*. Chicago: University of Chicago Press, 1986.
10. Gardi, Rene. *Indigenous African Architecture*. New York, Van Nostrand Reinhold, 1973.
11. Golany, Gideon S. *Chinese Earth-Sheltered Dwellings: Indigenous Lessons for Modern Urban Design*. Honolulu: University of Hawaii Press, 1992.
12. Howard, Ted. *Mud and Man: A History of Earth Building in Australia*. Melbourne: Earthbuild Publications, 1992.
13. Kahn, Lloyd. *Shelter,* 2nd ed. Bolinas, CA: Shelter Publications, 2000.
14. Kennedy, Joseph, ed. *Building Without Borders: Sustainable Construction for the Global Village*. Gabriola Island, British Columbia: New Society Publishers, 2004.
15. Kennedy, Joseph, Michael G. Smith, and Catherine Wankek, eds. *The Art of Natural Building: Design, Construction, Resources*. Gabriola Island, British Columbia: New Society Publishers, 2002.

16. Knevitt, Charles. *Shelter: Human Habitats from Around the World*. San Francisco: Pomegranate Art Books, 1994.

17. Lewcock, Ronald. *Traditional Architecture in Kuwait and the Northern Gulf*. London: AARP, 1978.

18. McGregor, Suzi Moore, and Nora Burba Trulsson. *Living Homes: Sustainable Architecture and Design*. San Francisco: Chronicle Books, 2001.

19. Nabokov, Peter, and Robert Easton. *Native American Architecture*. New York: Oxford University Press, 1989.

20. Oliver, Paul. *Dwellings: The House Across the World*. Austin: University of Texas Press, 1988.

21. _____, ed. *Shelter in Africa*. London: Barrie & Jenkins, 1971.

22. _____, ed. *The Encyclopedia of Vernacular Architecture of the World*. Cambridge: Cambridge University Press, 1997.

23. Pearson, David. *Earth to Spirit: In Search of Natural Architecture*. San Francisco: Chronicle Books, 1994.

24. Rapoport, Amos. *House, Form and Culture*. Englewood Cliffs, NJ: Prentice-Hall, 1969.

25. Roberts, Jennifer. *Good Green Homes*. Salt Lake City: Gibbs Smith, 2003.

26. Rudofsky, Bernard. *Architecture Without Architects: A Short Introduction to Nonpedigree Architecture*. Albuquerque: University of New Mexico, 1964.

27. _____. *The Prodigious Builders: Notes Toward a Natural History of Architecture*. New York: Harcourt Brace Jovanovich, 1977.

28. Scully, Vincent. *Pueblo: Mountain, Village, Dance*. New York: Viking, 1975.

29. Steen, Bill, Athena Steen, and Eiko Komatsu. *Built by Hand: Vernacular Buildings Around the World*. Salt Lake City: Gibbs Smith, 2003.

30. Taylor, James. *A Shelter Sketchbook*. White River Junction, VT: Chelsea Green Publishing, 1997.

31. Wodehouse, L. *Indigenous Architecture Worldwide: A Guide to Information Sources*. Detroit: Gale Research Co., 1980.

Construction Technologies

1. Bainbridge, David, Athena Swentzell Steen, and Bill Steen. *The Straw Bale House*. White River Junction, VT: Chelsea Green Publishing, 1994.

2. Bee, Becky. *The Cob Builders Handbook*. Murphy, OR: Groundworks Publishing, 1996.

3. Berlant, Steve. *The Natural Builder*, Vol. 1, *Creating Architecture from Earth*; Vol. 2, *Monolithic Adobe Known as English Cob*; Vol. 3, *Earth and Mineral Plasters*. Montrose, CO: Natural Builder Press, 1998.

4. Chappell, Steve. *Alternative Building Sourcebook: Traditional, Natural and Sustainable Building Products and Services*. Brownfield, ME: Fox Maple Press, 1998.

5. CRATerre, EAG. *Earth Building Materials and Techniques: Select Bibliography*. Eschborn, Germany: GATE, 1991 (annotated list of 305 publications in English, French, Spanish, Portuguese and German, dealing with all aspects of building with earth).

6. Cusack, Victor. *Bamboo Rediscovered: Growing and Using Non-invasive Bamboo*. White River Junction, VT: Chelsea Green Publishing, 1998.

7. DeBoer, Darrel. *Bamboo Building and Culture*. Alameda, CA: DeBoer Architects, 2004.

8. Easton, David. *The Rammed Earth House*. White River Junction, VT.: Chelsea Green Publishing, 1996.

9. Evans, Ianto, et al. *The Hand-Sculpted House: A Practical and Philosophical Guide to Building a Cob Cottage.* White River Junction, VT: Chelsea Green Publishing, 2002.

10. Farrelly, David. *The Book of Bamboo: A Comprehensive Guide to This Remarkable Plant, Its Uses and Its History.* San Francisco: Sierra Club Books, 1995.

11. Goldberg, Gale Beth. *Bamboo Style.* Layton, UT: Gibbs Smith Publisher, 2002.

12. Guelberth, Cedar Rose, and Dan Chiras. *The Natural Plaster* Book. Gabriola Island, British Columbia: New Society Publishers, 2003.

13. Haggard, Ken, and Scott Clark, eds. *Straw Bale Construction Details.* Angels Camp, CA: CASBA, 2000.

14. Hill, Neville R., with Stafford Holmes and David Mather. *Lime and Other Alternative Cements.* London: Intermediate Technology Publications, 1993.

15. Holms, Stafford, and Michael Wingate. *Building with Lime: A Practical Introduction.* London: Intermediate Technology Publications, 1997.

16. Houben, Hugo, and Hubert Guillaud. *Earth Construction: A Comprehensive Guide,* CRATerre-EAG. London: Intermediate Technology Publications, 1994.

17. Imhoff, Dan. *Building with Vision: Optimizing and Finding Alternatives to Wood.* Healdsburg, CA: Watershed Media, 2001.

18. Janssen, Jules J. A. *Building with Bamboo: A Handbook,* 2nd ed. London: Intermediate Technology Publications, 1995.

19. Jones, Barbara. *Building with Straw Bales.* Dartington, United Kingdom: Green Books, 2002. Distributed in the Americas by Chelsea Green Publishing.

20. Keable, Julian. *Rammed Earth Structures: A Code of Practice.* London: Intermediate Technology Publications, 1996.

21. Kennedy, Joseph F. *The Art of Natural Building.* Gabriola, British Columbia: New Society Publishers, 2002.

22. Khalili, Nader. *Ceramic Houses and Earth Architecture: How to Build Your Own.* Hesperia, CA: Cal-Earth Press, 1996.

23. Kiffmeyer, Doni, and Kaki Hunter. *Earthbag Building: The Tools, Tricks and Techniques.* Gabriola Island, British Columbia: New Society Publishers, 2005.

24. King, Bruce. *Buildings of Earth and Straw.* Sausalito, CA: Ecological Design Press, 1996.

25. Lanning, Bob, with Tom Greenwood and Paul Weiner. *Straw Bale Portfolio: A Collection of Sixteen Designs for Straw Bale Houses.* Tucson: Bob Lanning, 1996.

26. Lacinski, Paul, and Michel Bergeron. *Serious Straw Bale: A Home Construction Guide for All Climates.* White River Junction, VT: Chelsea Green, 2000.

27. LaPorte, Robert. *Mooseprints: A Holistic Home Building Guide.* Santa Fe: Natural House Building Center, 1993.

28. Lerner, Kelly, and Pamela Wadsworth Goode, eds. *The Building Official's Guide to Straw Bale Construction.* Angels Camp, CA: California Straw Bale Association, 2000.

29. Magwood, Chris, and Peter Mack. *Straw Bale Building.* Gabriola Island, British Columbia: New Society Publishers, 2000.

30. Magwood, Chris, and Chris Walker. *Strawbale Details: A Manual for Designers and Builders.* Gabriola Island, British Columbia: New Society Publishers, 2001.

31. McHenry, Paul Graham, Jr. *Adobe . . . Build It Yourself,* 2nd ed. Tucson: University of Arizona Press, 1985.

32. _____. *The Adobe Story: A Global Treaure*. Albuquerque: University of New Mexico Press, 1998.

33. Merrill, Robert. *Hybrid Construction: The Art of Building with Recycled and Indigenous Materials*. Dexter, OR: Lost Valley Publishing Company, 1994.

34. Myhrman, Matts, and S. O. MacDonald. *Build It with Bales: A Step-by-Step Guide to Straw Bale Construction,* 2nd ed. Tucson: Out on Bale, 1997.

35. Norton, John. *Building with Earth*. London: Intermediate Technology Publications, 1997.

36. Paschich, Ed, and Paula Hendricks. *The Tire House Book*. Santa Fe: Sunstone Press, 1995.

37. Reynolds, Michael. *Earthship, Volume 1*. Taos: Solar Survival Architecture, 1993.

38. Rigassi, Vincent. *Compressed Earth Blocks, Volume 1: Manual of Production*. Braunschweig-Wiesbaden: Vieweg and Sohn, 1995.

39. Roberts, Carolyn. *A House of Straw: A Natural Building Odyssey*. White River Junction, VT: Chelsea Green Publishing, 2002.

40. Romero, Orlando, and David Larkin (Michael Freeman, Photographer). *Adobe: Building and Living with Earth*. Boston: Houghton-Mifflin, 1994.

41. Roy, Rob. *Complete Book of Underground Houses: How to Build a Low-Cost Home.* New York: Sterling Publishers, 1994.

42. Scottish Lime Centre. *Lime Mortars Preparation and Use of Lime Mortar: An Introduction to the Principles of Using Lime Mortars*. Edinburgh: Scottish Lime Centre, 1995.

43. Seth, Sandra, and Laura Seth. *Adobe! Homes and Interiors of Taos, Santa Fe and the Southwest.* Stamford, CT; Architectural Book Publishing, 1998.

44. Steen, Athena, and Bill Steen. *The Beauty of Straw Bale Homes*. White River Junction, VT: Chelsea Green Publishing, 2000.

45. _____. *Earthen Floors*. Elgin, AZ: Canelo Project, 1997.

46. Steen, Athena, Bill Steen, and David Bainbridge. *The Straw Bale House*. White River Junction, VT: Chelsea Green Publishing, 1994.

47. Straw Bale Construction Association. *The New Mexico Engineering Tests and Building Code*. Santa Fe: Straw Bale Construction Association, 1996.

48. Stulz, Roland, and Kiran Mukerji. *Appropriate Building Materials: A Catalogue of Potential Solutions,* 3rd ed. St. Gallen, Switzerland: SKAT Publications, 1993.

49. Tibbets, Joseph M. *The Earthbuilder's Encyclopedia*. Bosque, NM: Southwest Solaradobe School, 1989.

50. VITRA Design Museum. *Grow Your Own House: Simon Velez and Bamboo Architecture.* Weil am Rein, Germany. Distributed in Americas by Chelsea Green Publishing, 2000.

51. Wanek, Catherine. *The New Straw Bale Home.* Layton, UT: Gibbs Smith Publisher, 2003.

52. Wells, Malcom. *The Earth-Sheltered House: An Architect's Sketchbook*. White River Junction, VT: Chelsea Green Publishing, 1998.

53. Wojciechowska, Paulina. *Building with Earth: A Guide to Flexible-Form Earthbag Construction*. Hopland, CA: Real Goods Solar Living Books, 2001.

54. Yoshida, Isao. *Building Bamboo Fences.* Tokyo: Graphic-Sha Publishing, 1999.

Ecological Practices and Design Guidelines

1. Alexander, Christopher. *The Timeless Way of Building*. New York: Oxford University Press, 1979.

2. _____. *The Production of Houses*. New York: Oxford University Press, 1995.

3. Alexander, Christopher, Sara Ishikawa, and Murray Silverstein. *A Pattern Language: Towns—Buildings—Construction*. New York: Oxford University Press, 1977.

4. Allen, T.F.H. and Thomas W. Hoekstra. *Toward a Unified Ecology*. New York: Columbia University Press, 1992.

5. Allard, Francis, ed. *Natural Ventilation in Buildings: A Design Handbook*. London: James & James, 1998.

6. Baker, Paula, Erica Elliot, and John Banta. *Prescriptions for a Healthy House: A Practical Guide for Architects, Builders, and Homeowners*. Santa Fe: Baker and Associates, 1997.

7. Barnett, Dianna Lopez, William D. Browning, and Rocky Mountain Institute, Department of Green Development Services. *A Primer on Sustainable Building*. Snowmass, CO: Rocky Mountain Institute, 1995.

8. Berry, Wendell. *The Unsettling of America: Culture and Agriculture*. San Francisco: Sierra Club Books, 1996.

9. Bower, John. *Healthy House Building: A Design and Construction Guide*. Bloomington, IN: Healthy House Institute, 1997.

10. Brand, Stewart. *How Buildings Learn*. New York: Viking/Penguin, 1994.

11. Brown, G. Z., and Mark DeKay. *Sun, Wind, and Light: Architectural Design Strategies*. Hoboken, NJ: John Wiley & Sons, 2000.

12. Canada Home Builders' Association. *Builders' Manual*. Ottawa, Ontario: Canada Home Builders' Association, 1994.

13. Chiras, Daniel. *The Natural House: A Complete Guide to Healthy, Energy-Efficient, Environmental Homes*. White River Junction, VT: Chelsea Green Publishing, 2000.

14. Clarke Snell, Clarke. *The Good House Book: A Common-Sense Guide to Alternative Homebuilding*. Asheville, NC: Lark Books, 2004.

15. Crosbie, Michael J. *Green Architecture: A Guide to Sustainable Design*. Washington, DC: American Institute of Architects Press, 1994.

16. Day, Christopher. *Places of the Soul: Architecture and Environmental Design*. New York: Elsevier Architectural Press, 2004.

17. Gauzin-Muller, Dominique, with Nicolas Favet. *Sustainable Architecture and Urbanism*. Boston: Birkhauser, 2002.

18. Grady, Wayne. *Green Home: Planning and Building the Environmentally Advanced House*. Buffalo, NY: Camden House Publishing, 1993.

19. Gottfried, David, and Annette Osso. *Sustainable Building Technical Manual; Green Building Design, Construction and Operations*. San Francisco: U.S. Green Building Council, and Washington, DC: Public Technology, 1996.

20. Guzowski, Mary. *Daylighting for Sustainable Design*. New York: McGraw-Hill, 1999.

21. Harland, Edward. *Eco-Renovation: The Ecological Home Improvement Guide*. White River Junction, VT: Chelsea Green Publishing, 1993.

22. Harwood, Barbara Bannon. *The Healing House*. Carlsbad, CA: Hay House, 1997.

23. Hunter, Robert J. *Simple Things Won't Save the Earth*. Austin: University of Texas Press, 1997.

24. Isaacs, Ken. *How to Build Your Own Living Structures*. New York: Harmony Books, Crown Publishers, 1974.

25. Javits, Tom, Helga Olkowski, Bill Olkowski, and Farallones Institute Staff. *The Integral Urban House: Self-Reliant Living in the City*. San Francisco: Sierra Club Books, 1979.

26. Jones, David Lloyd. *Architecture and the Environment: Bioclimatic Building Design*. New York: Overlook Press, 1998.

27. Klein, Kirsten, and Inger Klingenberg. *Recollection and Change: Examples of Ecological Architecture in Northern Europe, 1975–1995*. Denmark: Blaa Gaard, 1995.

28. Lawson, Bill, *Building Materials, Energy and the Environment: Towards Ecologically Sustainable Development*. Manuka: Royal Australian Institute of Architects, 1996.

29. Leibowitz Earley, Sandra. *Eco-Design and Building Schools: Guide to Green Design and Construction Education in the USA and Canada*. Oakland, CA: New Village Press, 2005.

30. Lyle, John Tillman. *Regenerative Design for Sustainable Development*. Hoboken, NJ: John Wiley & Sons, 1994.

31. Malin, Nadav, and Alex Wilson. *Building an Environmentally Friendly House*. Lincoln, MA: Massachusetts Audubon Society, 1991.

32. Marinelli, Janet, and Robert Kourik. *The Naturally Elegant Home: Environmental Style*. Waltham, MA: Little, Brown, & Co., 1992.

33. McClintock, Michael. *Alternative Housebuilding*. London: Sterling Publications, 1989.

34. McDonough, William, and Michael Braungart. *Cradle to Cradle: Remaking the Way We Make Things*. San Francisco: Northpoint Press, 2001.

35. McKenzie, Dorothy. *Design for the Environment*. New York: Rizzoli, 1991.

36. Mendler, Sandra. *The HOK Guidebook to Sustainable Design*. Hoboken, NJ: John Wiley & Sons, 2000.

37. Melet, Ed. *Sustainable Architecture: The Pursuit of a More Contrast-Rich Environment*. Rotterdam: Nai Publishers, 1999.

38. Mollison, Bill. *Permaculture: A Designer's Manual*. Tyalgum, New South Wales, Australia; Tagari Publications, 1998.

39. Moore, Fuller. *Environmental Design Systems: Heating, Cooling, Lighting*. New York: McGraw-Hill, 1992.

40. Mostaedi, Arian. *Sustainable Architecture: Lowtech Houses*. Carles Broto and Josop Ma Minguet, 2002.

41. Papanek, Victor. *The Green Imperative: Natural Design for the Real World*. London: Thames and Hudson, 1995.

42. Pearson, David. *The New Natural House Book*. New York: Fireside Books, 1998.

43. Piedmont-Palladino, Susan, and Mark Alden Branch. *Devil's Workshop: 25 Years of Jersey Devil Architecture*. Princeton, NJ: Princeton Architectural Press, 1997.

44. Roaf, Sue, and Manuel Thomas Fuentes. *Ecohouse 2: A Design Guide*. Oxford, United Kingdom: Butterworth-Heinemann, 2003.

45. Roodman, David Malin, and Nicholas Lenssen. *A Building Revolution: How Ecology and Health Concerns Are Transforming Construction*. Berkeley, CA: Siegel & Strain Architects, 1997.

46. Steele, James. *Sustainable Architecture*. New York: McGraw-Hill, 1981.

47. Stitt, Fred A. *Ecological Design Handbook*. New York: McGraw-Hill, 1999.

48. Strain, Larry. *Resourceful Specifications: Guideline Specifications for Environmentally Considered Building Materials and Construction Methods*. Berkeley, CA: Siegel & Strain Architects, 1997.

49. Swan, James A., ed. *Dialogues with the Living Earth: New Ideas on the Spirit of Place from Designers, Architects, and Innovators*. Wheaton, IL: Quest Books, 1996.

50. Talbott, John. *Simply Build Green: A Technical Guide to the Ecological Houses at the Findhorn Foundation*. Forres, Scotland: Findhorn Press, 1995.

51. Todd, Nancy Jack, and John Todd. *From Eco-Cities to Living Machines: Principles of Sustainable Design*. Pueblo, CO: U.S. Government Printing Office, 1993.

52. U.S. Department of the Interior, National Park Service. *Guiding Principles of Sustainable Design*. Pueblo, CO: U.S. Government Printing Office, 1993.

53. Van der Ryn, Sim, and Stuart Cowan. *Ecological Design*. Washington, DC: Island Press, 1996.

54. Wackernagel, Mathis, and William Rees. *Our Ecological Footprint*. Gabriola Island, British Columbia: New Society Publishers, 1996.

55. Wells, Malcolm. *Gentle Architecture*. New York: McGraw-Hill, 1981.

56. Woods, Charles G., and David Wright (Malcolm Wells, contributor). *Designing Your Natural House*. Hoboken, NJ: John Wiley & Sons, 1992.

57. Yeang, Ken. *Designing with Nature: The Ecological Basis for Architectural Design*. New York: McGraw-Hill, 1995.

58. Zeiher, Laura C. *The Ecology of Architecture: A Complete Guide to Creating the Environmentally Conscious Building*. New York: Whitney Library of Design/Imprint of Watson-Guptill, 1996.

59. Zelov, Chris, and Phil Cousineau, Phil. *Design Outlaws on the Ecological Frontier*. Philadelphia: Knossus Publishing, 1997.

Energy-Efficient Building

1. Anderson, Bruce, and Malcolm Wells. *Passive Solar Energy*. Amherst, MA: Brick House Publishing, 1994.

2. Balcomb, J. Douglas, ed. "Passive Solar Buildings" in *Solar Heat Technologies: Fundamentals and Applications,* Vol. 7. Cambridge, MA: MIT Press, 1992.

3. Cofaigh, Eoin O., John A. Olley, and J. Owen Lewis. *The Climatic Dwelling: An Introduction to Climate-Responsive Residential Architecture*. London: James & James, 1996.

4. Cole, Nancy, and P. J. Skerrett. *Renewables Are Ready*. White River Junction, VT: Chelsea Green Publishing, 1995.

5. Creech, Dennis B. *Homeowner's Guide to Energy Efficient and Passive Solar Homes*. Upland: Diane Publishing Co., 1996.

6. Crosbie, Michael J., ed. *The Passive Solar Design and Construction Handbook*. Hoboken, NJ: John Wiley & Sons, 1997.

7. Fong, Clay, and Alice Hubbard. *The Community Energy Workbook*. Snowmass, CO: Rocky Mountain Institute, 1995.

8. Freeman, Mark. *The Solar Home: How to Design and Build a House You Heat with the Sun*. Mechanicsburg, PA: Stackpole Books, 1994.

9. Givoni, Baruch. *Passive and Low Energy Cooling of Buildings*. Hoboken, NJ: John Wiley & Sons, 1994.

10. _____. *Climate Considerations in Building and Urban Design*. Hoboken, NJ: John Wiley & Sons, 1998.

11. Hawkes, Dean, and Wayne Forster. *Energy Efficient Buildings: Architecture, Engineering, and Environment*. New York: Norton, 2002.

12. Heede, Richard, and Rocky Mountain Institute Staff. *Homemade Money: How to Save Energy and Dollars in Your Home*. Amherst, MA: Brick House Publishing, 1995.

13. Johansson, Allan. *Clean Technology*. Boca Raton, FL: Lewis, 1992.

14. Johasson, Thomas B., H. Kelly, A.K.N. Reddy, R. H. Wiliams, and L. Burnham. *Renewable Energy: Sources for Fuels and Electricity*. Washington, DC: Island Press, 1993.

15. Johnson, Timothy E. *Low-E Glazing Design Guide*. Stoneham, MA: Butterworth-Heinemann, 1991.

16. Kachadorian, James. *The Passive Solar House*. White River Junction, VT: Chelsea Green Publishing, 1997.

17. Krigger, John T. *Your Home Cooling Energy Guide*. Helena, MT: Saturn Resources Management, 1992.

18. Lane, Tom. *Solar Hot Water Systems: Lessons Learned*. Gainesville, FL: Energy Conservation Services of North Florida, Inc., 2004.

19. Lechner, Norbert. *Heating, Cooling, Lighting: Design Methods for Architects*. Hoboken, NJ: John Wiley & Sons, 1991.

20. Lstiburek, Joe, and John Carmody. *Moisture Control Handbook*. Hoboken, NJ: John Wiley & Sons, 1991.

21. Mazria, Edward. *The Passive Solar Energy Book: A Complete Guide to Passive Solar Home, Greenhouse, and Building Design*. Emmaus, PA: Rodale Press, 1979.

22. Moore, Fuller. *Concepts and Practice of Architectural Daylighting*. Hoboken, NJ: John Wiley & Sons, 1991.

23. _____. *Environmental Control Systems*. New York: McGraw-Hill, 1993.

24. Olgyay, Victor. *Design with Climate: A Bioclimatic Approach to Architectural Regionalism*. Hoboken, NJ: John Wiley & Sons, 1992.

25. Potts, Michael. *The New Independent Home: People and Houses That Harvest the Sun, Wind, and Water*. White River Junction, VT: Chelsea Green Publishing, 1999.

26. Rocky Mountain Institute. *The Efficient House Sourcebook*. Snowmass, CO: Rocky Mountain Institute, 1992.

27. Schaffer, John. *A Place in the Sun: The Evolution of the Real Goods Solar Living Center*. White River Junction, VT: Chelsea Green Publishing, 1997.

28. Schaeffer, John, and Douglas Pratt. *Real Goods Solar Living Sourcebook: The Complete Guide to Renewable Energy Technologies and Sustainable Living*. Chelsea Green Publishing, distributor, 2001.

29. Solar Energy International. *Photovoltaics: Design and Installation Manual*. New Society Publishers, 2004.

30. Stein, Richard G. *Architecture and Energy*. Garden City, NY: Anchor Press/Doubleday, 1978.

31. Strong, Steven J. *The Solar Electric House: Energy for the Environmentally Responsive, Energy-Independent Home*. White River Junction, VT: Chelsea Green Publishing, 1993.

32. Watson, Donald, and Kenneth Labs. *Climactic Design: Energy-Efficient Building Principles and Practices*. New York: McGraw-Hill, 1993.

33. Wilson, Alex, and John Morrill. *Consumer Guide to Home Energy Savings.* Berkeley, CA: American Council for an Energy-Efficient Economy, 1995.

34. Steven Winter Associates. *The Passive Solar Design and Construction Handbook.* Hoboken, NJ: John Wiley & Sons, 1997.

Ecological Interiors

1. Anderson, Nina, and Albert Benoit. *Your Health and Your House.* New Canaan, CT: Keats Publishing, 1994.

2. Bower, John. *The Healthy House: How to Buy One, How to Cure a Sick One, How to Build One,* 3rd ed. Bloomington, IN: Healthy House Institute, 1997.

3. Bower, Lynn Marie. *The Healthy Household.* Bloomington, IN: Healthy House Institute, 1995.

4. Canada Mortgage and Housing Corporation. *Building Materials for the Environmentally Hypersensitive.* Toronto, Canada: CMHC, 1997.

5. Dadd, Debra Lynn. *The Nontoxic Home: Protecting Yourself and Your Family from Everyday Toxins and Health Hazards.* Los Angeles: Jeremy P. Tarcher, 1986.

6. _____. *Home Safe Home.* New York: Tarcher/Putnam, 1997.

7. Marinelli, Janet, and Paul Bierman-Lytle. *Your Natural Home: The Complete Source Book and Design Manual for Creating a Healthy, Beautiful and Environmentally Sensitive House.* Boston: Little, Brown, & Co., 1995.

8. Pearson, David. *The Natural House Catalog.* New York: Simon & Schuster, 1996.

9. Rosseau, David, and James Wasley. *Healthy by Design: Building and Remodeling Solutions for Creating Healthy Homes.* Point Roberts, WA: Hartley & Marks Publishers, 1997.

10. Venolia, Carol. *Healing Environments: Your Guide to Indoor Well-Being.* Berkeley, CA: Celestial Arts News, 1993.

Periodicals

1. *Adobe Builder Magazine,* Joe Tibbets, ed. Southwest Solar Adobe School, Bosque, NM (quarterly).

2. *Building for a Future,* Keith Hall, ed. Association for Environment-Conscious Building, Coaley, Gloucestershire, England (quarterly).

3. *Designer Builder: New Mexico's Magazine of Architecture and Design,* Jerilou Hammett, ed. Fine Editions, Inc., Santa Fe.

4. *Environmental Building News: A Monthly Newsletter on Environmentally Responsible Design and Construction,* Alex Wilson and Nadav Malin, eds. E Build, Inc., Brattleboro, VT.

5. *Environmental Design and Construction,* John Sailer, ed. Business News Publishing, Saddle Brook, NJ (bimonthly).

6. *Jointers' Quarterly: The Journal of Timber Framing and Traditional Building,* Steve Chappell, ed. Fox Maple Press, Brownfield, ME.

7. *The Journal of Light Construction,* Sal Alfano, ed. Richmond, VT. (monthly).

8. *The Last Straw: The Journal of Straw Bale Construction,* Catherine Wanek and Mark Piepkorn, eds. Hillsboro, NM (quarterly).

9. *Natural Home,* Laurel Lund, ed. Natural Home, LLC, Loveland, CO. (bimonthly).

10. *The Permaculture Activist,* Peter Bane, pub. Black Mountain, NC.

Energy Conservation and Energy Efficiency Periodicals

1. *Home Energy: The Magazine of Residential Energy Conservation*, Alan Meier, ed. Energy Auditor & Retrofitter, Inc., Berkeley, CA (bimonthly).
2. *Home Power: The Hands-On Journal of Home-Made Power*. Home Power, Inc., Ashland, OR (bimonthly).
3. *Solar Today*, Maureen McIntyre, ed. American Solar Energy Society, Boulder, CO (bimonthly).

Resource Guides

1. *Architectural Resource Guide*, David Kibbey, ed. Northern California ADPSR, Berkeley, CA (searchable CD-ROM database and printed guide), 2000.
2. *Environmental Resource Guide*, American Institute of Architects, Joseph Demkin, ed. John Wiley & Sons, Hoboken, NJ, 1999.
3. *Green Building Materials: A Guide to Product Selection and Specification*. Ross Spiegel and Dru Meadows. John Wiley & Sons, Hoboken, NJ, 1999.
4. *Green Building Resource Guide*, John Hermannsson, ed. Taunton Press, Newton, CT, 1997.
5. *GreenSpec: The Environmental Building News Product Directory and Guideline Specifications*, Dwight Holmes, Larry Holmes, Alex Wilson, and Sandra Leibowitz. E Build, Inc., Brattleboro, VT, 1999.
6. *HOK Database for Healthy and Sustainable Building Materials*. Hellmuth, Obata + Kassabaum, Washington, DC, 1997.
7. *REDI: Resources for Environmental Design Index*. Iris Communications, Bend, OR. http://oikos.com/redi
8. *Resourceful Specifications: Guideline Specifications for Environmentally Considered Building Materials and Construction Methods*, Larry Strain. Strain Publishing, Emeryville, CA, 1997.
9. *Sustainable Building Sourcebook: Supplement to the Green Builder Program*. City of Austin, Environment and Conversation, Austin, TX, 1993. http://www.greenbuilder.com/general/BuildingSources.html

Videos

1. *At Home with Mother Earth* (40 min.). Feat of Clay Productions, Los Angeles, CA, 1995.
2. *Building with the Earth: Oregon's Cob Cottage Company* (28 min.). Cottage Grove, OR.
3. *Building with Straw*, Vol. 1, *A Straw Bale Workshop* (73 min.); Vol. 2, *A Straw Bale Home Tour* (60 min.); Vol. 3, *Straw Bale Code Testing* (40 min.). Black Range Films, Kingston, NM, 1994–1996.
4. *Creating Healing Environments for Holistic Living* (30 min.). The Natural House Building Center, Santa Fe, 1996.
5. *How to Build Your Elegant Home with Straw Bales* (90 min.). Video and manual. Sustainable Systems Support, Bisbee, AZ, 1996.
6. *Rammed Earth Basics* (50 min). EarthWright Institute, Napa, CA, 1996.
7. *The Rammed Earth Renaissance* (31 min.). Lyceum Productions, Napa, CA, 1996.
8. *The Straw Bale Solution* (30 min.). NetWorks Productions, Kingston, NM, 1998.

Appendix B

Alternative Construction Resource Centers

Natural Building Education Centers and Green Building Resource Centers

These are centers that offer hands-on training and technical assistance, and ecological construction information.

United States and Canada

Apeiron Institute
451 Hammet Rd.
Coventry, RI 02816-5008
401-397-3430
www.apeiron.org
info@apeiron.org

Aprovecho Research Center
80574 Hazelton Rd.
Cottage Grove, OR 97424
541-942-8198
http://www.aprovecho.net/
apro@efn.org

Arcosanti, A Project of the Cosanti Foundation
HC 74, Box 4136
Mayer, AZ 86333
928-632-6233
www.arcosanti.org
info@arcosanti.org

Autonomous and Sustainable Housing, Inc.
9211 Scurfield Dr. N.W.
Calgary, AB T3L 1V9
Canada
403-239-1882
www.ecobuildings.net
jdo@ecobuildings.net

Builders Without Borders
119 Main St.
Kingston, NM 88042
505-895-5400
www.builderswithoutborders.org
mail@builderswithoutborders.org

**California Earth Art and Architecture Institute
   (Cal-Earth Institute)**
10376 Shangri-La Avenue
Hesperia, CA 92345
760-244-0614
www.calearth.org
calearth@aol.com

California Straw Building Association
P.O. Box 1293
Angels Camp, CA 93522-1293
209-785-7077
www.strawbuilding.org
casba@strawbuilding.org

The Canelo Project
HC1 Box 324
Elgin, AZ 85611
520-455-5548
www.caneloproject.com
absteen@dakotacom.net

Center for Maximum Potential Building Systems
8604 FM 969
Austin, TX 78724
512-928-4786
www.cmpbs.org
center@cmpbs.org

Cleveland Green Building Coalition
Cleveland Environmental Center
3500 Lorain Avenue, Ste. 200
Cleveland, OH 44113
216-961-8850
www.clevelandgbc.org
melanie@clevelandgbc.org

Cob Cottage Company's North American School of Natural Building
P.O. Box 123
Cottage Grove, OR 97424
541-942-2005
www.cobcottage.com
cobcottage@hotmail.com

Cobworks
R.P.#1
Mayne Island, BC V0N 2T0
Canada
250-539-5253
www.cobworks.com
pat@cobworks.com

Colorado Straw Bale Association (COSBA)
P.O. Box 7398
Boulder, CO 80306
303-415-0638
www.coloradostrawbale.org
sfrancis@ecentral.com

Culture's Edge at Earthaven Eco-Village
1025 Camp Elliott Rd.
Black Mountain, NC 28711
828-669-3937
www.earthaven.org
culturesedge@earthaven.org

DAWN SouthWest
6570 W. Illinois St.
Tucson, AZ 85735
520-624-1673
www.greenbuilder.com/dawn
dawnaz@earthlink.net

Development Center for Appropriate Technology
P.O. Box 27513
Tucson AZ 85726-7513
520-624-6628
www.dcat.net
info@dcat.net

Earthwood Building School
366 Murtagh Hill Rd.
West Chazy, NY 12992
518-493-7744
www.cordwoodmasonry.com
robandjaki@yahoo.com

East Coast Alternative Building Center
3400 Eastern Blvd., Ste. A-15
York, PA 17402
717-840-6502
home.ptd.net/~zantar79/
ecabc@juno.com

Ecological Building Network
209 Caledonia St.
Sausalito, CA 94965-1926
415-331-7630
www.ecobuildnetwork.org
ecobruce@sbcglobal.net

Econest Building Company
P.O. Box 864
Tesuque, NM 87574
505-984-2928
or 505-989-1813
www.econest.com/
paula@bakerlaporte.com

Ecosa Institute
212B South Marina St.
Prescott, AZ 86303
928-541-1002
www.ecosainstitute.org
info@ecosainstitute.org

Ecoveristy
2639 Agua Fria
Santa Fe, NM 87505
505-424-9797
www.ecoversity.org
info@ecoversity.org

Eco-Village Training Center
189 Schoolhouse Rd.
P.O. Box 90
Summertown, TN 38483
931-964-4324
www.thefarm.org/etc/
ecovillage@thefarm.org

Emerald Earth Workshops
P.O. Box 764
Boonville, CA 95415
707-895-3302
www.emeraldearth.org
lorax@mail.ap.net

Everdale Environmental Learning Centre
Box 29
Hillsburgh, ON N0B 1Z0
Canada
519-855-4859
www.everdale.org
info@everdale.org

Fox Maple School of Traditional Building
P.O. Box 249
65 Corn Hill Rd.
Brownfield, ME 04010
207-935-3720
www.foxmaple.com
info@foxmaple.com

Goshen Timber Frames
37 Phillips St.
Franklin, NC 28734
828-524-8662
www.goshenframes.com
bonnie@goshenframes.com

Green Advantage
12606 Trillium Glen Lane
Lovettsville, VA 20180
540-822-9449
http://www.greenadvantage.org
gorear@greenadvantage.org

Green Building Resource Center
2218 Main St.
Santa Monica, CA 90405
310-452-7677
www.globalgreen.org/gbrc
gbrc@globalgreen.org

Groundworks
P.O. Box 381
Murphy, OR 97533
541-471-3470
www.cpros.com/~sequoia/
cobalot@cpros.com

Heartwood School for the Homebuilding Crafts
Johnson Hill Rd.
Washington, MA 01223
413-623-6677
www.heartwoodschool.com
willb@heartwoodschool.com

House Alive
7540 Griffin Lane
Jacksonville, OR 97530
541-889-3751
www.housealive.org
Coenraad@housealive.org

*International Institute for Bau-biologie and Ecology, Inc.
  (IBE)*
P.O. Box 387
Clearwater, FL 33757
727-461-4371
www.buildingbiology.net
baubiologie@earthlink.net

Island School of Building Arts
3199 Coast Rd.
Gabriola Island, BC V0R 1X7
250-247-8922
Canada
www.logandtimberschool.com
Info@logandtimberschool.com

Joslyn Castle Institute
3902 Davenport St.
Omaha, NE 68131
402-595-1902
http://www.ecospheres.com
ccdahlin@sustainabledesign.org

Kortright Centre
9550 Pine Valley Dr.
Woodbridge, ON L4L IAG
Canada
905-832-2289
http://www.kortright.org
KCC@look.ca

Kupono Natural Builders
P.O. Box 828
Bella Vista, CA 96008
530-275-3623
www.kuponobuilders.com
kuponobuilders@snowcrest.net

Natural Building Resources
119 Main St.
Kingston, NM 88042
505-895-3389
www.strawbalecentral.com
resources@strawbalecentral.com

Occidental Arts and Ecology Center
15290 Coleman Valley Rd.
Occidental, CA 95465
707-874-1557
www.oaec.org
inquiry@oaec.org

OK OK OK Productions
256 East 100 South
Moab, Utah 84532
435-259-8378
www.ok-ok-ok.com
okokok@frontiernet.ne

One United Resource Ecovillage
Box 530
Shawnigan Lake, BC V0R 2W0
Canada
250-743-3067
www.ourecovillage.org
our@pacificcoast.net

Pacific Gas & Electric's Pacific Energy Center
851 Howard St.
San Francisco, CA 94103
415-973-2277
www.pge.com/pec
bwc8@pge.com

The Pangea Partnership
239 Bradley Ave.
Ottawa, ON K1L 7E8
Canada
613-747-9185
www.pangeapartnership.org
rstone@PangeaPartnership.org

Powell Center for Construction & Environment
University of Florida
Rinker Hall, Room 304
Gainesville, FL 33711-5703
www.cce.ufl.edu/
ckibert@ufl.edu

Rocky Mountain Institute
Green Development Services
1739 Snowmass Creek Rd.
Snowmass, CO 81654-9199
970-927-3851
www.rmi.org

Seven Generations Natural Builders
SGNB, Attn. Sasha Rabin
P.O. Box 735
Bolinas, CA 94924
415-310-7460
www.sgnb.com
tim@sgnb.com

Shelter Institute
873 Route One
Woolwich, ME 04579
207-442-7938
www.shelterinstitute.com
info@shelterinstitute.com

Solar Energy International
P.O. Box 715,
76 S. 2nd St., Ste. B
Carbondale, CO 81623
970-963-8855
www.solarenergy.org
sei@solarenergy.org

Solar Living Institute
P.O. Box 836
13771 S. Highway 101
Hopland, CA 95449
707-744-2100
www.solarliving.org
sli@solarliving.org

Southface Energy Institute
241 Pine St. NE
Atlanta, GA 30308
404-872-3549
www.southface.org
info@southface.org

Southwest Solar Adobe School
P.O. Box 153
Bosque, NM 87006
505-861-2287
www.adobebuilder.com
adobebuilder@juno.com

U.S. Green Building Council
1015 18th St. NW, Ste. 508
Washington, DC 20036
202-828-7422
www.usgbc.org
leedinfo@usgbc.org

The United States Partnership for the Decade of
    Education for Sustainable Development
c/o ULSF
2100 "L" St., NW
Washington, DC 20037
www.uspartnership.org
info@uspartnership.org

Yestermorrow Design/Build School
189 VT Route 100
Warren, VT 05674
802-496-5545
www.yestermorrow.org
designbuild@yestermorrow.org

International Building Technology Centers

**Association for Environment Conscious Building
    (AECB)**
P.O. Box 32
Llandysul, Wales, UK SA44 5ZA
Tel: 01-559-370908
www.aecb.net
info@aecb.net

Auroville Earth Institute
Auroshilpam, Auroville 605 101 - T.N. India
Tel: +91 (0) 413 - 262 3064 / 262 3330
earth-institute@auroville.org.in
www.earth-auroville.com

**BASIN (Building Advisory Service and
    Information Network)**
www.gtz.de/basin

The BASIN Partners:

CRATerre
Maison Levrat (Parc Fallavier)
Rue de la Buthière - B.P. 53
F-38092 Villefontaine-CEDEX
France

Tel: +33-474-954391
www.craterre.archi.fr
craterre@club-internet.fr

The Centro Experimental de la Vivienda Económica
    (CEVE)
Igualdad 3585—B° Villa Siburu
5003 Córdoba, Argentina
Tel: +54-351-489-4442
www.ceve.org.ar
basin@ceve.org.ar

Development Alternatives
B-32 Tara Crescent
Qutab Institutional Area
New Delhi—110 016, India
Tel: +91-11-66-5370; 696-7938; 685-1158
www.devalt.org
tara@sdalt.ernet.in

EcoSouth: The Network for an Ecologically and
    Economically Sustainable Habitat
EcoSur
Apdo 107
Jinotepe, Nicaragua

Tel/Fax: +505-4223325
www.ecosur.org
ecosur@ibw.com.ni

Ecosouth
Schatzgutstrasse 9
8750 Glarus, Switzerland

The German Appropriate Technology Exchange (gate) of the Deutsche Gesellschaft für Technische Zusammenarbeit (GTZ)
P.O. Box 5180
D-65726 Eschborn, Germany
Tel: +49-6196-79-4212
www5.gtz.de/gate/
gate-basin@gtz.de

Intermediate Technology Development Group (ITDG) Schumacher Centre for Technology and Development
Bourton Hall, Bourton-on-Dunsmore
Warwickshire CV23 9QZ
United Kingdom
Tel: +44-1926-634400
www.itdg.org
luckyl@itdg.org.uk

Pagtambayayong—A Foundation for Mutual Aid, Inc.
102 P. del Rosario Ext.
Cebu City, 6000 Philippines
Tel: +63-32-2537974
pagtamba@cnms.net

Shelter Forum—East Africa
P.O Box 9202,00100,
Nairobi, Kenya
Tel: +254-(-020)-3753181/2
www.shelterforum.org
shelter@shelterforum.or.ke

The Swiss Centre for Development Cooperation in Technology Management (SKAT)
Vadianstr. 42
CH-9000 St. Gallen, Switzerland
Tel: +41-71-2285454
www.skat.ch
info@skat.ch

Earth Hands & Houses
18 the Willows, Byfleet
Surrey KT14-7QY, England
United Kingdom
Tel: +44 (0)1932 352129
www.earthhandsandhouses.org
enquiries@earthhandsandhouses.org

International Council for Research and Innovation in Building Construction
Postbox 1837
3000 BV Rotterdam, The Netherlands
Tel: 31 10 411 0240
www.cibworld.nl
secretariat@cibworld.nl

Lund Centre for Habitat Studies (LCHS)
Lund University
Box 118
SE-221 00 LUND, Sweden
Tel: +46-46 222 97 61
www.ark3.lth.se
Bl@lchs.lth.se

United Nations Centre for Human Settlements (UN-Habitat)
P.O. Box 30030
Nairobi, Kenya
Tel: (254-2) 623151
www.unchs.org
habitat.press@unchs.org

About the Contributors

Cassandra Adams became fascinated with environmentally conscious architecture in the mid-1970s when she helped to develop an early natural daylighting design software program for office buildings, and assisted friends in building a sod house on the Alaskan taiga. She spent the 1980s constructing large energy-intensive projects and state-of-the-art, energy-conserving buildings. Since the early 1990s Cassandra has been researching the environmental implications of construction materials' resource utilization and has written and co-written numerous articles based on this work. She is co-author with Frank Ching of the most recent edition of *Building Construction Illustrated* and maintains a private architecture and consulting practice.

Frank Andresen has been involved in the professional use of clay for building structures in Germany since the early 1980s. He has been applying traditional and modern techniques in Europe and America for both historical and new buildings. He teaches European modular light-clay building techniques for blocks, walls, and ceiling panels at various natural building workshops in the United States, such as those offered by the Fox Maple School of Traditional Building in Maine and the Canelo Project in Arizona.

Karl Bareis lived in Japan for nine years, completing a traditional apprenticeship in timber-frame architecture and fine finish carpentry. Mr. Bareis travels extensively in Southeast Asia, China, and Japan, researching and documenting uses of bamboo in building. He is past president of the Northern California Chapter of the American Bamboo Society, and has been International Coordinator of the International Association of Bamboo Societies for the past seven years. Mr. Bareis teaches bamboo architecture and propagation through the University of California at Santa Cruz Extension.

Bob Berkebile, FAIA, is a principal of Berkebile Nelson Immenschuh McDowell Architects (BNIM) of Kansas City, Missouri. One of the leading advocates of ecological building methods, he helped found the American Institute of Architect's Committee on the Environment. He is also a founding member of the Union of International Architects' Road from Rio Working Group. Berkebile is currently working with the U.S. National Parks Service, Department of Energy, and Department of Defense, and with Canadian provincial governments and the U.S. National Science Foundation to develop sustainable guidelines for their projects.

Tom Bender is an architect in Nehalem, Oregon. Top award winner in the 1993 International Sustainable Community Solutions and 1981 California Affordable Housing competitions, he is former editor of *RAIN: Journal of Appropriate Technology* and author of *Environmental Design Primer, The Heart of Place*, and *Silence, Song & Shadows,* as well as numerous articles. As a consultant to private and public organizations, he has been developing technical as well as spiritual dimensions of ecological design and sustainable communities.

Carole Crews learned about mud while growing up in a place and time when it was still being used—Taos, New Mexico, in the 1950s. She received an art degree from the University of Texas–Austin, then returned home to a life of building, raising three daughters, and creating various objects, the most recent being an adobe dome with cob extensions. Her love of mica, clay, and color are expressed in the practical wall finishes of her company, "Gourmet Adobe," as well as in sculpted and painted bas-relief panels.

Darrel DeBoer is a San Francisco Bay Area architect and furniture maker. He is a coauthor of the *Resource Guide to Resource-Efficient Building Materials* and the book *Building Less Waste*. He has consulted with several municipalities about the incorporation of recycled and less toxic building materials. He teaches workshops on bamboo construction and travels to China, Costa Rica, and Colombia to research local use of bamboo.

David Easton is a widely recognized authority on rammed earth. His 25-year career includes the construction of well over 100 structures and authorship of several books on monolithic earth wall technologies. He is a director, along with his wife and partner Cynthia Wright, of the EarthWright Institute, a nonprofit educational foundation that provides training programs for earth construction projects in developing nations. The couple live in what passes for a 300-year-old French farmhouse, which they built in Napa Valley, California.

David Eisenberg cofounded and directs the Development Center for Appropriate Technology in Tucson, Arizona. He has more than 20 years' construction experience and his many projects range from troubleshooting the high-tech steel and glass cover of Biosphere 2 in Oracle, Arizona, to building with adobe, rammed earth, and straw-bale. Currently, he is leading a national effort to create a sustainable context for building code development, modification, and enforcement. He is a coauthor of *The Straw Bale House* and helped write the first load-bearing straw-bale construction building code for the City of Tucson and Pima County, Arizona.

Lynne Elizabeth is an advocate of sustainable community development and appropriate building technologies. She has helped to organize educational programs, resource centers, and demonstration projects for the nonprofit organization Architects/Designer/Planners for Social Responsibility (ADPSR) since 1982. She currently directs New Village Press, a publishing project of ADPSR, and resides in Temescal Commons, a solar-powered co-housing community in Oakland, California.

Ken Haggard, Polly Cooper, and Jennifer Rennick all work at San Luis Sustainability Group, a green architecture firm near San Luis Obispo, California. There they run their own straw-bale office and residential complex, the "Old Trout Farm," with passive solar, off-grid systems. Besides practicing architecture and teaching at California Polytechnic State University, they are involved in the politics of sustainability in San Luis Obispo County. Ken Haggard and Polly Cooper were awarded the Passive Pioneer Award by the American Solar Energy Society in 1996.

Lou Host-Jablonski, **AIA**, is a father and an architect with Design Coalition Inc., a community design center, working in the Madison, Wisconsin, area since 1972. His work with the nonprofit firm emphasizes socially and ecologically conscious design and planning. He has designed multifamily housing, child care centers, cohousing, new homes and additions, community-built projects, community centers and playgrounds, and environments for persons with disabilities. Designs for two eco-villages are currently on the boards.

Dominic Howes is a home builder, a consultant, and musician who currently lives and works in Minneapolis, Minnesota. Mr. Howes began his alternative home construction education at the age of 7, while helping his family construct their passive solar home in rural Virginia. More recently he has been gifted by the wonderful guidance of many accomplished builders and artists during his world travels.

Kaki Hunter and Doni Kiffmeyer, principals of OK OK OK Productions of Moab, Utah, are both avid thespians, home builders, and freedom-promoting anarchists. They have brought a new level of precision to earthbag construction, developing tools and techniques to streamline and refine this nascent construction process. They have produced instructional manuals and videos, and recently authored *Earthbag Building: The Tools, Tricks, and Techniques*.

Steve Kemble and Carol Escott have been involved in education and promotion of ecological building since 1991 with their partnership, Sustainable Systems Support. Both are certified and experienced Permaculture designers. Together they lead wall-raising workshops and have produced two popular videos on straw-bale construction. Carol Escott has designed and built her own passive solar adobe home, as well as a number of straw-bale houses. Steve Kemble, a licensed engineer has designed and built numerous straw-bale homes.

Joseph F. Kennedy consults for natural building and ecological design projects around the world, including Sardinia, South Africa, Mexico, Canada, and the Czech Republic. He spent 1999 teaching natural building at ecovillage sites in Argentina, Europe, and South Africa. He received degrees in architecture from the University of California at Berkeley and the Southern California Institute of Architecture. Mr. Kennedy organized the 1996 and 1997 Southwest Natural Building Colloquia. He is also a Permaculture designer and a published author and illustrator.

Bruce King is a structural engineer with a private consulting practice in Sausalito, California. He is a member of the Structural Engineers Association of Northern California, the International Conference of Building Officials, and the American Society of Civil Engineers. Mr. King received a bachelor of science degree in Architectural Engineering from the University of Colorado and is the author of *Buildings of Earth and Straw—Structural Design for Rammed Earth and Straw Bale Architecture*.

Robert Laporte, a principal of EcoNest Building Company and the Natural House Building Center, has been designing and building environmentally sensitive homes for the past 20 years in Canada and the United States. He is a leading expert in earth, straw, and timber structures. He has researched natural building extensively in Europe and has been a major influence in the introduction of light-straw-clay building to North America. Mr. Laporte and the EcoNest Team recently returned to Europe to teach these building techniques throughout Scandinavia.

Kelly Lerner designs and builds energy-efficient, ecologically based buildings in the western United States and Asia. In her work as advisor to the United Nations Development Programme, she has introduced straw-bale construction technologies to Mongolia and designed the first straw-bale buildings in China. She received her master of architecture degree from the University of Oregon. She is a founding member and past director of the California Straw Building Association (CASBA) and coauthor of the *California Guide to Straw-Bale Construction for Building Code Officials*.

Michael Moquin has been working as a craftsman in historic preservation of earthen structures since 1968, specializing in landmark adobe buildings in the Albuquerque area. He organized a statewide preservation group to stop the needless demolition of historic adobe churches and served as its first president. He is the founding editor of the *Adobe Journal,* documenting earthen-based construction methods from around the world. He recently designed a 3,200-square-foot passive-solar adobe home, built in Albuquerque.

C. Wayne Nelson, Associate Director of Appropriate Technology for Habitat for Humanity International's Construction and Environmental Resources, has assisted in providing training and construction information to leaders in 1,700 Habitat affiliates in more than 60 countries. He began building with sustainable local materials while helping affiliates in places such as Ghana, West Africa, the Philippines, Guatemala, Honduras, Bolivia, Brazil, and the Dominican Republic. Habitat uses volunteers for much of the home construction, along with sweat equity contributed by the family who will occupy the home. The loan for materials is without interest, in keeping with the biblical principle.

Michael Reynolds, architect, has been working the past 32 years in Taos, New Mexico, where he has developed an ecologically integrated, whole-house system called the *Earthship*. His vision of sustainable architecture has led him to pioneer methods of building with recycled materials and independent heating, cooling, electrical, waste management, and water catchment systems.

Dan Smith, an architect in Berkeley for 20 years, has been adapting straw-bale construction for use in seismic California since 1993 and helped to found the California Straw Building Association. He is a principal of DSA Architects, along with associates Bob Theis, Dietmar Lorenz, and, previously, Kelly Lerner. The firm has been in the forefront of straw-bale revival in California, with more than 11 bale projects completed, 9 in construction, and 8 more in design. He earned a B.A. in History of Art with Vincent Scully at Yale University and an M.Arch. at the University of California–Berkeley with Chris Alexander and Sim Van der Ryn.

Michael Smith has a background in environmental engineering, international development, and social forestry. In 1993 he helped found the Cob Cottage Company in Oregon and was a director and workshop instructor there for four years. He wrote *The Cobber's Companion: How to Build Your Own Earthen Home* and is coauthor of *The Cob Cottage*. Mr. Smith currently lives in northern California, where he is a freelance workshop teacher, writer, journal editor, and design consultant, focusing on hybrid natural building and related social and environmental concerns.

Athena and Bill Steen are co-authors of the book *The Straw Bale House*, as well as publications on earthen baking ovens and earthen floors and plasters. They are directors of the Canelo Project of Elgin, Arizona, which brings together people from around the world to share specialized skills and cultural traditions. Current projects include construction of straw-earthen housing in Mexico, training women and children in the traditional livelihood of earth building, and ongoing workshops on natural building and other living arts.

Bob Theis obtained his B.Arch. from Pratt Institute and specialized in the adaptive reuse of historic buildings in and around New York City. He moved to California to work as a construction supervisor and design assistant for Christopher Alexander, earning a M.Arch. in the process. The popular success of straw bale construction, which he introduced to DSA Architects, has encouraged the firm's inclusion of related interests, such as Permaculture, building ornament, and the design of neighborhoods.

Sim Van der Ryn's leadership in architecture, planning, and design education has advanced the integration of ecological principles and practices into the design professions and made them real and timely solutions for the built environment. He is Professor Emeritus of Architecture, University of California–Berkeley, California State Architect Emeritus, and President of Van der Ryn Architects and the Ecological Design Institute, Sausalito, California. His books include *The Toilet Papers, Sustainable Communities, Ecological Design*, and the forthcoming *The Geometry of Hope.*

Marci Webster-Mannison is an architect and Director of Design at Charles Sturt University, Australia. She is responsible for site planning and building design for all campuses, including the ecologically innovative Thurgoona campus under construction in New South Wales. Ms. Webster-Mannison has designed a wide range of commercial, laboratory, residential, and institutional buildings. Her doctoral thesis formulates strategies for architecture that responds to environmental conditions. She has won design awards for water management and landscape, as well as a National Trust Heritage Award for historic renovation.

Paulina Wojciechowska was born in Poland and spent her childhood there as well as in Afghanistan. She received her architectural degree from Kingston University in England, and worked in London architectural practices for several years. She has traveled widely to study original and natural building techniques in other parts of the world, and now specializes in design and construction of ecological houses from local materials. Paulina is the author of *Building with Earth: A Guide to Flexible-Form Earthbag Construction,* and she is the founder and director of Earth Hands & Houses, an appropriate building technology center based in England.

Index